Applied Modeling Techniques and Data Analysis 1

Big Data, Artificial Intelligence and Data Analysis Set

coordinated by
Jacques Janssen

Volume 7

Applied Modeling Techniques and Data Analysis 1

Computational Data Analysis Methods and Tools

Yannis Dimotikalis
Alex Karagrigoriou
Christina Parpoula
Christos H Skiadas

iSTE

WILEY

First published 2021 in Great Britain and the United States by ISTE Ltd and John Wiley & Sons, Inc.

ISTE Ltd
27-37 St George's Road
London SW19 4EU
UK

www.iste.co.uk

John Wiley & Sons, Inc.
111 River Street
Hoboken, NJ 07030
USA

www.wiley.com

Library of Congress Control Number: 2020951002

British Library Cataloguing-in-Publication Data
A CIP record for this book is available from the British Library
ISBN 978-1-78630-673-9

Contents

Preface xi
Yannis DIMOTIKALIS, Alex KARAGRIGORIOU, Christina PARPOULA and Christos H. SKIADAS

Part 1. Computational Data Analysis 1

Chapter 1. A Variant of Updating PageRank in Evolving Tree Graphs 3
Benard ABOLA, Pitos Seleka BIGANDA, Christopher ENGSTRÖM, John Magero MANGO, Godwin KAKUBA and Sergei SILVESTROV

1.1. Introduction 3
1.2. Notations and definitions 5
1.3. Updating the transition matrix 5
1.4. Updating the PageRank of a tree graph 10
1.4.1. Updating the PageRank of tree graph when a batch of edges changes 12
1.4.2. An example of updating the PageRank of a tree 15
1.5. Maintaining the levels of vertices in a changing tree graph 17
1.6. Conclusion 21
1.7. Acknowledgments 21
1.8. References 21

Chapter 2. Nonlinearly Perturbed Markov Chains and Information Networks 23
Benard ABOLA, Pitos Seleka BIGANDA, Sergei SILVESTROV, Dmitrii SILVESTROV, Christopher ENGSTRÖM, John Magero MANGO and Godwin KAKUBA

2.1. Introduction 23
2.2. Stationary distributions for Markov chains with damping component 26

2.2.1. Stationary distributions for Markov chains with damping component 26
2.2.2. The stationary distribution of the Markov chain $X_{0,n}$ 28
2.3. A perturbation analysis for stationary distributions of Markov chains with damping component 29
2.3.1. Continuity property for stationary probabilities 29
2.3.2. Rate of convergence for stationary distributions 29
2.3.3. Asymptotic expansions for stationary distributions 30
2.3.4. Results of numerical experiments 32
2.4. Coupling and ergodic theorems for perturbed Markov chains with damping component 39
2.4.1. Coupling for regularly perturbed Markov chains with damping component 39
2.4.2. Coupling for singularly perturbed Markov chains with damping component 41
2.4.3. Ergodic theorems for perturbed Markov chains with damping component in the triangular array mode 42
2.4.4. Numerical examples 43
2.5. Acknowledgments 51
2.6. References 51

Chapter 3. PageRank and Perturbed Markov Chains 57
Pitos Seleka BIGANDA, Benard ABOLA, Christopher ENGSTRÖM, Sergei SILVESTROV, Godwin KAKUBA and John Magero MANGO

3.1. Introduction 57
3.2. PageRank of the first-order perturbed Markov chain 59
3.3. PageRank of the second-order perturbed Markov chain 60
3.4. Rates of convergence of PageRanks of first- and second-order perturbed Markov chains 70
3.5. Conclusion 72
3.6. Acknowledgments 72
3.7. References 72

Chapter 4. Doubly Robust Data-driven Distributionally Robust Optimization 75
Jose BLANCHET, Yang KANG, Fan ZHANG, Fei HE and Zhangyi HU

4.1. Introduction 75
4.2. DD-DRO, optimal transport and supervised machine learning 79
4.2.1. Optimal transport distances and discrepancies 80
4.3. Data-driven selection of optimal transport cost function 81
4.3.1. Data-driven cost functions via metric learning procedures 81
4.4. Robust optimization for metric learning 83

4.4.1. Robust optimization for relative metric learning 83
4.4.2. Robust optimization for absolute metric learning 86
4.5. Numerical experiments 88
4.6. Discussion and conclusion 89
4.7. References 89

Chapter 5. A Comparison of Graph Centrality Measures Based on Lazy Random Walks 91
Collins ANGUZU, Christopher ENGSTRÖM and Sergei SILVESTROV

5.1. Introduction 91
5.1.1. Notations and abbreviations 93
5.1.2. Linear systems and the Neumann series 94
5.2. Review on some centrality measures 95
5.2.1. Degree centrality 95
5.2.2. Katz status and β-centralities 95
5.2.3. Eigenvector and cumulative nomination centralities 96
5.2.4. Alpha centrality 97
5.2.5. PageRank centrality 98
5.2.6. Summary of the centrality measures as steady state, shifted and power series 99
5.3. Generalizations of centrality measures 99
5.3.1. Priors to centrality measures 99
5.3.2. Lazy variants of centrality measures 100
5.3.3. Lazy α-centrality 100
5.3.4. Lazy Katz centrality 102
5.3.5. Lazy cumulative nomination centrality 103
5.4. Experimental results 104
5.5. Discussion 106
5.6. Conclusion 109
5.7. Acknowledgments 109
5.8. References 110

Chapter 6. Error Detection in Sequential Laser Sensor Input 113
Gwenael GATTO and Olympia HADJILIADIS

6.1. Introduction 113
6.2. Data description 114
6.3. Algorithms 116
6.3.1. Algorithm for consecutive changes in mean 118
6.3.2. Algorithm for burst detection 120
6.4. Results 125
6.5. Acknowledgments 127
6.6. References 127

Chapter 7. Diagnostics and Visualization of Point Process Models for Event Times on a Social Network 129
Jing WU, Anna L. SMITH and Tian ZHENG

7.1. Introduction 129
7.2. Background 131
7.2.1. Univariate point processes 131
7.2.2. Network point processes 132
7.3. Model checking for time heterogeneity 134
7.3.1. Time rescaling theorem 134
7.3.2. Residual process 136
7.4. Model checking for network heterogeneity and structure 138
7.4.1. Kolmogorov–Smirnov test 138
7.4.2. Structure score based on the Pearson residual matrix 141
7.5. Summary 143
7.6. Acknowledgments 144
7.7. References 144

Part 2. Data Analysis Methods and Tools 147

Chapter 8. Exploring the Distribution of Conditional Quantile Estimates: An Application to Specific Costs of Pig Production in the European Union 149
Dominique DESBOIS

8.1. Introduction 150
8.2. Conceptual framework and methodological aspects 150
8.2.1. The empirical model for estimating the specific production costs 151
8.2.2. The procedures for estimating and testing conditional quantiles . . 152
8.2.3. Symbolic PCA of the specific cost distributions 154
8.2.4. Symbolic clustering analysis of the specific cost distributions . . . 162
8.3. Results 165
8.3.1. The SO-PCA of specific cost estimates 167
8.3.2. The divisive hierarchy of specific cost estimates 170
8.4. Conclusion 171
8.5. References 172

Chapter 9. Maximization Problem Subject to Constraint of Availability in Semi-Markov Model of Operation 175
Franciszek GRABSKI

9.1. Introduction 175
9.2. Semi-Markov decision process 176
9.3. Semi-Markov decision model of operation 177

9.3.1. Description and assumptions 177
9.3.2. Model construction 177
9.4. Optimization problem 178
9.4.1. Linear programming method 179
9.5. Numerical example 182
9.6. Conclusion 184
9.7. References 185

Chapter 10. The Impact of Multicollinearity on Big Data Multivariate Analysis Modeling 187
Kimon NTOTSIS and Alex KARAGRIGORIOU

10.1. Introduction 187
10.2. Multicollinearity 188
10.3. Dimension reduction techniques 191
10.3.1. Beale *et al.* 192
10.3.2. Principal component analysis 192
10.4. Application 194
10.4.1. The modeling of PPE 194
10.4.2. Concluding remarks 200
10.5. Acknowledgments 200
10.6. References 200

Chapter 11. Weak Signals in High-Dimensional Poisson Regression Models 203
Orawan REANGSEPHET, Supranee LISAWADI and Syed Ejaz AHMED

11.1. Introduction 203
11.2. Statistical background 204
11.3. Methodologies 205
11.3.1. Predictor screening methods 205
11.3.2. Post-screening parameter estimation methods 206
11.4. Numerical studies 208
11.4.1. Simulation settings and performance criteria 208
11.4.2. Results 209
11.5. Conclusion 217
11.6. Acknowledgments 218
11.7. References 218

Chapter 12. Groundwater Level Forecasting for Water Resource Management 221
Andrea ZIRULIA, Alessio BARBAGLI and Enrico GUASTALDI

12.1. Introduction 221
12.2. Materials and methods 222

12.2.1. Study area 222
12.2.2. Forecast method 222
12.3. Results 224
12.4. Conclusion 230
12.5. References 230

Chapter 13. Phase I Non-parametric Control Charts for Individual Observations: A Selective Review and Some Results 233
Christina PARPOULA

13.1. Introduction 234
13.1.1. Background 234
13.1.2. Univariate non-parametric process monitoring 235
13.2. Problem formulation 237
13.3. A comparative study 239
13.3.1. The existing methodologies 239
13.3.2. Simulation settings 240
13.3.3. Simulation-study results 242
13.4. Concluding remarks 247
13.5. References 247

Chapter 14. On Divergence and Dissimilarity Measures for Multiple Time Series 249
Konstantinos MAKRIS, Alex KARAGRIGORIOU and Ilia VONTA

14.1. Introduction 249
14.2. Classical measures 250
14.3. Divergence measures 252
14.4. Dissimilarity measures for ordered data 254
14.4.1. Standard dissimilarity measures 254
14.4.2. Advanced dissimilarity measures 256
14.5. Conclusion 259
14.6. References 259

List of Authors 261

Index 265

Preface

Data analysis as an area of importance has grown exponentially, especially during the past couple of decades. This can be attributed to a rapidly growing technology industry and the wide applicability of computational techniques, in conjunction with new advances in analytic tools. Modeling enables analysts to apply various statistical models to the data they are investigating, to identify relationships between variables, to make predictions about future sets of data, as well as to understand, interpret and visualize the extracted information more strategically. Many new research results have recently been developed and published and many more are developing and in progress at the present time. The topic is also widely presented at many international scientific conferences and workshops. This being the case, the need for the literature that addresses this is self-evident. This book includes the most recent advances on the topic. As a result, on one hand, it unifies in a single volume all new theoretical and methodological issues and, on the other, introduces new directions in the field of applied data analysis and modeling, which are expected to further grow the applicability of data analysis methods and modeling techniques.

This book is a collective work by a number of leading scientists, analysts, engineers, mathematicians and statisticians, who have been working on the front end of data analysis. The chapters included in this collective volume represent a cross-section of current concerns and research interests in the above-mentioned scientific areas. This volume is divided into two parts with a total of 14 chapters in a form that provides the reader with both theoretical and applied information on data analysis methods, models and techniques, along with appropriate applications.

Part 1 focuses on computational data analysis and includes seven chapters: Chapter 1, "A Variant of Updating PageRank in Evolving Tree Graphs", by *Benard Abola, Pitos Seleka Biganda, Christopher Engström, John Magero Mango, Godwin Kakuba and Sergei Silvestrov*; Chapter 2, "Nonlinearly Perturbed Markov Chains and Information Networks", by *Benard Abola, Pitos Seleka Biganda, Sergei*

Silvestrov, Dmitrii Silvestrov, Christopher Engström, John Magero Mango and Godwin Kakuba; Chapter 3, "PageRank and Perturbed Markov Chains", by *Pitos Seleka Biganda, Benard Abola, Christopher Engström, Sergei Silvestrov, Godwin Kakuba and John Magero Mango*; Chapter 4, "Doubly Robust Data-driven Distributionally Robust Optimization", by *Jose Blanchet, Yang Kang, Fan Zhang, Fei He and Zhangyi Hu*; Chapter 5, "A Comparison of Graph Centrality Measures Based on Lazy Random Walks", by *Collins Anguzu, Christopher Engström and Sergei Silvestrov*; Chapter 6, "Error Detection in Sequential Laser Sensor Input", by *Gwenael Gatto and Olympia Hadjiliadis*; Chapter 7, "Diagnostics and Visualization of Point Process Models for Event Times on a Social Network", by *Jing Wu, Anna L. Smith and Tian Zheng.*

Part 2 covers the area of data analysis methods and tools and comprises seven chapters: Chapter 8, "Exploring the Distribution of Conditional Quantile Estimates: An Application to Specific Costs of Pig Production in the European Union", by *Dominique Desbois*; Chapter 9 "Maximization Problem Subject to Constraint of Availability in Semi-Markov Model of Operation", by *Franciszek Grabski*; Chapter 10, "The Impact of Multicollinearity on Big Data Multivariate Analysis Modeling", by *Kimon Ntotsis and Alex Karagrigoriou*; Chapter 11, "Weak Signals in High-Dimensional Poisson Regression Models", by *Orawan Reangsephet, Supranee Lisawadi and Syed Ejaz Ahmed*; Chapter 12, "Groundwater Level Forecasting for Water Resource Management", by *Andrea Zirulia, Alessio Barbagli and Enrico Guastaldi*; Chapter 13, "Phase I Non-parametric Control Charts for Individual Observations: A Selective Review and Some Results", by *Christina Parpoula*; Chapter 14, "On Divergence and Dissimilarity Measures for Multiple Time Series", by *Konstantinos Makris, Alex Karagrigoriou and Ilia Vonta.*

We wish to thank all the authors for their insights and excellent contributions to this book. We would like to acknowledge the assistance of all those involved in the reviewing process of this book, without whose support this could not have been successfully completed. Finally, we wish to express our thanks to the secretariat and, of course, the publishers. It was a great pleasure to work with them in bringing to life this collective volume.

Yannis DIMOTIKALIS
Crete, Greece
Alex KARAGRIGORIOU
Samos, Greece
Christina PARPOULA
Athens, Greece
Christos H. SKIADAS
Athens, Greece
December 2020

PART 1

Computational Data Analysis

1

A Variant of Updating PageRank in Evolving Tree Graphs

A PageRank update refers to the process of computing new PageRank values after a change(s) (addition or removal of links/vertices) has occurred in real-life networks. The purpose of updating is to avoid re-calculating the values from scratch. To efficiently carry out the update, we consider PageRank to be the expected number of visits to a target vertex if multiple random walks are performed, starting at each vertex once and weighing each of these walks by a weight value. Hence, it might be looked at as updating a non-normalized PageRank. We focus on networks of tree graphs and propose an approach to sequentially update a scaled adjacency matrix after every change, as well as the levels of the vertices. In this way, we can update the PageRank of affected vertices by their corresponding levels.

1.1. Introduction

Most real-world networks are continuously changing, and this phenomenon has come with challenges to the known data mining algorithms that assume the static form of datasets (Bahmani *et al.* 2012). Besides, it is a primary aim of network analysts to keep track of critical nodes.

Since Brin and Page (1998) pioneered PageRank centrality more than two decades ago, the centrality measure has found its applications in many disciplines (Gleich 2015). Recently, the study of PageRank of evolving graphs has drawn considerable attention and several computational models have been proposed.

Chapter written by Benard ABOLA, Pitos Seleka BIGANDA, Christopher ENGSTRÖM, John Magero MANGO, Godwin KAKUBA and Sergei SILVESTROV.

For instance, Langville and Meyer (2006) proposed an algorithm to update PageRank using an aggregation/disaggregation concept. In Bahmani *et al.* (2012), an algorithm that estimates a PageRank vector by crawling a small portion of the graph has been proposed, while Ohsaka *et al.* (2015) proposed an updating method for personalized PageRanks. In addition, the idea of partitioning information networks into connected acyclic and strongly connected components and then applying iterative methods of solving linear systems – while keeping an eye on the component(s) that has evolved – is proposed in Engström (2016) and Engström and Silvestrov (2016). In fact, all of these authors treat PageRank computation using matrix–vector multiplication.

There is evidence that tree graphs have applications in science and engineering. For instance, in Han *et al.* (2016), a directed acyclic graph (DAG) has been used in optimization problems, while Shandilya *et al.* (2014) applies the graph model in security systems.

Furthermore, tree graphs have been used to investigate ecological systems and functional similarity in biological networks (Ulanowicz 2004; Rocchi *et al.* 2017). For the case of an ecosystem, analysis of extinction risk or identifying important species in such a network is of significance (Allesina and Bodini 2004; Allesina and Pascual 2009). The notion of modeling an ecosystem as a tree graph is the result of mutual dependence between interacting species, for instance, dominator tree and food web structures (Ulanowicz 2004). In fact, dependence interaction can lead to interesting problems of network evolution. For example, if an organism Y requires nutrients from an organism X, in short, $X \to Y$, then lack of nutrients from X may lead to the extinction of Y unless it switches to another form of resource by addition of a new link. This is one of the primary roles of ecosystem-based management, as mentioned in Rocchi *et al.* (2017).

According to Wang *et al.* (2010), it has been found that a gene's functions are closely associated with different diseases, and such a dependence relationship requires a tree graph model. In view of Wang *et al.* (2010) and Han *et al.* (2016), a graph model improves computation cost and conceptualizes large networks.

The existing methods are virtually rooted in the iterative techniques which might be demanding when a localized change only occurs in a large network. Furthermore, the methods pay little attention to specific kinds of networks such as tree and strongly connected graphs. Computing PageRank of such components can lead to a great reduction in time complexity. For example, vertices of a strongly connected graph can be reordered by a feedback vertex set. In light of all of the above, we focus on the algorithm that updates PageRank when vertices are reordered by their levels, specifically for evolving directed tree graphs.

The remaining sections of this chapter are organized as follows. In section 1.2, we propose a technique for updating transition matrices when an edge is added or removed. Section 1.3 presents a single vertex update of PageRank when an edge is inserted or removed. Furthermore, we demonstrate that refinement iterative formation of linear systems fits in a single vertex update. Section 1.4 is devoted to maintaining the levels of vertices after some changes, and the conclusion is given in section 1.5.

1.2. Notations and definitions

For convenience, we define some notations that will be used throughout this chapter.

– Let us denote a directed graph by $G = (V, E)$, where V and E are sets of vertices and edges, respectively. The number of vertices in graph G will be denoted by $|V|$ or n. Furthermore, $G \cup \Delta G$ will denote a new graph after some changes of the edges or vertices.

– $\mathbf{A} \in \mathbb{R}^{n \times n}$ represents a matrix derived from G, and $\mathbf{A}_i$ means the ith row of $\mathbf{A}$.

– $\vec{u} \in \mathbb{R}^{n \times 1}$ denotes a column vector, and $\vec{u}^\top$ represents a row vector. We denote $\vec{e}_i$ as a vector with 1 in the ith entry and 0 elsewhere.

– $\vec{\omega}$ is the personalization vector of the vertices, which can be thought of as preference assigned to each of the vertices, and c is the damping parameter, usually taken to be 0.85.

DEFINITION 1.1.– *A directed tree graph G is a digraph without cycles.*

DEFINITION 1.2.– *The PageRank of a vertex v_i, denoted by π_i, is defined as*

$$\pi_i = (1 - c)\omega_i + c \sum_{u \in N_o} \frac{\pi_u}{deg(u)}, \qquad [1.1]$$

where $deg(u)$ is the degree of u and N_o is the out-neighbor of u. If v_i is a root, then $\pi_i = (1 - c)\omega_i$.

1.3. Updating the transition matrix

In this section, we present how to update the transition matrix when an edge(s) is added or removed. Two cases are considered:

(1) a source vertex v_i has at most one outgoing edge and at least one edge is added;

(2) a source vertex v_i has at least one outgoing edge and at least one edge is removed.

Each case will be handled separately.

Case 1: Addition of an edge to a source vertex with and without outgoing edges.

LEMMA 1.1.– *Let* $\mathbf{A}^{(1)}$ *and* $\mathbf{A}^{(2)}$ *be transition matrices of a tree graph* G *and* G∪ΔG, *respectively. Let* $deg(v_i)$ *be the out-degree of vertex* v_i. *Suppose an edge* $v_i \to v_j$ *is added, then a new matrix* $\mathbf{A}^{(2)}$ *can be updated as*

(i) $\mathbf{A}^{(2)} = \mathbf{A}^{(1)} + \vec{e}_i\vec{e}_j^\top$, *if* $deg(v_i) = 0$;

(ii) $\mathbf{A}^{(2)} = \left(\mathbf{I} - \frac{1}{deg(v_i)+1}\vec{e}_i\vec{e}_i^\top\right)\mathbf{A} + \frac{1}{deg(v_i)+1}\vec{e}_i\vec{e}_j^\top$, *if* $deg(v_i) \geq 1$.

We remark that a matrix $\mathbf{A}$ in (ii) is a sub-matrix of unaffected vertices in the graph G.

PROOF.– (i) Let $\vec{e}_i$ be a column vector with 1 at the ith entry and 0 elsewhere. If there was no edge between vertex v_i and v_j, then the only edge is the new one, $v_i \to v_j$. To write the new transition matrix $A^{(2)}$, we apply the well-known shearing matrix properties: If $\mathbf{A}^{(2)} = \left[a_{i,j}^{(2)}\right]$, for $i, j = 1, 2, \ldots, n$, then

$$\begin{aligned}\mathbf{A}^{(2)} &= \sum_{i,j} \vec{e}_i\vec{e}_j^\top a_{i,j}^{(2)},\\ &= \mathbf{A}^{(1)} + \vec{e}_i\vec{e}_j^\top a_{i,j}^{(2)},\end{aligned}$$

where $\mathbf{A}^{(1)} = \sum_{i,k\in V\setminus\{v_i\to v_j\}} \vec{e}_i\vec{e}_k^\top a_{i,k}^{(2)}$. Upon adding $v_i \to v_j$ the entry $a_{i,j}^{(2)} = 1$, hence $\mathbf{A}^{(2)} = \mathbf{A}^{(1)} + \vec{e}_i\vec{e}_j^\top$.

(ii) Let us define an operation on identity matrix $\mathbf{I}$, at an entry (i, i) as in Lancaster and Tismenetsky (1985), then

$$\mathbf{E}_{i,i} = \begin{matrix} \\ \\ \\ i \\ \\ \\ \end{matrix}\overset{\begin{matrix} & & & i & & \end{matrix}}{\begin{pmatrix} 1 & 0 & \cdots & \cdots & \cdots & 0 \\ 0 & 1 & \cdots & \cdots & \cdots & 0 \\ \vdots & \vdots & \ddots & \vdots & \vdots & \vdots \\ \cdot & \cdot & \cdot & \mu & \cdot & \cdot \\ \vdots & \vdots & \vdots & \vdots & \ddots & \vdots \\ 0 & \cdots & \cdots & \cdots & \cdots & 1 \end{pmatrix}} = \mathbf{I} - (1-\mu)\vec{e}_i\vec{e}_i^\top.$$

Then, the result of multiplying the ith row of $A^{(1)}$ by a non-zero scalar μ yields

$$\mathbf{E}_{i,i}\mathbf{A}^{(1)} = \mathbf{A}^{(1)} - (1-\mu)\vec{e}_i\vec{e}_i^\top\mathbf{A}^{(1)}, \qquad [1.2]$$

Using Lemma 1.1 (i) and substituting $a_{i,j}^{(1)} = \frac{1}{\deg(v_i)}$, we write

$$\mathbf{A}^{(1)} = \mathbf{A} + \frac{1}{\deg(v_i)}\vec{e}_i\vec{e}_j^\top,$$

where $\mathbf{A} = \sum_{i,k\in V\setminus\{v_i\to v_j\}} \vec{e}_i\vec{e}_k^\top a_{i,k}^{(2)}$.

Adding an edge $v_i \to v_k$, for $k \neq j$ to the source vertex v_i, the weight of the outgoing edges changes to $\frac{1}{\deg(v_i)+1}$. Let $\mu = \frac{\deg(v_i)}{\deg(v_i)+1}$, and by applying relation [1.2], we get

$$\begin{aligned}
\mathbf{A}^{(2)} &= \mathbf{A}^{(1)} - (1-\mu)\vec{e}_i\vec{e}_i^{\top}\left(\mathbf{A} + \frac{1}{\deg(v_i)}\vec{e}_i\vec{e}_j^{\top}\right), \\
&= \mathbf{A} + \frac{1}{\deg(v_i)}\vec{e}_i\vec{e}_j^{\top} - (1-\mu)\vec{e}_i\vec{e}_i^{\top}\left(\mathbf{A} + \frac{1}{\deg(v_i)}\vec{e}_i\vec{e}_j^{\top}\right), \\
&= \mathbf{A} + \frac{1}{\deg(v_i)}\vec{e}_i\vec{e}_j^{\top} - \frac{1}{\deg(v_i)+1}\vec{e}_i\vec{e}_i^{\top}\mathbf{A} - \frac{1}{\deg(v_i)}\cdot\frac{1}{\deg(v_i)+1}\vec{e}_i\vec{e}_j^{\top}, \\
&= \mathbf{A} + \frac{1}{\deg(v_i)+1}\vec{e}_i\vec{e}_j^{\top} - \frac{1}{\deg(v_i)+1}\vec{e}_i\vec{e}_i^{\top}\mathbf{A}, \qquad [1.3]
\end{aligned}$$

rearranging [1.3] completes the proof. □

Let V_0 be a set of vertices without outgoing edges and suppose $v_h \in V_0$. Then, a generalization of Lemma 1.1(i) when multiple edges, say k edges, are added takes the form

$$\mathbf{A}^{(2)} = \mathbf{A} + \sum_{h=1}^{k} \vec{e}_h\vec{e}_{j_h}^{\top}, \; j_h \in V \backslash V_0. \qquad [1.4]$$

Without loss of generality, we define the transition matrix $\mathbf{A}^{(2)}$ when the source vertex v_i has the degree $deg(v_i)$ and k edges are added to distinct target vertices as

$$\mathbf{A}^{(2)} = \mathbf{A} + \frac{1}{\deg(v_i)+k}\vec{e}_i\left(\vec{e}_j^{\top} - \mathbf{A}_i\right), \qquad [1.5]$$

where $\vec{e}_j^{\top}$ is a row vector with 1 in all the k target vertices and 0 elsewhere.

Case 2: Removal of edge(s) from a source vertex with at least one outgoing edge.

LEMMA 1.2.– *Let* $\mathbf{A}^{(1)}$ *and* $\mathbf{A}^{(2)}$ *be transition matrices of tree graphs* G *and* G∪ΔG, *respectively. Suppose an edge* $v_i \to v_j$ *is removed, then* $\mathbf{A}^{(2)}$ *can be updated as*

(i) $\mathbf{A}^{(2)} = \mathbf{A}^{(1)} - \vec{e}_i\vec{e}_j^{\top}$, *if* $\deg(v_i) = 1$;

(ii) $\mathbf{A}^{(2)} = \left(\mathbf{I} + \frac{1}{deg(v_i)-1}\vec{e}_i\vec{e}_i^{\top}\right)\mathbf{A} - \frac{1}{deg(v_i)-1}\vec{e}_i\vec{e}_j^{\top}$, *if* $deg(v_i) > 1$.

PROOF.– The proof of part (i) is trivial, so we prove part (ii) only. Recall that we do the reverse of Lemma 1.1 (ii), hence we use the property that the inverse

$\mathbf{E}_{i,i}^{-1} = \mathbf{I} + (1-\mu)\vec{e}_i\vec{e}_i^{\top}$. Supposing v_i has the degree $\deg(v_i)$ and then removing a single outgoing edge from v_i, the transition matrix $\mathbf{A}^{(2)}$ can be expressed as

$$\begin{aligned}\mathbf{A}^{(2)} &= \mathbf{A}^{(1)} + (1-\mu)\vec{e}_i\vec{e}_i^{\top}\mathbf{A}^{(1)},\\ &= \left(\mathbf{A} - \frac{1}{\deg(v_i)}\vec{e}_i\vec{e}_j^{\top}\right) + (1-\mu)\vec{e}_i\vec{e}_i^{\top}\left(\mathbf{A} - \frac{1}{\deg(v_i)}\vec{e}_i\vec{e}_j^{\top}\right), \qquad [1.6]\\ &= \mathbf{A} - \frac{1}{\deg(v_i)}\vec{e}_i\vec{e}_j^{\top} + \frac{1}{\deg(v_i)-1}\vec{e}_i\vec{e}_i^{\top}\mathbf{A} - \frac{1}{\deg(v_i)}\cdot\frac{1}{\deg(v_i)-1}\vec{e}_i\vec{e}_j^{\top},\\ &= \mathbf{A} + \frac{1}{\deg(v_i)-1}\vec{e}_i\vec{e}_i^{\top}\mathbf{A} - \frac{1}{\deg(v_i)-1}\vec{e}_i\vec{e}_j^{\top}, \qquad [1.7]\end{aligned}$$

rearranging [1.7] completes the proof. □

At implementation level, it is important to simplify relation [1.7] to

$$\mathbf{A}^{(2)} = \mathbf{A} - \frac{1}{\deg(v_i)-1}\vec{e}_i\left(\vec{e}_j^{\top} - \mathbf{A}_i\right). \qquad [1.8]$$

Consequently, if k edges are removed, then

$$\mathbf{A}^{(2)} = \mathbf{A} - \frac{1}{\deg(v_i)-k}\vec{e}_i\left(\vec{e}_j^{\top} - \mathbf{A}_i\right). \qquad [1.9]$$

In this case, we have out-vector multiplication, which is relatively easy to handle since the vectors are sparse.

For clarity, we demonstrate Lemmas 1.1 (i) and 1.2 (ii) accordingly.

EXAMPLE 1.1.– *Consider a graph $G := (V, E)$, where $V = \{v_1, v_2, v_3\}$, as shown in Figure 1.1. Suppose a new edge $v_2 \to v_3$ is added, then*

$$\begin{aligned}\mathbf{A}^{(2)} &= \mathbf{A}^{(1)} + \vec{e}_2\vec{e}_3^{\top},\\ &= \begin{pmatrix} 0 & \frac{1}{2} & \frac{1}{2}\\ 0 & 0 & 0\\ 0 & 0 & 0\end{pmatrix} + \begin{pmatrix}0\\1\\0\end{pmatrix}\begin{pmatrix}0 & 0 & 1\end{pmatrix},\\ &= \begin{pmatrix} 0 & \frac{1}{2} & \frac{1}{2}\\ 0 & 0 & 1\\ 0 & 0 & 0\end{pmatrix}.\end{aligned}$$

EXAMPLE 1.2.– *Let us consider a case where an edge, $v_2 \to v_5$ is removed, as indicated in Figure 1.2. Then, by Lemma 1.2 (ii), we get*

$$\begin{aligned}\mathbf{A}^{(2)} &= \mathbf{A} + \frac{1}{deg(v_2)-1}\vec{e}_2\mathbf{A}_2 - \frac{1}{deg(v_2)-1}\vec{e}_2\vec{e}_5^{\top},\\ &= \mathbf{A} + \frac{1}{2}\vec{e}_2\mathbf{A}_2 - \frac{1}{2}\vec{e}_2\vec{e}_5^{\top},\end{aligned}$$

$$= \begin{pmatrix} 0 & \frac{1}{2} & 0 & 0 & \frac{1}{2} \\ 0 & 0 & \frac{1}{3} & \frac{1}{3} & \frac{1}{3} \\ 0 & 0 & 0 & 0 & 0 \\ 0 & 0 & 0 & 0 & 1 \\ 0 & 0 & 0 & 0 & 0 \end{pmatrix} + \frac{1}{2} \begin{pmatrix} 0 \\ 1 \\ 0 \\ 0 \\ 0 \end{pmatrix} \left(0\,0\,\tfrac{1}{3}\,\tfrac{1}{3}\,\tfrac{1}{3} \right) - \frac{1}{2} \begin{pmatrix} 0 \\ 1 \\ 0 \\ 0 \\ 0 \end{pmatrix} \left(0\,0\,0\,0\,1 \right),$$

$$= \begin{pmatrix} 0 & \frac{1}{2} & 0 & 0 & \frac{1}{2} \\ 0 & 0 & \frac{1}{3} & \frac{1}{3} & \frac{1}{3} \\ 0 & 0 & 0 & 0 & 0 \\ 0 & 0 & 0 & 0 & 1 \\ 0 & 0 & 0 & 0 & 0 \end{pmatrix} + \begin{pmatrix} 0 & 0 & 0 & 0 & 0 \\ 0 & 0 & \frac{1}{6} & \frac{1}{6} & -\frac{1}{3} \\ 0 & 0 & 0 & 0 & 0 \\ 0 & 0 & 0 & 0 & 0 \\ 0 & 0 & 0 & 0 & 0 \end{pmatrix} = \begin{pmatrix} 0 & \frac{1}{2} & 0 & 0 & \frac{1}{2} \\ 0 & 0 & \frac{1}{2} & \frac{1}{2} & 0 \\ 0 & 0 & 0 & 0 & 0 \\ 0 & 0 & 0 & 0 & 1 \\ 0 & 0 & 0 & 0 & 0 \end{pmatrix}.$$

The lemma below will be essential in the subsequent section.

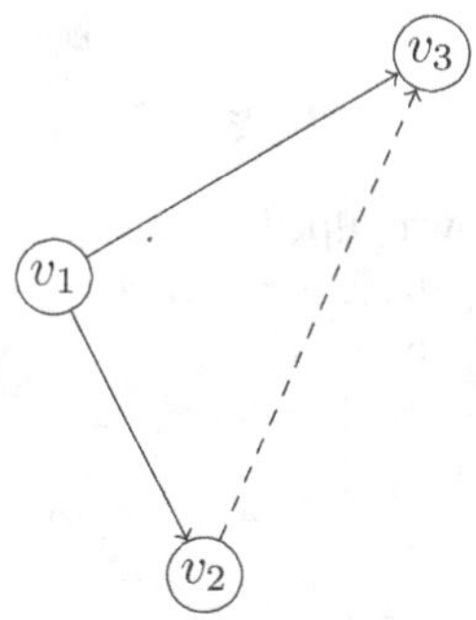

Figure 1.1. *A tree graph with an edge added to vertex v_2 without outgoing edge*

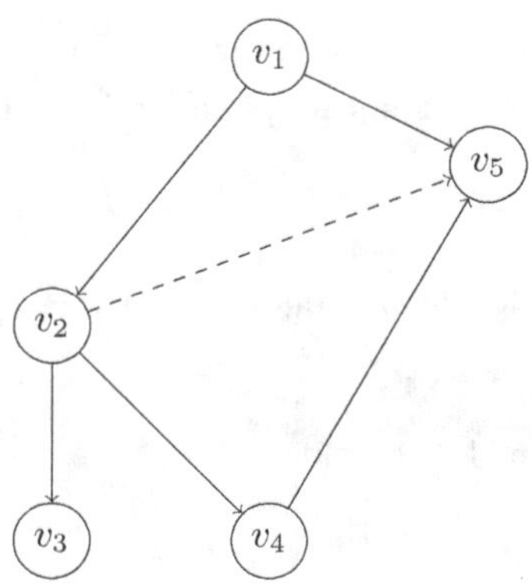

Figure 1.2. *A tree graph with the edge $v_2 \to v_5$ removed (dashed line)*

LEMMA 1.3.– *Let* $\mathbf{L}_{ij}$, $i, j = 1, 2$, *be a block lower triangular matrix. Then, the solution to the linear system*

$$\begin{bmatrix} \mathbf{L}_{11} & 0 \\ \mathbf{L}_{21} & \mathbf{L}_{22} \end{bmatrix} \begin{bmatrix} \vec{x}_1 \\ \vec{x}_2 \end{bmatrix} = \begin{bmatrix} \vec{b}_1 \\ \vec{b}_2 \end{bmatrix}$$

takes the form

$$\begin{bmatrix} \vec{x}_1 \\ \vec{x}_2 \end{bmatrix} = \begin{bmatrix} \mathbf{L}_{11}^{-1}\vec{b}_1 \\ -\mathbf{L}_{22}^{-1}\mathbf{L}_{21}\mathbf{L}_{11}^{-1}\vec{b}_1 + \mathbf{L}_{22}^{-1}\vec{b}_2 \end{bmatrix}. \qquad [1.10]$$

PROOF.– The proof of this lemma follows the fact that the inverse of a lower triangular matrix is also a lower triangular matrix, and it becomes straightforward by applying the properties of the inverse of block matrices (Lancaster and Tismenetsky 1985). □

1.4. Updating the PageRank of a tree graph

In this section, we will look at how to update the PageRank of the affected vertices, i.e. descendants of source vertex v_i (these are vertices that can be reached from a source vertex). The subsequent lemma allows us to update a PageRank of a target vertex when an edge is added to it.

LEMMA 1.4.– *Suppose* $\mathbf{A}^{(1)}$ *and* $\mathbf{A}^{(2)}$ *are transition matrices of tree graphs* G *and* $\mathrm{G} \cup \Delta\mathrm{G}$, *respectively. Consider a source vertex* v_i *without outgoing edges. Then, after addition of an edge* $v_i \to v_j$ *that does not create a cycle, the new PageRank of* v_j *with respect to* $\mathbf{A}^{(2)}$ *is expressed as*

$$\pi_j^{(2)} = \pi_j^{(1)} + c\pi_i^{(1)} \qquad [1.11]$$

PROOF.– Note that $\left(\mathbf{I} - c\mathbf{A}^{(2)^\top}\right)$ is a lower triangular matrix and is non-singular. Define the PageRank problem as a linear system of the form

$$\left(\mathbf{I} - c\mathbf{A}^{(2)^\top}\right)\vec{\pi}^{(2)} = (1-c)\vec{u}. \qquad [1.12]$$

Since an edge, $v_i \to v_j$ is added at the source vertex v_i, $\mathbf{A}^{(2)^\top} = \mathbf{A}^{(1)^\top} + \vec{e}_j\vec{e}_i^\top$ by Lemma 1.1.

Substituting for $\mathbf{A}^{(2)^\top}$ in [1.12] yields

$$\left(\left(\mathbf{I} - c\mathbf{A}^{(1)^\top}\right) - c\vec{e}_j\vec{e}_i^\top\right)\vec{\pi}^{(2)} = (1-c)\vec{u}.$$

Therefore

$$\vec{\pi}^{(2)} = (1-c)\left(\left(\mathbf{I} - c\mathbf{A}^{(1)^\top}\right) - c\vec{e}_j\vec{e}_i^\top\right)^{-1}\vec{u}. \qquad [1.13]$$

Set $\mathbf{C} = \left(\mathbf{I} - c\mathbf{A}^{(1)^\top}\right)$, $p = c\vec{e}_j$ and $q^\top = \vec{e}_i^\top$. By the Sherman–Morrison formula (Lancaster and Tismenetsky 1985), the inverse of $(\mathbf{C} - pq^\top)$ is expressed as

$$(\mathbf{C} - pq^\top)^{-1} = \mathbf{C}^{-1} + \frac{\mathbf{C}^{-1}pq^\top\mathbf{C}^{-1}}{1 - q^\top\mathbf{C}^{-1}p} \qquad [1.14]$$

Consequently, relation [1.13] becomes

$$\vec{\pi}^{(2)} = (1-c)\left(I - cA^{(1)^\top}\right)^{-1}\vec{u} + \frac{(1-c)c\mathbf{C}^{-1}\vec{e}_j\vec{e}_i^\top\mathbf{C}^{-1}}{1 - c\vec{e}_i^\top\mathbf{C}^{-1}\vec{e}_j}\vec{u}. \qquad [1.15]$$

The quantity $\vec{\pi}^{(1)}$ is the old PageRank vector before addition of an edge. Since $\vec{\pi}^{(1)} = (1-c)\left(\mathbf{I} - c\mathbf{A}^{(1)^\top}\right)^{-1}\vec{u}$, relation [1.15] becomes

$$\vec{\pi}^{(2)} = \vec{\pi}^{(1)} + \frac{c\mathbf{C}^{-1}\vec{e}_j\vec{e}_i^\top\vec{\pi}^{(1)}}{1 - c\vec{e}_i^\top\mathbf{C}^{-1}\vec{e}_j}. \qquad [1.16]$$

By writing $\left(\mathbf{I} - c\mathbf{A}^{(1)^\top}\right)^{-1}$ as power series, the denominator of the second term of [1.15] can be simplified as follows:

$$\begin{aligned}
1 - c\vec{e}_i^\top\mathbf{C}^{-1}\vec{e}_j &= 1 - c\vec{e}_i^\top\left(I - c\mathbf{A}^{(1)^\top}\right)^{-1}\vec{e}, \\
&= 1 - c\sum_{k=0}^{\infty}\vec{e}_i^\top\left(c\mathbf{A}^{(1)^\top}\right)^k\vec{e}_j, \\
&= 1 - c\underbrace{\left(\vec{e}_i^\top\vec{e}_j + c\vec{e}_i^\top\mathbf{A}^{(1)^\top}\vec{e}_j + \cdots\right)}_{=0}, \\
&= 1.
\end{aligned}$$

Thus, relation [1.16] simplifies to

$$\begin{aligned}
\vec{\pi}^{(2)} &= \vec{\pi}^{(1)} + c\mathbf{C}^{-1}\vec{e}_j\vec{e}_i^\top\vec{\pi}^{(1)}, \\
&= \vec{\pi}^{(1)} + c\sum_{k=0}^{\infty}\left(c\mathbf{A}^{(1)^\top}\right)^k\vec{e}_j\vec{e}_i^\top\vec{\pi}^{(1)}, \\
&= \vec{\pi}^{(1)} + c\left(\mathbf{I} + c\mathbf{A}^{(1)^\top} + c^2(\mathbf{A}^{(1)^\top})^2 + \cdots\right)\vec{e}_j\vec{e}_i^\top\vec{\pi}^{(1)}, \\
&= \vec{\pi}^{(1)} + c\left(\vec{e}_j\vec{e}_i^\top\vec{\pi}^{(1)} + c\mathbf{A}^{(1)^\top}\vec{e}_j\vec{e}_i^\top\vec{\pi}^{(1)} + \cdots\right).
\end{aligned} \qquad [1.17]$$

Simplifying further, we get $\vec{e}_j\vec{e}_i^\top\vec{\pi}^{(1)} = \begin{pmatrix}\vec{0}\\ \pi_i\end{pmatrix}$ and $\mathbf{A}^{(1)^\top}\vec{e}_j = \vec{0}$ before an edge $v_i \to v_j$ is added. Therefore, the PageRank of vertex v_j can be updated as $\pi_j^{(2)} = \pi_j^{(1)} + c\pi_i$. □

Similarly, if an outgoing edge from $v_i \to v_j$ is deleted, then

$$\pi_j^{(2)} = \pi_j^{(1)} - c\pi_i. \qquad [1.18]$$

Lemma 1.4 shows us that to update $\vec{\pi}^{(2)}$, we need to access the incoming edges of v_j. This seems to be effective for updates but may have a drawback if we have a batch of edges changing in a graph. In fact, it may be difficult to keep track of source and target vertices. We consider this case in section 1.4.1.

LEMMA 1.5.– *Let $\vec{\pi}^{(1)}$ be the PageRank of the transition matrix $\mathbf{A}^{(1)}$, and let $deg^{(1)}(v_i)$ and $deg^{(2)}(v_i)$ be the degrees of v_i before and after some edges are added to it, respectively. Suppose k new edges from v_i to $v_j = \{v_{j_1}, \cdots, v_{j_k}\}$ are added without a cycle being formed, the new PageRank of the target vertex $\pi_j^{(2)}$ can be expressed as*

$$\pi_j^{(2)} = \pi_j^{(1)} + \frac{p(i,j)}{deg^{(2)}(v_i)}\pi_i^{(1)}, \qquad [1.19]$$

where $p(i,j) = c$ is the probability to follow the edge from vertex v_i to v_j,

If, before adding the k edges, vertex v_i was linked to v_b, then the new PageRank of v_b is expressed as

$$\pi_b^{(2)} = \pi_b^{(1)} - \frac{kp(i,j)}{deg^{(1)}(v_i)deg^{(2)}(v_i)}\pi_i^{(1)}. \qquad [1.20]$$

PROOF.– This result is a consequence of a theorem (Abola *et al.* 2020); hence, we will not give the proof. □

1.4.1. *Updating the PageRank of tree graph when a batch of edges changes*

Here, we do not assume any knowledge of the vertex that might be affected by the evolution of a graph. The main emphasis is on how to update the PageRank of affected vertices effectively while exploiting local changes in the network. The update of the PageRank of a graph $G \cup \Delta G$ is based on the following result:

LEMMA 1.6.– *Let $\mathbf{A}^{(1)}$ and $\mathbf{A}^{(2)}$ be the old and new transition matrices of graphs G and $G \cup \Delta G$, respectively. Suppose at some $(k-1)^{th}$ change, the PageRank* *and residual vector of* $\mathbf{A}^{(1)}$ *are* $\vec{\pi}_{k-1}^{(1)}$ *and* $\vec{r}_{k-1}^{(1)}$*, respectively. Then, the PageRank of transition matrix* $\mathbf{A}^{(2)}$ *is updated as*

$$\vec{r}_k = \vec{r}_{k-1}^{(1)} + c\delta\mathbf{A}^{(1)^\top}\vec{\pi}_{k-1}^{(1)}, \qquad [1.21]$$

$$\vec{\pi}_k^{(2)} = \vec{\pi}_{k-1}^{(1)} + \delta\vec{\pi}_{k-1}^{(1)}, \qquad [1.22]$$

where $\delta\mathbf{A}^{(1)^\top} = \mathbf{A}^{(2)^\top} - \mathbf{A}^{(1)^\top}$ *and* $\delta\vec{\pi}_{k-1}^{(1)} = \left(\mathbf{I} - c\mathbf{A}^{(2)^\top}\right)^{-1}\vec{r}_k$.

PROOF.– Let $\mathbf{A}^{(2)} = \mathbf{A}^{(1)} + \delta\mathbf{A}^{(1)^\top}$ also allow us to use information available at moment $(k-1)^{\text{th}}$ to compute the PageRank at the kth evolution. Hence, the residual at $(k-1)^{\text{th}}$ approximates $\vec{r}_k$ as

$$\begin{aligned}\vec{r}_k &= (1-c)\vec{u} - \left(\mathbf{I} - c(\mathbf{A}^{(1)^\top} + \delta\mathbf{A}^{(1)^\top})\right)\vec{\pi}_{k-1}^{(1)}, &[1.23]\\ &= (1-c)\vec{u} - \vec{\pi}_{k-1}^{(1)} + c\mathbf{A}^{(1)^\top}\vec{\pi}_{k-1}^{(1)} + c\delta\mathbf{A}^{(1)^\top}\vec{\pi}_{k-1}^{(1)}, \\ &= (1-c)\vec{u} - \left(\mathbf{I} - c\mathbf{A}^{(1)^\top}\right)\vec{\pi}_{k-1}^{(1)} + c\delta\mathbf{A}^{(1)^\top}\vec{\pi}_{k-1}^{(1)}, \\ &= \vec{r}_{k-1}^{(1)} + c\delta\mathbf{A}^{(1)^\top}\vec{\pi}_{k-1}^{(1)}. &[1.24]\end{aligned}$$

Let $\vec{b} = (1-c)\vec{u}$, then [1.23] becomes $\vec{r}_k = \vec{b} - \left(\mathbf{I} - c\mathbf{A}^{(2)^\top}\right)\vec{\pi}_{k-1}^{(1)}$. Multiplying both sides by $\left(\mathbf{I} - c\mathbf{A}^{(2)^\top}\right)^{-1}$, we get

$$\left(\mathbf{I} - c\mathbf{A}^{(2)^\top}\right)^{-1}\vec{r}_k = \left(\mathbf{I} - c\mathbf{A}^{(2)^\top}\right)^{-1}\vec{b} - \vec{\pi}_{k-1}^{(1)},$$

Thus, $\vec{\pi}_k^{(2)} = \vec{\pi}_{k-1}^{(1)} + \delta\vec{\pi}_{k-1}^{(1)}$. □

From Lemma 1.6, we estimate the new PageRank indirectly, i.e. we need to solve for $\delta\vec{\pi}_{k-1}^{(1)}$ instead of $\vec{\pi}_k^{(2)}$. In the case where $\mathbf{A}^{(2)^\top}$ is dense, an iterative technique is recommended. It is known from numerical methods that if $\delta\vec{\pi}_{k-1}^{(1)}$ can be obtained with higher accuracy, then $\vec{\pi}_k^{(2)}$ will have minimum error (Golub and Van Loan 2013; Erin and Higham 2017). Next, we check the convergence of the linear system $\left(\mathbf{I} - c\mathbf{A}^{(2)^\top}\right)\delta\vec{\pi}_{k-1}^{(1)} = \vec{r}_k$.

LEMMA 1.7.– *If* $\vec{\pi}_k^{(2)}$ *is the PageRank of* $\mathbf{A}^{(2)^\top}$ *corresponding to graph* $G \cup \Delta G$*, then* $\delta\vec{\pi}_{k-1}^{(1)}$ *converges.*

PROOF.– The following facts are known for the matrix $\left(\mathbf{I} - c\mathbf{A}^{(2)^\top}\right)$. It is M-matrix, non-singular and the column sums are less than or equal to $1-c$ (Andersson and

Silvestrov 2008; Langville and Meyer 2012). Using norm-1, i.e. $\|\vec{z}\|_1 = \sum_{i=1}^{n} z_i$, we get

$$\|\delta\vec{\pi}_{k-1}^{(1)}\|_1 = \| \left(\mathbf{I} - c\mathbf{A}^{(2)^\top}\right)^{-1} \vec{r}_k\|_1, \qquad [1.25]$$

$$\leq \| \left(\mathbf{I} - c\mathbf{A}^{(2)^\top}\right)^{-1} \|_\infty \|\vec{r}_k\|_1,$$

$$\leq \frac{1}{1 - \|c\mathbf{A}^{(2)^\top}\|_1},$$

$$= \frac{1}{1-c}. \qquad [1.26]$$

Hence, it is bounded and, by continuity of the normed space, it converges. □

Let us introduce some notations that will be useful in the subsequent section. The subscripts $\bar{u}$ and $\bar{u}'$ represent the sets of unaffected and affected vertices of the graph $\mathrm{G} \cup \Delta\mathrm{G}$, respectively.

The following lemma is immediate from Lemma 1.3.

LEMMA 1.8.– *Suppose a triangular matrix $\mathbf{A}^{(2)^\top}$ is partitioned into two blocks of unchanged and changed vertices. If Lemma 1.6 holds, such that $\left(I - c\mathbf{A}^{(2)^\top}\right)\delta\vec{\pi}_{k-1}^{(1)} = \vec{r}_k$ has a unique solution, then*

$$\begin{bmatrix} \delta\vec{\pi}_{\bar{u},k-1}^{(1)} \\ \delta\vec{\pi}_{\bar{u}',k-1}^{(1)} \end{bmatrix} = \begin{bmatrix} \mathbf{L}_{11}^{-1}\vec{r}_{\bar{u},1} \\ -\mathbf{L}_{22}^{-1}\mathbf{L}_{21}\mathbf{L}_{11}^{-1}\vec{r}_{\bar{u},1} + \mathbf{L}_{22}^{-1}\vec{r}_{\bar{u}',2} \end{bmatrix} \approx \begin{bmatrix} \mathbf{0} \\ \mathbf{L}_{22}^{-1}\vec{r}_{\bar{u}',2} \end{bmatrix} \qquad [1.27]$$

and the PageRank vector of $A^{(2)^\top}$ is updated as

$$\begin{bmatrix} \vec{\pi}_{\bar{u},k-1}^{(2)} \\ \vec{\pi}_{\bar{u}',k-1}^{(2)} \end{bmatrix} = \begin{bmatrix} \vec{\pi}_{\bar{u},k-1}^{(1)} \\ \vec{\pi}_{\bar{u}',k-1}^{(1)} \end{bmatrix} + \begin{bmatrix} \mathbf{0} \\ \mathbf{L}_{22}^{-1}\vec{r}_{\bar{u}',2} \end{bmatrix}, \qquad [1.28]$$

where $\mathbf{L}_{22}^{-1}$ consists of target vertices and the corresponding neighbors (descendants).

PROOF.– The proof is based on Lemma 1.6; hence, it is straightforward. □

Following relation [1.28], our intention is to avoid finding the inverse of $\mathbf{L}_{22}$. This can be achieved by considering $\vec{r}_{\bar{u}',2}$ as a residual weight vector. For consistency, let us denote $\vec{r}_{\bar{u}',2}$ by $\hat{\vec{w}}$ such that $\hat{w}_i$ is the residual weight of v_i and $\hat{\vec{w}} = \delta\mathbf{A}^{(1)^\top}\vec{\pi}_{\bar{u}'}^{(1)}$. Furthermore, suppose $\delta\hat{\pi}_i^{(1)}$ is the residual PageRank of v_i after some changes, then

$$\delta\hat{\pi}_i^{(1)} = \hat{w}_i + c \sum_{v_j \in V_v | v_j \to v_i} \frac{\delta\hat{\pi}_j^{(1)}}{deg^{(2)}(v_j)} \qquad [1.29]$$

and $\delta\hat{\pi}_i^{(1)} = \hat{w}_i$ if v_i is a root. Importantly, we need to carry out the matrix–vector product once to determine $\hat{\vec{w}}$, and since $\delta\mathbf{A}^{(1)^\top}$ is very sparse, the multiplication is quite fast. We demonstrate how this procedure can be realized using an example.

To this end, the PageRank of any vertex affected by the changes is expressed as

$$\pi_i^{(2)} = \pi_i^{(1)} + \delta\hat{\pi}_i^{(1)}. \qquad [1.30]$$

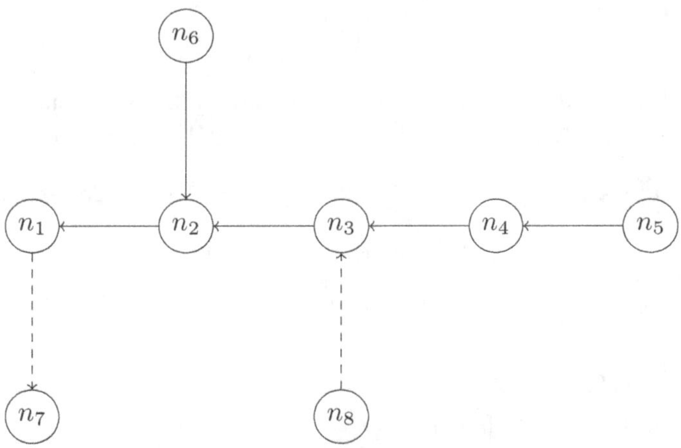

Figure 1.3. *A tree graph with two edges added simultaneously, as indicated by dashed lines*

1.4.2. *An example of updating the PageRank of a tree*

Let $\vec{\pi}^{(1)}$ be the PageRank before addition of any edge to graph G and $V = \{n_1, n_2, \ldots, n_8\}$ be a set of vertices. Suppose the vertices n_7 and n_8 are connected simultaneously, as shown in Figure 1.3, then the PageRank of $\mathbf{A}^{(2)^\top}$ is computed as follows. It is essential to note that updating matrix $\mathbf{A}^{(2)^\top}$ using Lemma 3.1 is challenging; hence, it is worth applying relation [1.28]. First, we re-order vertices of $\mathbf{A}^{(2)^\top}$ using a depth first search (DFS) starting at vertex n_5; obviously, any vertex can work. Then, we write $\mathbf{A}^{(2)^\top}$, $\mathbf{A}^{(1)^\top}$ and $\delta\mathbf{A}^{(1)^\top}$ as

$$\mathbf{A}^{(2)^\top} = \begin{array}{c} \begin{matrix} n_5 & n_6 & n_4 & n_8 & n_3 & n_2 & n_1 & n_7 \end{matrix} \\ \begin{pmatrix} 0 & 0 & 0 & 0 & 0 & 0 & 0 & 0 \\ 0 & 0 & 0 & 0 & 0 & 0 & 0 & 0 \\ 1 & 0 & 0 & 0 & 0 & 0 & 0 & 0 \\ 0 & 0 & 0 & 0 & 0 & 0 & 0 & 0 \\ 0 & 0 & 1 & 1 & 0 & 0 & 0 & 0 \\ 0 & 1 & 0 & 0 & 1 & 0 & 0 & 0 \\ 0 & 0 & 0 & 0 & 0 & 1 & 0 & 0 \\ 0 & 0 & 0 & 0 & 0 & 0 & 1 & 0 \end{pmatrix} \end{array},$$

$$\mathbf{A}^{(1)^\top} = \begin{array}{c} \begin{array}{cccccccc} n_5 & n_6 & n_4 & n_8 & n_3 & n_2 & n_1 & n_7 \end{array} \\ \begin{pmatrix} 0 & 0 & 0 & 0 & 0 & 0 & 0 & 0 \\ 0 & 0 & 0 & 0 & 0 & 0 & 0 & 0 \\ 1 & 0 & 0 & 0 & 0 & 0 & 0 & 0 \\ 0 & 0 & 0 & 0 & 0 & 0 & 0 & 0 \\ 0 & 0 & 1 & 0 & 0 & 0 & 0 & 0 \\ 0 & 1 & 0 & 0 & 1 & 0 & 0 & 0 \\ 0 & 0 & 0 & 0 & 0 & 1 & 0 & 0 \\ 0 & 0 & 0 & 0 & 0 & 0 & 0 & 0 \end{pmatrix} \end{array}; \ \delta\mathbf{A}^{(1)^\top} = \vec{e}_{n_3}\vec{e}_{n_8}^{\top} + \vec{e}_{n_7}\vec{e}_{n_1}^{\top}.$$

Set $k = 1$, and following Engström and Silvestrov (2016), Biganda *et al.* (2017) and Abola *et al.* (2018), the PageRank of vertices of $A^{(1)^\top}$ is expressed as

$$\vec{\pi}_0^{(1)} = \left(1,\, 1,\, 1+c,\, 1,\, 1+c+c^2,\, 1+2c+c^2+c^3,\, 1+c+2c^2+c^3+c^4,\, 1\right)^\top.$$

Since $\vec{\pi}_0^{(1)}$ is the exact PageRank of the transition matrix $\mathbf{A}^{(1)^\top}$, $\vec{r}_0 = \vec{0}$. Therefore,

$$\begin{aligned} \vec{r}_1 &= c\delta\mathbf{A}^{(1)^\top}\vec{\pi}_0^{(1)}, \\ &= c(\vec{e}_{n_8}\vec{e}_{n_3}^{\top} + \vec{e}_{n_1}\vec{e}_{n_7}^{\top})^\top \vec{\pi}_0^{(1)}, \\ &= \left(0,\, 0,\, 0,\, 0,\, c,\, 0,\, 0,\, c(1+c+2c^2+c^3+c^4)\right)^\top \end{aligned}$$

and $\delta\vec{\pi}_0^{(1)} = \left(\mathbf{I} - c\mathbf{A}^{(2)^\top}\right)^{-1} \vec{r}^{(1)} = \left(0,\, 0,\, 0,\, 0,\, c,\, c^2,\, c^3,\, \underline{c}\right)^\top$, where $\underline{c} = c + c^2 + 2c^3 + 2c^4 + c^5$. Hence,

$$\begin{aligned} \vec{\pi}^{(2)} &= \vec{\pi}_0^{(1)} + \delta\vec{\pi}_0^{(1)}. \\ &= \left(1, 1, 1+c, \bar{c}, 1+2c+2c^2+c^3, 1+c+2c^2+2c^3+c^4, 1+\underline{c}\right)^\top, \end{aligned} \quad [1.31]$$

where $\bar{c} = 1 + c + c^2$.

Following relation [1.29], let us find $\delta\hat{\pi}_i^{(1)}$, for $n_i \in \{n_3, n_2, n_1, n_7\}$ as in Figure 1.3, and recall that the weight vector $\hat{\vec{w}} = c\delta\mathbf{A}^{(1)^\top}\vec{\pi}_{\bar{u}'}$. We get $\hat{\vec{w}} = [c, 0, 0, c + c^2 + 2c^3 + c^4 + c^5]^\top$. Using relation [1.29], the quantity $\delta\hat{\pi}_{n_8}^{(1)} = 0$, since it is not pointed at by any new incoming edge. The quantity $\delta\hat{\pi}_{n_3}^{(1)}$ of vertex v_3 is expressed as

$$\begin{aligned} \delta\hat{\pi}_{n_3}^{(1)} &= \hat{w}_{n_3} + c \sum_{v_j \in V_v | v_j \to n_3} \frac{\delta\hat{\pi}_j^{(1)}}{deg^{(2)}(v_j)}, \\ &= c + c.\frac{\delta\hat{\pi}_{n_8}^{(1)}}{1}, \\ &= c. \end{aligned} \quad [1.32]$$

For vertex n_2, we use relation [1.29], i.e.

$$\begin{aligned}\delta\hat{\pi}_{n_2}^{(1)} &= \hat{w}_{n_2} + c \sum_{v_j \in V_v | v_j \to n_2} \frac{\delta\hat{\pi}_j^{(1)}}{deg^{(2)}(v_j)},\\ &= 0 + c.\frac{\delta\hat{\pi}_{n_3}^{(1)}}{1},\\ &= c^2.\end{aligned} \quad [1.33]$$

Similarity, the residual PageRanks of n_1 and n_7 are $\delta\hat{\pi}_{n_1}^{(1)} = c^3$ and $\delta\hat{\pi}_{n_7}^{(1)} = c + c^2 + 2c^3 + 2c^4 + c^5$, respectively. In sequel, the new PageRank of a set of vertices $\{n_3, n_2, n_1, n_7\}$ is updated using relation [1.30].

Therefore, updating the PageRank of the graph associated with matrix $\mathbf{A}^{(2)\top}$ using relation [1.15], we obtain the result as in [1.31].

1.5. Maintaining the levels of vertices in a changing tree graph

We describe how to keep track of levels of vertices in a changing tree graph. We focus on changes associated with the addition and removal of edges. It is assumed that this can be extended to the deletion or addition of vertices, although it might require slight modification. For instance, adding a vertex with an edge to G is the same as first adding a vertex, then an edge later. Therefore, it involves creating a row and column with entries equal to zero, and an edge is added, as previously mentioned in Lemma 1.1. We will not go into the details of that for the time being.

For our case, one essential process $\mathcal{P}1$ is to check whether the addition of an edge between v_i and v_j creates a cycle. This is done by traversing a path from the target to a source using a depth first search. We stop scanning and abort the addition of an edge if a new component will be formed. This is important because we are interested only in a directed acyclic graph. In fact, $\mathcal{P}1$ requires linear time.

The proposed Algorithm 1.1 is described as follows. Start with an arbitrary number of vertices, then allow either the addition or removal of an edge. The algorithm sequentially updates adjacency matrix $\mathbf{A}$, as well as the levels of the vertices $V \in G$ denoted as $\ell_{\mathbf{A}}$. Suppose (v_i, v_j) is added to G, we check if a cycle is formed. If it is true, the process stops (line 3). However, if it is false and the degree of v_i is zero, $\mathbf{A}^{(2)}$ is computed as in line 4. If $deg(v_i) > 0$, we determine $\mathbf{A}^{(2)}$ as in line 7. Without loss of generality, when an edge (v_i, v_j) is to be removed, we need to check for existence of the degree of vertex $deg(v_i)$, i.e. if $deg(v_i) = 1$. If the test is true, then $\mathbf{A}^{(2)}$ is computed as in line 9. This also verifies whether removal of an edge creates two tree graphs. However, it is fortunate that if it is true that two graphs are created, then the two graphs can be handled in parallel.

Algorithm 1.1 Adjacency matrix and vertex level updates

INPUT: The old adjacency matrix $\mathbf{A}$, out-degree $deg(v_i)$, levels of vertices $\ell_{\mathbf{A}}$ and the edge (v_i, v_j)

if (v_i, v_j) does not create cycle & $deg(v_i) = 0$. % Adding an edge **then**
  $A^{(2)} \leftarrow \mathbf{A} + \vec{e}_i \vec{e}_j^\top$
else
  $\vec{u} \leftarrow \frac{\vec{e}_i}{deg(v_i)+1}, \vec{v}^\top \leftarrow (\vec{e}_j^\top - \mathbf{A}_i)$
  $A^{(2)} \leftarrow \mathbf{A} + \vec{u}\vec{v}^\top$
  if (v_i, v_j) is removed for $deg(v_i) = 1$. **then**
    $\mathbf{A}^{(2)} \leftarrow \mathbf{A} - \vec{e}_i \vec{e}_j^\top$
  else
    $\vec{u} \leftarrow \frac{\vec{e}_i}{deg(v_i)-1}, \vec{v}^\top \leftarrow (\vec{e}_j^\top - \mathbf{A}_i)$
    $\mathbf{A}^{(2)} \leftarrow \mathbf{A} + \vec{u}\vec{v}^\top$
  end if
end if
OUTPUT: $\mathbf{A}^2$

Finally, let us look at maintaining the levels of vertices after a change. The following are essential to note. First, the lowest levels are for leaves (negative labeling of a level is allowed). For instance, let us assume that the vertices of $G = (V, E)$ are partitioned in such a way that

$$V = L(v_{-h}) \cup L(v_{-h+1}) \cup \cdots \cup L(v_k),$$

for $h, k > 0$ and $L(v_t) \cap L(v_s) = \emptyset$ for all $t, s \in I$ (set of indexed vertices). Suppose the vertices are partitioned in such a way that the top level $L(v_k)$ consists of vertices without incoming edges and $L(v_{-h})$ represents nodes (leaves) without outgoing edges. Recall that the vertices on $L(v_k)$ point to vertices on $L(v_{k-1})$ or lower.

Second, maintain partial order, but not the unique partial order with a minimum number of levels (see Figure 1.4).

Third, only updates for added edges are outlined. The update of the levels consists of one main function, denoted by *newLevel* (Algorithm 1.2), and two sub-functions *DfsLevel* and *checkForCycle*; see Algorithms 1.3 and 1.4, respectively. To avoid repetition, we will not describe the pseudo-code for the sub-functions.

We start by checking whether the level of v_i differs from v_j. Assuming that they are different and $L(v_i) > L(v_j)$, no update of the levels is required. However, the effect of the change will obviously propagate downward and affect PageRank values below v_i. If, before addition of an edge, the two vertices were at the same level, i.e. $L(v_i) = L(v_j)$, then $L(v_j)$ drops by a unit and the recursive process of updating levels of the children continue (lines 3–8). We need also to ensure that a cycle is not formed (line 9). $L(v_j)$ is updated only if *checkForCycle*$(v_i, v_j) = 0$, otherwise a warning pop-up (lines 10–14). The levels of the vertices unaffected by the change remain the same. These include $L(v_{i+})$, or v_i is not its parent. The process terminates when new levels are updated.

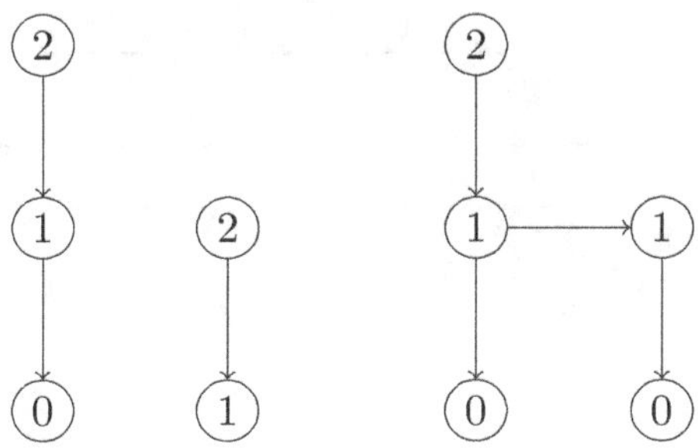

Figure 1.4. *Maintaining partial order before (first two on the left) and after (right) addition of an edge. The numbers in the circle indicate the levels of sub-graphs*

Algorithm 1.2 Maintaining levels of vertices

INPUT: The new adjacency matrix $A^{(2)}$ and an edge (v_i, v_j)

3: **if** $L(v_i) > L(v_j)$ **then**
    % stay put
else
6:   **if** $L(v_i) = L(v_j)$ **then**
      $L(v_j) = L(v_i) - 1$
      DfsLevel(v_j)
9:   **else**
      if *checkForCycle*$(v_i, v_j) = 0$ **then**
        $L(v_j) = L(v_i) - 1$
12:       *DfsLevel*(v_j)
      else
        % warning a cycle is created
15:     **end if**
    end if
end if
18: OUTPUT: (V, L)

Algorithm 1.3 *DfsLevel*(v_j)

INPUT: Starting vertex (v_i)
for each neighbor v_j
if $L(v_i) \leq L(v_j)$ **then**
  $L(v_j) = L(v_i) - 1$
  DfsLevel(v_j)
end if

Algorithm 1.4 *checkForCycle*(v_i, v_j)

INPUT: Insert an edge (v_i, v_j)
for each neighbor v_k of v_j
if $v_k = v_j$ **then**
  return 1
else
  if $L(v_i) \leq L(v_j)$ **then**
    % stay put
  else
    $r = \textit{checkForCycle}(v_i, v_j)$
    if $r = 1$ **then**
      return 1
    end if
  end if
  return 0
end if

To update the PageRank after doing some changes (say, the addition of an edge) on a tree graph when the vertices are partitioned into levels, the following adjustments are necessary. Adjust the old adjacency matrix to $\mathbf{A}^{(2)\top}$ and the corresponding levels. Reorder $\mathbf{A}^{(2)\top}$ starting with the top levels, and also maintain both the old PageRank and the current level. Then, finally perform a single matrix–vector product $c\delta\mathbf{A}^{(1)^\top}\vec{\pi}_{\bar{u}'}$ to obtain what we term its weight vector $\hat{\vec{w}}$. It can readily be seen that a change in the higher level propagates downward but not the reverse because we try to avoid the back edge. Thus, starting from the top levels, for each level, compute the PageRank depending on where the level of the source vertex is located as follows. The PageRank of vertices at level $L(v_i)$ takes the form

$$\vec{\pi}^{(2)}_{L(v_{i-1})} = \vec{\pi}^{(1)}_{L(v_{i-1})} + c\mathbf{A}^{(2)\top}_{L(v_i+),L(v_{i-1})}\vec{\pi}^{(1)}_{L(v_i+)}, \qquad [1.34]$$

where $\mathbf{A}^{(2)\top}_{L(v_i+),L(v_{i-1})}$ is the sub-matrix of scaled adjacency matrix $\mathbf{A}^{(2)\top}$ with rows at level $L(v_{i-1})$ and all columns of level $L(v_{i-1})$ or higher.

1.6. Conclusion

In this chapter, we have focused on updating the scaled adjacency matrix, maintaining levels and calculating the PageRank of a tree graph after some changes. Importantly, we observed that if the change is either the addition or removal of edges, then updating the PageRank of a graph depends on the location of the source vertices. This is equivalent to performing a random walk from the source (parent) vertices to all its children. In addition, the approach seems to improve the accuracy as well as speed up computation of the PageRank values as compared to the ordinary matrix–vector products.

1.7. Acknowledgments

This research was supported by the Swedish International Development Cooperation Agency (Sida), the International Science Programme (ISP) (namely the International Programme in the Mathematical Sciences (IPMS)) and Sida Bilateral Research Programmes for research and education capacity development in Mathematics in Uganda and Tanzania. The authors are also grateful to the research environment Mathematics and Applied Mathematics (MAM), Division of Applied Mathematics, Mälardalen University, for providing an excellent and inspiring environment for research in Mathematics.

1.8. References

Abola, B., Biganda, P.S., Engström, C., Mango, J.M., Kakuba, G., Silvestrov, S. (2018). PageRank in evolving tree graphs. In *Stochastic Processes and Applications*, Silvestrov, S., Rančić, M., Malyarenko, A. (eds). Springer, Cham.

Abola, B., Biganda, P.S., Engström, C., Mango, J.M., Kakuba, G., Silvestrov, S. (2020). Updating of PageRank in evolving tree graphs. In *Data Analysis and Application 3: Computational, Classification, Financial, Statistical and Stochastic Methods*, Makrides, A., Karagrigoriou, A., Skiadas, C.H. (eds). ISTE Ltd, London, and Wiley, New York.

Allesina, S. and Bodini, A. (2004). Who dominates whom in the ecosystem? Energy flow bottlenecks and cascading extinctions. *J. Theor. Biol.*, 230(3), 351–358.

Allesina, S. and Pascual, M. (2009). Googling food webs: Can an eigenvector measure species' importance for coextinctions? *PLoS Comput. Biol.*, 5(9), e1000494.

Andersson, F.K. and Silvestrov, S.D. (2008). The mathematics of internet search engines. *Acta Appl. Math.*, 104(2), 211–242.

Bahmani, B., Kumar, R., Mahdian, M., Upfal, E. (2012). PageRank on an evolving graph. *Proceedings of the 18th ACM SIGKDD International Conference on Knowledge Discovery and Data Mining*. Beijing.

Biganda, P.S., Abola, B., Engström, C., Silvestrov, S. (2017). PageRank, connecting a line of nodes with multiple complete graphs. In *Proceedings of the 17th Applied Stochastic Models and Data Analysis International Conference with the 6th Demographics Workshop*. June 6–9, London.

Brin, S. and Page, L. (1998). The anatomy of a large-scale hypertextual Web search engine. *Comput. Netw. ISDN Syst.*, 30(1–7), 107–117.

Engström, C. (2016). PageRank in evolving networks and applications of graphs in natural language processing and biology. Doctoral Dissertation. Mälardalen University, Västerås.

Engström, C. and Silvestrov, S. (2016). PageRank, a look at small changes in a line of nodes and the complete graph. In *Engineering Mathematics II. Algebraic, Stochastic and Analysis Structures for Networks, Data Classification and Optimization*, Silvestrov, S. and Rančić, M. (eds). Springer, Cham.

Engström, C. and Silvestrov, S. (2017). PageRank for networks, graphs, and Markov chains. *Theor. Probab. Math. Statist.*, 96, 59–82 (2018).

Erin, C. and Higham. N.J. (2017). A new analysis of iterative refinement and its application to accurate solution of ill-conditioned sparse linear systems. *SIAM J. Sci. Comput.*, 396, A2834–A2856.

Gleich, D.G. (2015). PageRank beyond the Web. *SIAM Rev.*, 57(3), 321–363.

Golub, G.H. and Van Loan, F.C. (2013). *Matrix Computations*, 4th edition. Johns Hopkins University Press, Baltimore.

Han, S.W., Chen, G., Cheon, M.S., Zhong, H. (2016). Estimation of directed acyclic graphs through two-stage adaptive lasso for gene network inference. *J. Am. Stat. Assoc.*, 111(515), 1004–1019.

Lancaster, P. and Tismenetsky, M. (1985). *The Theory of Matrices*, 2nd edition. Academic Press, Orlando.

Langville, A.N. and Meyer, C.D. (2006). Updating Markov chains with an eye on Google's PageRank. *SIAM J. Matrix Anal. Appl.*, 27(4), 968–987.

Langville, A.N. and Meyer, C.D. (2012). *Google's PageRank and Beyond: The Science of Search Engine Rankings*. Princeton University Press, Princeton.

Ohsaka, N., Maehara, T., Kawarabayashi, K.I. (2015). Efficient PageRank tracking in evolving networks. *Proceedings of the 21st ACM SIGKDD International Conference on Knowledge Discovery and Data Mining*, 875–884.

Rocchi, M., Scotti, M., Micheli, F., Bodini, A. (2017). Key species and impact of fishery through food web analysis: A case study from Baja California Sur, Mexico. *J. Mar. Syst.*, 165, 92–102.

Shandilya, V., Simmons, C.B., Shiva, S. (2014). Use of attack graphs in security systems. *J. Comput. Networks Comm.*, vol. 2014.

Ulanowicz, R.E. (2004). Quantitative methods for ecological network analysis. *Comput. Biol. Chem.*, 28(5–6), 321–339.

Wang, D., Wang, J., Lu, M., Song, F., Cui, Q. (2010). Inferring the human microRNA functional similarity and functional network based on microRNA-associated diseases. *Bioinformatics*, 26(13), 1644–1650.

Yonina, E.C. and Needell, D. (2011). Acceleration of randomized Kaczmarz method via the Johnson-Lindenstrauss lemma. *Numerical Algorithms*, 58(2), 163–177.

2

Nonlinearly Perturbed Markov Chains and Information Networks

This chapter is devoted to studies of perturbed Markov chains, commonly used for the description of information networks. In such models, the matrix of transition probabilities for the corresponding Markov chain is usually regularized by adding a special damping matrix, multiplied by a small damping (perturbation) parameter ε. In this chapter, we present the results of detailed perturbation analysis of Markov chains with damping component and numerical experiments supporting and illustrating the results of this perturbation analysis.

2.1. Introduction

Perturbed Markov chains is a popular and important aspect in the study of the theory of Markov processes and their applications to stochastic networks, queuing and reliability models, bio-stochastic systems and many other stochastic models.

We refer here to some recent books and papers devoted to perturbation problems for Markov-type processes: Stewart (1994, 1998, 2001), Hartfiel and Meyer (1998), Korolyuk and Korolyuk (1999), Englund (2001); Konstantinov *et al.* (2003), Korolyuk and Limnios (2005), Mitrophanov (2005), Bini *et al.* (2005), Yin and Zhang (2005), Gambini *et al.* (2008), Gyllenberg and Silvestrov (2008), Ni *et al.* (2008), Ni (2011), Avrachenkov *et al.* (2013, 2018), Silvestrov and Petersson (2014), Petersson (2016), Silvestrov and Silvestrov (2016, 2017a, b, c, d), Silvestrov *et al.* (2018) and Yin and Zhang (2013). In particular, we would like to mention works by Gyllenberg and Silvestrov (2008), Avrachenkov *et al.* (2013), Silvestrov and Silvestrov (2016) and

Chapter written by Benard ABOLA, Pitos Seleka BIGANDA, Sergei SILVESTROV, Dmitrii SILVESTROV, Christopher ENGSTRÖM, John Magero MANGO and Godwin KAKUBA.

Silvestrov and Silvestrov (2017a), where the extended bibliographies of works in the area and the corresponding methodological and historical remarks can be found.

We are especially interested in models of Markov chains commonly used for the description of information networks. In such models, an information network is represented by the Markov chain associated with the corresponding node link graph. Stationary distributions and other related characteristics of information Markov chains usually serve as basic tools for the ranking of nodes in information networks. The ranking problem may be complicated by singularity of the corresponding information Markov chain, where its phase space is split into several weakly or completely non-communicating groups of states. In such models, the matrix of transition probabilities $\mathbf{P}_0$ of the information Markov chain is usually regularized and approximated by matrix $\mathbf{P}_\varepsilon = (1 - \varepsilon)\mathbf{P}_0 + \varepsilon\mathbf{D}$, where $\mathbf{D}$ is a so-called damping stochastic matrix with identical rows and all positive elements, while $\varepsilon \in [0, 1]$ is a damping (perturbation) parameter.

The power method is often used to approximate the corresponding stationary distribution $\bar{\pi}_\varepsilon$ by rows of matrix $\mathbf{P}^n_\varepsilon$. The damping parameter ε should be chosen neither too small nor too large. In the first case, where ε takes too small a value, the damping effect will not work against absorbing and pseudo-absorbing effects, since the second eigenvalue for such matrices (determining the rate of convergence in the above-mentioned ergodic approximation) take values approaching 1. In the second case, the ranking information (accumulated by matrix $\mathbf{P}_0$ via the corresponding stationary distribution) may be partly lost, due to the deviation of matrix $\mathbf{P}_\varepsilon$ from matrix $\mathbf{P}_0$. This actualizes the problem of construction of asymptotic expansions for perturbed stationary distribution $\bar{\pi}_\varepsilon$, with respect to damping parameter ε, as well as studies of asymptotic behavior of matrices $\mathbf{P}^n_\varepsilon$ in triangular array mode, where $\varepsilon \to 0$ and $n \to \infty$, simultaneously.

Real-world systems consist of interacting units or components. These components constitute what is termed information networks. With recent advances in technology, filtering information has become a challenge in such systems. Moreover, their significance is visible as they find their applications in Internet search engines, biological, financial, transport, queuing networks and many others (Brin and Page (1998), Andersson and Silvestrov (2008), Avrachenkov *et al.* (2013), Engström (2016), Biganda *et al.* (2017), Langville and Meyer (2012), Sun and Han (2013), Gleich (2015) and Engström and Silvestrov (2017)).

PageRank is the link-based criterion that captures the importance of web pages and provides rankings of the pages in the search engine Google (Brin and Page (1998), Avrachenkov *et al.* (2013), Biganda *et al.* (2017), Engström (2016), Langville and Meyer (2012) and Engström and Silvestrov (2017)). The transition matrix (also called Google matrix $\mathbf{G}$) of a Markov chain in a PageRank problem is defined in Haveliwala and Kamvar (2003), Andersson and Silvestrov (2008) and Langville and Meyer (2012) as $\mathbf{G} = c\mathbf{P}+(1-c)\mathbf{E}$, where $\mathbf{P}$ is an $n \times n$ row-stochastic matrix (also called hyperlink matrix), $\mathbf{E}$ (the damping matrix) is the $n \times n$ rank-one stochastic matrix and $c \in (0, 1)$ is the damping parameter.

The fundamental concept of PageRank is to use the stationary distribution of the Markov chain on the network to rank web pages. However, other algorithms similar to PageRank exist in the literature, for instance, the EigenTrust algorithm (Kamvar *et al.* (2003)) and the DeptRank algorithm (Battiston *et al.* (2012)). In addition, variants of PageRank in relation to some specific networks have been studied, for example, in Biganda *et al.* (2018, 2020), and also updating PageRank due to changes in some network is in the literature, for instance, in Abola *et al.* (2018, 2020).

The parameter c is very important in the PageRank definition, because it regulates the level of uniform noise introduced to the system (Langville and Meyer 2012; Avrachenkov *et al.* 2013). If $c = 1$, there are several absorbing states for the random walk defined by $\mathbf{P}$. However, if $0 < c < 1$, the Markov chain induced by matrix $\mathbf{G}$ is ergodic (Avrachenkov *et al.* 2013). Langville and Meyer (2012) argued that parameter c controls the asymptotic rate of convergence of the power method algorithm. Similar arguments were given in Andersson and Silvestrov (2008), where it is pointed out that the choice of c is crucial. It may result in convergence and stability problems, if not carefully chosen.

The damping factor c may be denoted and interpreted differently depending on the model being studied. For instance, a model of Markov chain with restart is considered in Avrachenkov *et al.* (2018), where parameter p is the probability to restart the move and $1-p$ is the probability to follow the link according to the corresponding transition probability of the above Markov chain. Hence, we can argue that parameter p has the same interpretation as the damping factor in Google's PageRank problem.

Our representation of perturbed Markov chains is traditional for perturbed stochastic processes. In fact, PageRank is the stationary distribution of the singularly perturbed Markov chain with perturbation parameter $\varepsilon = 1 - c$. Hence, we wish to point out here that representation of the information network model by a Markov chain with matrix of transition probabilities $\mathbf{P}_\varepsilon = (1 - \varepsilon)\mathbf{P}_0 + \varepsilon\mathbf{D}$ should not create any confusion to the reader. We perform asymptotic analysis of such Markov chains, in particular, under the assumption that $\varepsilon \to 0$.

This chapter includes five sections. In section 2.2, we give a special series representation for stationary distributions of Markov chains with a damping component. In section 2.3, we describe continuity properties of stationary distributions with respect to perturbation (damping) parameter, give explicit upper bounds for rates of convergence in approximations of the stationary distributions for Markov chain with damping component and present asymptotic expansions for stationary distribution of Markov chains with damping component. We also present numerical examples, which support and illustrate theoretical results presented in this section. In section 2.4, we get explicit coupling-type upper bounds for the rate of convergence in ergodic theorems for Markov chain with damping component and present ergodic theorems for Markov chains with damping component in a triangular array mode, where time tends to infinity and perturbation parameter tends to zero, simultaneously. Also results of numerical experiments, which show how the results presented in this section can be interpreted and useful in studies of information networks, are presented.

In this chapter, we concentrate our attention on experimental results supporting the outcomes of perturbation analysis of Markov chains with damping component. We refer to the recent comprehensive preprint (Abola *et al.* 2019) for the proofs of theorems and lemmas presented in this chapter as well as some additional comments.

2.2. Stationary distributions for Markov chains with damping component

In this section, we introduce the model of Markov chains with damping component. We explain how such Markov chains can be considered as a special case of regularly or singularly perturbed Markov chains and present a special series representation for stationary distributions of such Markov chains. This representation plays the key role in the following perturbation analysis. We also give a formula for the stationary distribution of the corresponding limiting (unperturbed) Markov chain.

2.2.1. *Stationary distributions for Markov chains with damping component*

Let (a) $\mathbb{X} = \{1, 2, \dots, m\}$ be a finite space, (b) $\bar{p} = \langle p_1, \dots, p_m \rangle$, $\bar{d} = \langle d_1, \dots, d_m \rangle$, be two discrete probability distributions, (c) $\mathbf{P}_0 = \|p_{0,ij}\|$ be an $m \times m$ stochastic matrix, (d) $\mathbf{D} = \|d_{ij}\|$ be an $m \times m$ damping stochastic matrix with elements $d_{ij} = d_j > 0, i, j = 1, \dots, m$ and (e) $\mathbf{P}_\varepsilon = \|p_{\varepsilon,ij}\| = (1 - \varepsilon)\mathbf{P}_0 + \varepsilon \mathbf{D}$ is a stochastic matrix with elements $p_{\varepsilon,ij} = (1 - \varepsilon)p_{0,ij} + \varepsilon d_j, i, j = 1, \dots, m$, where $\varepsilon \in [0, 1]$.

We refer to a Markov chain $X_{\varepsilon,n}, n = 0, 1, \dots$, with the phase space $\mathbb{X}$, an initial distribution $\bar{p}$ and the matrix of transition probabilities $\mathbf{P}_\varepsilon$ as a Markov chain with damping component.

We denote by $\mathcal{P}_m$ the class of all initial distributions $\bar{p} = \langle p_1, \dots, p_m \rangle$.

Let $p_{\varepsilon,ij}(n) = \mathsf{P}\{X_{\varepsilon,n} = j / X_{\varepsilon,0} = i\}, i, j \in \mathbb{X}$, $n = 0, 1, \dots$ be the transition probabilities for the Markov chain $X_{\varepsilon,n}$. Obviously, $p_{\varepsilon,ij}(0) = \mathrm{I}(i = j), i, j \in \mathbb{X}$ and $p_{\varepsilon,ij}(1) = p_{\varepsilon,ij}, i, j \in \mathbb{X}$. Let, also, $p_{\varepsilon,\bar{p},j}(n) = \mathsf{P}_{\bar{p}}\{X_{\varepsilon,n} = j\} = \sum_{i \in \mathbb{X}} p_i p_{\varepsilon,ij}(n), \bar{p} \in \mathcal{P}_m, j \in \mathbb{X}$, $n = 0, 1, \dots$. Obviously, $p_{\varepsilon,\bar{p},j}(0) = p_j, j \in \mathbb{X}$.

Here and henceforth, the symbols $\mathsf{P}_{\bar{p}}$ and $\mathsf{E}_{\bar{p}}$ are used for probabilities and expectations related to a Markov chain with an initial distribution $\bar{p}$. When the initial distribution is concentrated in a state i, the above symbols take the form P_i and E_i, respectively.

The phase space $\mathbb{X}$ for the Markov chain $X_{\varepsilon,n}$ is one aperiodic class of communicative states, for every $\varepsilon \in (0, 1]$.

The following theorem gives a series representation for stationary distribution of Markov chain $X_{\varepsilon,n}$. This representation plays the key role in the following perturbation analysis.

THEOREM 2.1.– *The following ergodic relation takes place for any $\varepsilon \in (0, 1]$, an initial distribution $\bar{p} \in \mathcal{P}_m$ and $j \in \mathbb{X}$:*

$$p_{\varepsilon,\bar{p},j}(n) \to \pi_{\varepsilon,j} = \varepsilon \sum_{l=0}^{\infty} p_{0,\bar{d},j}(l)(1-\varepsilon)^l \text{ as } n \to \infty. \qquad [2.1]$$

The proof of this theorem is based on the special procedure of embedding Markov chain $X_{\varepsilon,n}$ in the model of discrete time regenerative processes with additional binary damping component, which controls the choice between two possible variants of transitions with probabilities taken from matrix $\mathbf{P}_0$ or matrix $\mathbf{D}$. This embedding lets us write down the following renewal type equation, for any $\bar{p} \in \mathcal{P}_m$ and $j \in \mathbb{X}$,

$$p_{\varepsilon,\bar{p},j}(n) = p_{0,\bar{p},j}(n)(1-\varepsilon)^n + \varepsilon \sum_{l=0}^{n} p_{0,\bar{d},j}(l)(1-\varepsilon)^l, n \geq 0. \qquad [2.2]$$

and then obtain the ergodic relation [2.1] using the classical renewal theorem.

The phase space $\mathbb{X}$ is one aperiodic class of communicative states for the Markov chain $X_{\varepsilon,n}$, for every $\varepsilon \in (0, 1]$. In this case, its stationary distribution $\bar{\pi}_\varepsilon = \langle \pi_{\varepsilon,j}, j \in \mathbb{X} \rangle$ is the unique positive solution for the system of linear equations,

$$\sum_{i \in \mathbb{X}} \pi_{\varepsilon,i} p_{\varepsilon,ij} = \pi_{\varepsilon,j}, j \in \mathbb{X}, \sum_{j \in \mathbb{X}} \pi_{\varepsilon,j} = 1. \qquad [2.3]$$

Also, the stationary probabilities $\pi_{\varepsilon,j}$ can be represented in the form $\pi_{\varepsilon,j} = e_{\varepsilon,j}^{-1},, j \in \mathbb{X}$, via the expected return times $e_{\varepsilon,j}$, with the use of the regeneration property of the Markov chain $X_{\varepsilon,n}$ at moments of return in state j. The series representation [2.1] for the stationary distribution of Markov chain $X_{\varepsilon,n}$ is based on the use of alternative damping regeneration times. This representation is, in our opinion, a more effective tool for performing asymptotic perturbation analysis for Markov chains with damping component.

2.2.2. *The stationary distribution of the Markov chain* $X_{0,n}$

In the present chapter, we assume that the following condition holds for some $h \geq 1$:

$\mathbf{A}_{h,1}$: The phase space $\mathbb{X} = \cup_{g=1}^{h} \mathbb{X}^{(g)}$, where: **(a)** $\mathbb{X}^{(g)}, g = 1, \dots, h$ are non intersecting subsets of $\mathbb{X}$ and **(b)** $\mathbb{X}^{(g)}, g = 1, \dots, h$ are non-empty, closed aperiodic classes of communicative states for the Markov chain $X_{0,n}$.

Matrix $\| p_{\varepsilon,ij}(n) \| = \mathbf{P}_{\varepsilon}^{n}$, for $n \geq 0$. Therefore, $p_{\varepsilon,ij}(n) \to p_{0,ij}(n)$ as $\varepsilon \to 0$, for $i, j \in \mathbb{X}, n \geq 0$, This relation allows us to consider the Markov chain $X_{\varepsilon,n}$, for $\varepsilon \in (0, 1]$, as a perturbed version of the Markov chain $X_{0,n}$ and to interpret the damping parameter ε as a perturbation parameter.

The simplest case is, where $h = 1$, i.e. the phase space $\mathbb{X}^{(1)} = \mathbb{X}$ is one aperiodic class of communicative states for the Markov chain $X_{0,n}$. This case relates to regularly perturbed models. The case, where $h > 1$, i.e. the phase space $\mathbb{X}$ splits into several closed classes of communicative states, relates to singularly perturbed models.

If the initial distribution of the Markov chain $X_{0,n}$ is concentrated at the set $\mathbb{X}^{(g)}$, for some $g = 1, \dots, h$, then $X_{0,n} = X_{0,n}^{(g)}, n = 0, 1, \dots$ can be considered as a Markov chain with the reduced phase space $\mathbb{X}^{(g)}$ and the matrix of transition probabilities $\mathbf{P}_{0,g} = \| p_{0,rk} \|_{k,r \in \mathbb{X}^{(g)}} = \| p_{0,rk}^{(g)} \|_{k,r \in \mathbb{X}^{(g)}}$. Let, also, $\mathbf{P}_{0,g}^{n} = \| p_{0,rk}^{(g)}(n) \|_{r,k \in \mathbb{X}^{(g)}}, n = 0, 1, \dots$ be matrices of n-step transition probabilities for the Markov chain $X_{0,n}^{(g)}$, for $g = 1, \dots, h$. According to condition $\mathbf{A}_{h,1}$, the Markov chain $X_{0,n}^{(g)}$ is ergodic, for every $g = 1, \dots, h$. This means that n-step transition probabilities $p_{0,rk}^{(g)}(n) \to \pi_{0,k}^{(g)}$ as $n \to \infty$, for $r, k \in \mathbb{X}^{(g)}$, where $\bar{\pi}_{0}^{(g)} = \langle \pi_{0,k}^{(g)}, k \in \mathbb{X}^{(g)} \rangle$ is the stationary distribution of the Markov chain $X_{0,n}^{(g)}$. This distribution is, for every $g = 1, \dots, h$, the unique solution of the system of linear equations [2.3] written for the Markov chain $X_{0,n}^{(g)}$. Let us also introduce probabilities $f_{\bar{p}}^{(g)} = \sum_{i \in \mathbb{X}^{(g)}} p_i$, for $\bar{p} \in \mathcal{P}_m, g = 1, \dots, h$.

LEMMA 2.1.– *Let condition* $\mathbf{A}_{h,1}$ *hold. Then, the following ergodic relation takes place, for* $\bar{p} \in \mathcal{P}_m$, $k \in \mathbb{X}^{(g)}, g = 1, \dots, h$,

$$p_{0,\bar{p},k}(n) \to \pi_{0,\bar{p},k} = f_{\bar{p}}^{(g)} \pi_{0,k}^{(g)} \text{ as } n \to \infty. \qquad [2.4]$$

Ergodic relation [2.4] shows that in the singular case, where condition $\mathbf{A}_{h,1}$ holds, for some $h > 1$, the stationary probabilities $\pi_{0,\bar{p},k}$ may depend on the initial distribution. It coincides with the stationary distribution $\pi_{0,\bar{d},k}$, if probabilities

$f_{\bar{p}}^{(g)} = f_{\bar{d}}^{(g)}, g = 1, \ldots, h$. In particular, these relations hold for any initial distribution $\bar{p} \in \mathcal{P}_m$, in the regular case, where $h = 1$.

2.3. A perturbation analysis for stationary distributions of Markov chains with damping component

In this section, we present results concerned with continuity of stationary distributions $\bar{\pi}_\varepsilon$ with respect to damping (perturbation) parameter $\varepsilon \to 0$. We also give explicit upper bounds for the rate of convergence in the above continuity convergence relations for stationary probabilities, and the corresponding asymptotic expansions with respect to perturbation parameter ε.

2.3.1. *Continuity property for stationary probabilities*

In what follows, relation $\varepsilon \to 0$ is a reduced version of relation $0 < \varepsilon \to 0$.

THEOREM 2.2.– *Let condition* $\mathbf{A}_{h,1}$ *hold. Then, the following asymptotic relation holds, for* $k \in \mathbb{X}$,

$$\pi_{\varepsilon,k} \to \pi_{0,\bar{d},k} \text{ as } \varepsilon \to 0. \quad [2.5]$$

The proof of this theorem is based on the representation of stationary probabilities $\pi_{\varepsilon,j}, i \in \mathbb{X}$, in the form, $\pi_{\varepsilon,j} = \mathsf{E}p_{0,\bar{d},j}(\nu_\varepsilon - 1)$, where ν_ε is a geometrically distributed random variable with parameter ε, and the following use of the corresponding variant of the Lebesgue theorem.

Theorem 2.2 implies that, in the case where condition $\mathbf{A}_{h,1}$ holds, the continuity property for stationary distributions $\bar{\pi}_\varepsilon$, as $\varepsilon \to 0$, takes place under the additional assumption that place if $f_{\bar{p}}^{(g)} = f_{\bar{d}}^{(g)}, g = 1, \ldots, h$. In particular it is so, for the regular case, where $h = 1$.

2.3.2. *Rate of convergence for stationary distributions*

Let us assume that condition $\mathbf{A}_{h.1}$ holds. In this case, the reduced Markov chain $X_{0,n}^{(g)}$ with the phase space $\mathbb{X}^{(g)}$ and the matrix of transition probabilities $\mathbf{P}_{0,g}$ is, for every $g = 1, \ldots, h$, exponentially ergodic and the following inequalities take place, for $k \in \mathbb{X}^{(g)}, g = 1, \ldots, h$ and $n \geq 1$,

$$\max_{r,k \in \mathbb{X}^{(g)}} |p_{0,rk}^{(g)}(n) - \pi_{0,k}^{(g)}| \leq C_g \lambda_g^n, \quad [2.6]$$

with some constants $C_g = C_g(\mathbf{P}_{0,g}) \in [0, \infty), \lambda_g = \lambda_g(\mathbf{P}_{0,g}) \in [0, 1), g = 1, \dots, h$ and distributions $\bar{\pi}_0^{(g)} = \langle \pi_{0,k}^{(g)}, k \in \mathbb{X}^{(g)} \rangle, g = 1, \dots, h$, with all positive components.

According to the Perron–Frobenius theorem, the role of λ_g can be played, for every $g = 1, \dots, h$, by the absolute value of the second (by absolute value) eigenvalue for matrix $\mathbf{P}_{0,g}$, and C_g is the constant, which can be computed using the algorithm described, for example, in Feller (1968).

Condition $\mathbf{A}_{h,1}$ is, in fact, equivalent to the following condition:

$\mathbf{A}_{h,2}$: The phase space $\mathbb{X} = \cup_{g=1}^{h} \mathbb{X}^{(g)}$, where: **(a)** $\mathbb{X}^{(g)}, g = 1, \dots, h$ are non-intersecting subsets of $\mathbb{X}$ and **(b)** $\mathbb{X}^{(g)}, g = 1, \dots, h$ are non-empty, closed classes of states for the Markov chain $X_{0,n}$ such that inequalities [2.6] hold.

THEOREM 2.3.– *Let condition $\mathbf{A}_{h,2}$ hold. Then, the following relation holds, for $\varepsilon \in (0, 1]$ and $k \in \mathbb{X}^{(g)}, g = 1, \dots, h$,*

$$|\pi_{\varepsilon,k} - \pi_{0,\bar{d},k}| \leq \varepsilon(|d_k - \pi_{0,\bar{d},k}| + F_g), \text{ where } F_g = \frac{f_{\bar{d}}^{(g)} C_g \lambda_g}{1 - \lambda_g}. \quad [2.7]$$

The proof of this theorem is based on the substitution of estimates given in inequalities [2.6] in the series representation for stationary probabilities $\pi_{\varepsilon,k}$ given in relation [2.1].

2.3.3. *Asymptotic expansions for stationary distributions*

Let us now consider the case where condition $\mathbf{A}_{h,1}$ holds. We can assume that class $\mathbb{X}^{(g)}$ includes $m_g \geq 1$ states, for $g = 1, \dots, h$.

Let us denote by $\rho_{g,1}, \dots, \rho_{g,m_g}$ the eigenvalues of the stochastic matrices $\mathbf{P}_{0,g}$, $g = 1, \dots, h$.

Condition $\mathbf{A}_{h,1}$ is, due to the Perron–Frobenius theorem, equivalent to the following condition:

$\mathbf{A}_{h,3}$: The phase space $\mathbb{X} = \cup_{g=1}^{h} \mathbb{X}^{(g)}$, where: **(a)** $\mathbb{X}^{(g)}$, $g = 1, \dots, h$ are non-intersecting subsets of $\mathbb{X}$, **(b)** $\mathbb{X}^{(g)}, g = 1, \dots, h$ are non-empty, closed classes of states for the Markov chain $X_{0,n}$ and **(c)** inequalities $\rho_{g,1} = 1 > |\rho_{g,2}| \geq \dots \geq |\rho_{g,m_g}|$, $g = 1, \dots, h$, hold.

Condition $\mathbf{A}_{h,3}$ implies that the following eigenvalue decomposition representations take place, for $r, k \in \mathbb{X}^{(g)}, g = 1, \dots, h$ and $n \geq 1$,

$$p_{0,rk}^{(g)}(n) = \pi_{0,k}^{(g)} + \rho_{g,2}^{n} \pi_{0,rk}^{(g)}[2] + \dots + \rho_{g,m_g}^{n} \pi_{0,rk}^{(g)}[m_g], \quad [2.8]$$

where: (a) $\bar{\pi}_0^{(g)} = \langle \pi_{0,k}^{(g)}, k \in \mathbb{X}^{(g)} \rangle$ is a distribution with all positive components, for $g = 1, \dots, h$, and (b) $\pi_{0,rk}^{(g)}[l], r, k \in \mathbb{X}^{(g)}, l = 2, \dots, m_g, g = 1, \dots, h$ are some real- or complex-valued coefficients.

It is appropriate to mention that relation [2.8] implies that, for $n \geq 1$,

$$\max_{r,k \in \mathbb{X}^{(g)}} |p_{0,rk}^{(g)}(n) - \pi_{0,k}^{(g)}| = |\rho_{g,2}|^n \max_{r,k \in \mathbb{X}^{(g)}} |\sum_{l=2}^{m_g} \pi_{0,rk}^{(g)}[l] \frac{\rho_{g,l}^n}{\rho_{g,2}^n}|$$

$$\leq |\rho_{g,2}|^n \max_{r,k \in \mathbb{X}^{(g)}} \sum_{l=2}^{m_g} |\pi_{0,rk}^{(g)}[l]| \, |\frac{\rho_{g,l}}{\rho_{g,2}}|. \qquad [2.9]$$

and, thus, inequality [2.6] holds, with parameters, which take the following form, for $g = 1, \dots, h.$,

$$\lambda_g = |\rho_{g,2}|, \; C_g = \max_{r,k \in \mathbb{X}^{(g)}} \sum_{l=2}^{m_g} |\pi_{0,rk}^{(g)}[l]| \, |\frac{\rho_{g,l}}{\rho_{g,2}}|, \; F_g = \frac{f_{\bar{d}}^{(g)} C_g \lambda_g}{1 - \lambda_g}. \qquad [2.10]$$

We again refer to Feller (1968) for the description of an effective algorithm for finding matrices $\mathbf{\Pi}_l^{(g)} = \|\pi_{0,rk}^{(g)}[l]\|, l = 2, \dots, m_g, g = 1, \dots, h$.

Relation [2.8] implies, in this case, that the following relation holds, for any $k \in \mathbb{X}^{(g)}, g = 1, \dots, h$ and $n \geq 1$,

$$p_{0,\bar{d},k}(n) = \pi_{0,\bar{d},k} + \rho_{g,2}^n \pi_{0,\bar{d},k}^{(g)}[2] + \cdots + \rho_{g,m_g}^n \pi_{0,\bar{d},k}^{(g)}[m_g], \qquad [2.11]$$

where $\pi_{0,\bar{d},k}^{(g)}[l] = \sum_{r \in \mathbb{X}^{(g)}} d_r \pi_{0,rk}^{(g)}[l]$, for $k \in \mathbb{X}^{(g)}, l = 2, \dots, m_g, g = 1, \dots, h$,

Let us also define coefficients, for $k \in \mathbb{X}^{(g)}, g = 1, \dots, h, n \geq 1$,

$$\tilde{\pi}_{0,\bar{d},k}^{(g)}[n] = \begin{cases} d_k - \pi_{0,\bar{d},k} + \sum_{l=2}^{m_g} \pi_{0,\bar{d},k}^{(g)}[l] \frac{\rho_{g,l}}{1-\rho_{g,l}} & \text{for } n = 1, \\ (-1)^{n-1} \sum_{l=2}^{m_g} \pi_{0,\bar{d},k}^{(g)}[l] \frac{\rho_{g,l}^{n-1}}{(1-\rho_{g,l})^n} & \text{for } n > 1. \end{cases} \qquad [2.12]$$

Below, symbol $O(\varepsilon^n)$ is used for quantities such that $O(\varepsilon^n)/\varepsilon^n$ is bounded as a function of $\varepsilon \in (0, 1]$.

The following theorem takes place.

THEOREM 2.4.– *Let condition* $\mathbf{A}_{h,3}$ *hold. Then, the following asymptotic expansion takes place, for* $k \in \mathbb{X}^{(g)}, g = 1, \dots, h, n \geq 1$, *and* $\varepsilon \to 0$,

$$\pi_{\varepsilon,k} = \pi_{0,\bar{d},k} + \tilde{\pi}_{0,\bar{d},k}^{(g)}[1]\varepsilon + \cdots + \tilde{\pi}_{0,\bar{d},k}^{(g)}[n]\varepsilon^n + O(\varepsilon^{n+1}). \qquad [2.13]$$

The proof of this theorem is based on the substitution of the eigenvalue decomposition representations [2.11] in the series representation for stationary probabilities $\pi_{\varepsilon,k}$ given in relation [2.1].

It is worth noting that some of the eigenvalues $\rho_{g,l}$ and coefficients $\pi^{(g)}_{0,rk}[l]$ can be complex numbers. Despite this, coefficients $\tilde{\pi}^{(g)}_{0,\bar{d},k}[n], n \geq 1$ in the expansions given in relation [2.13] are real numbers. Indeed, $\pi_{\varepsilon,k}, \varepsilon \in (0,1]$ and $\pi_{0,\bar{d},k}$ are positive numbers. Relation [2.13] implies that $(\pi_{\varepsilon,k} - \pi_{0,\bar{d},k})\varepsilon^{-1} \to \tilde{\pi}^{(g)}_{0,\bar{d},k}[1]$ as $\varepsilon \to 0$. Thus, $\tilde{\pi}^{(g)}_{0,\bar{d},k}[1]$ is a real number. In a similar way, the above proposition can be proved for all coefficients in expansions [2.13]. This implies that the remainders of these expansions $O(\varepsilon^{n+1})$ also are real-valued functions of ε. Moreover, since $\bar{\pi}_\varepsilon = \langle \pi_{\varepsilon,k}, k \in \mathbb{X}\rangle, \varepsilon \in (0,1]$ and $\bar{\pi}_{0,\bar{d}} = \langle \pi_{0,\bar{d},k}, k \in \mathbb{X}\rangle$ are probability distributions, the following equalities connect coefficients in the asymptotic expansions [2.13], for $n \geq 1$,

$$\sum_{g=1}^{h} \sum_{k \in \mathbb{X}^{(g)}} \tilde{\pi}^{(g)}_{0,\bar{d},k}[n] = 0. \qquad [2.14]$$

2.3.4. *Results of numerical experiments*

In this section, we present some numerical examples supporting and illustrating the theoretical results given in section 2.3. The first example demonstrates results associated with regularly perturbed Markov chains. The second example focuses on the use of higher-order terms of asymptotic expansion in regularly perturbed Markov chains.

Let us consider the Markov chain associated with a simple information network with five nodes and a complete link graph. We also restrict our attention to the model with the simplest damping matrix with all elements being equal. The matrices $\mathbf{P}_0$ and $\mathbf{D}$ have the following forms:

$$\mathbf{P}_0 = \begin{Vmatrix} \frac{1}{5} & \frac{1}{5} & \frac{1}{5} & \frac{1}{5} & \frac{1}{5} \\ \frac{1}{4} & 0 & \frac{1}{4} & \frac{1}{4} & \frac{1}{4} \\ 0 & \frac{1}{3} & 0 & \frac{1}{3} & \frac{1}{3} \\ 0 & \frac{1}{3} & \frac{1}{3} & 0 & \frac{1}{3} \\ 0 & \frac{1}{3} & \frac{1}{3} & \frac{1}{3} & 0 \end{Vmatrix}, \quad \mathbf{D} = \begin{Vmatrix} \frac{1}{5} & \frac{1}{5} & \frac{1}{5} & \frac{1}{5} & \frac{1}{5} \\ \frac{1}{5} & \frac{1}{5} & \frac{1}{5} & \frac{1}{5} & \frac{1}{5} \\ \frac{1}{5} & \frac{1}{5} & \frac{1}{5} & \frac{1}{5} & \frac{1}{5} \\ \frac{1}{5} & \frac{1}{5} & \frac{1}{5} & \frac{1}{5} & \frac{1}{5} \\ \frac{1}{5} & \frac{1}{5} & \frac{1}{5} & \frac{1}{5} & \frac{1}{5} \end{Vmatrix}. \qquad [2.15]$$

The corresponding link graph is presented in Figure 2.1.

In this case, condition $\mathbf{A}_{1,1}$ holds.

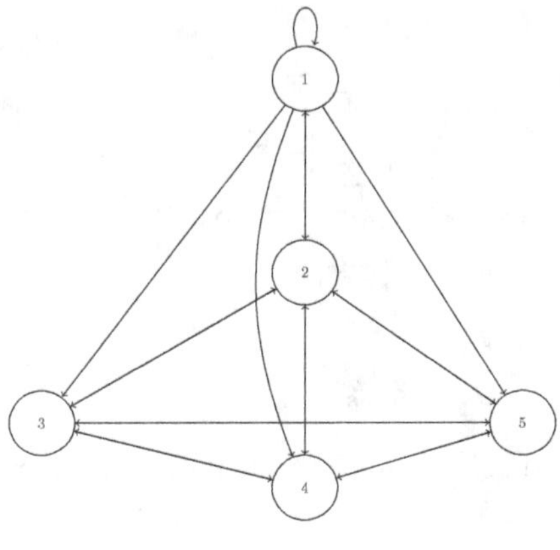

Figure 2.1. *A link graph of a simple information network*

The eigenvalues $\rho_k = \rho_{1,k}, k = 1, \ldots, 5$ of matrix $\mathbf{P}_0$ (computed by solving equation, $\det(\rho\mathbf{I} - \mathbf{P}_0) = 0$), are

$$\rho_1 = 1, \ \rho_2 = \rho_3 = -\frac{1}{3}, \ \rho_4 = -\frac{1}{15} - \frac{\sqrt{34}}{30}, \ \rho_5 = -\frac{1}{15} + \frac{\sqrt{34}}{30}. \qquad [2.16]$$

The stationary projector matrix $\mathbf{\Pi} = \|\pi_{0,rk}\|$, with elements $\pi_{0,rk} = \pi_{0,\bar{d},k}, r, k = 1, \ldots, 5$ and matrices $\mathbf{\Pi}_l = \mathbf{\Pi}_l^{(1)} = \|\pi_{rk}^{(1)}[l]\|, l = 2, \ldots, 5$ (computed with the use of the algorithm described in Feller (1968)), take the following forms:

$$\mathbf{\Pi} = \left\| \begin{matrix} \frac{5}{66} & \frac{8}{33} & \frac{5}{22} & \frac{5}{22} & \frac{5}{22} \\ \frac{5}{66} & \frac{8}{33} & \frac{5}{22} & \frac{5}{22} & \frac{5}{22} \\ \frac{5}{66} & \frac{8}{33} & \frac{5}{22} & \frac{5}{22} & \frac{5}{22} \\ \frac{5}{66} & \frac{8}{33} & \frac{5}{22} & \frac{5}{22} & \frac{5}{22} \\ \frac{5}{66} & \frac{8}{33} & \frac{5}{22} & \frac{5}{22} & \frac{5}{22} \end{matrix} \right\|, \ \mathbf{\Pi}_2 = \mathbf{\Pi}_3 = \left\| \begin{matrix} 0 & 0 & 0 & 0 & 0 \\ 0 & 0 & 0 & 0 & 0 \\ 0 & 0 & \frac{1}{3} & -\frac{1}{6} & -\frac{1}{6} \\ 0 & 0 & -\frac{1}{6} & \frac{1}{3} & -\frac{1}{6} \\ 0 & 0 & -\frac{1}{6} & -\frac{1}{6} & \frac{1}{3} \end{matrix} \right\|, \qquad [2.17]$$

and

$$\mathbf{\Pi}_4 = \left\| \begin{matrix} -a_1 & -b_1 & c_1 & c_1 & c_1 \\ a_2 & b_2 & -c_2 & -c_2 & -c_2 \\ -a_3 & -b_3 & c_3 & c_3 & c_3 \\ -a_3 & -b_3 & c_3 & c_3 & c_3 \\ -a_3 & -b_3 & c_3 & c_3 & c_3 \end{matrix} \right\|, \mathbf{\Pi}_5 = \left\| \begin{matrix} e_1 & f_1 & -g_1 & -g_1 & -g_1 \\ e_2 & -f_2 & g_2 & g_2 & g_2 \\ -e_3 & -f_3 & g_3 & g_3 & g_3 \\ -e_3 & -f_3 & g_3 & g_3 & g_3 \\ -e_3 & -f_3 & g_3 & g_3 & g_3 \end{matrix} \right\|, \qquad [2.18]$$

where

$$\begin{aligned} a_1 &= \tfrac{46\sqrt{34}}{561} - \tfrac{61}{132},\ a_2 = -\tfrac{335\sqrt{34}}{4488} - \tfrac{5}{132},\ a_3 = -\tfrac{20\sqrt{34}}{561} + \tfrac{5}{132}, \\ b_1 &= -\tfrac{29\sqrt{34}}{1122} + \tfrac{4}{33},\ b_2 = \tfrac{95\sqrt{34}}{1122} + \tfrac{25}{66},\quad b_3 = \tfrac{37\sqrt{34}}{1122} + \tfrac{4}{33}, \\ c_1 &= \tfrac{7\sqrt{34}}{374} - \tfrac{5}{44},\quad c_2 = \tfrac{5\sqrt{34}}{1496} + \tfrac{5}{44},\quad c_3 = -\tfrac{\sqrt{34}}{1122} + \tfrac{7}{132}, \\ e_1 &= \tfrac{46\sqrt{34}}{561} + \tfrac{61}{132},\ e_2 = \tfrac{67\sqrt{34}}{4488} - \tfrac{1}{132},\quad e_3 = \tfrac{20\sqrt{34}}{561} + \tfrac{5}{132}, \\ f_1 &= -\tfrac{29\sqrt{34}}{1122} - \tfrac{4}{33},\ f_2 = \tfrac{95\sqrt{34}}{1122} - \tfrac{25}{66},\quad f_3 = -\tfrac{37\sqrt{34}}{1122} + \tfrac{4}{33}, \\ g_1 &= \tfrac{7\sqrt{34}}{374} + \tfrac{5}{44},\quad g_2 = \tfrac{5\sqrt{34}}{1496} - \tfrac{5}{44},\quad g_3 = \tfrac{\sqrt{34}}{1122} + \tfrac{7}{132}. \end{aligned} \quad [2.19]$$

Respectively, the matrix eigenvalue representation based on relation [2.8] takes the following form, for $n \geq 1$:

$$\begin{aligned} \mathbf{P}_0^n &= \mathbf{\Pi} + 2(-\frac{1}{3})^n \mathbf{\Pi}_2 + (-\frac{1}{15} - \frac{1}{30}\sqrt{34})^n \mathbf{\Pi}_4 \\ &+ (-\frac{1}{15} + \frac{1}{30}\sqrt{34})^n \mathbf{\Pi}_5. \end{aligned} \quad [2.20]$$

Relation [2.10] yields, in this case, the values, $\lambda_1 = \frac{1}{3}$ and $F_1 = \frac{67}{4488}\sqrt{34} + \frac{49}{132}$, and the inequality [2.7] takes the following form, for $\varepsilon \in (0, 1]$ and $k = 1, \ldots, 5$:

$$|\pi_{\varepsilon,k} - \pi_{0,\bar{d},k}| \leq \varepsilon(|\frac{1}{5} - \pi_{0,\bar{d},k}| + \frac{67}{4488}\sqrt{34} + \frac{49}{132}), \quad [2.21]$$

where stationary probabilities, $\pi_{0,\bar{d},1} = \frac{5}{66}, \pi_{0,\bar{d},2} = \frac{8}{33}, \pi_{0,\bar{d},r} = \frac{5}{22}, r = 3, 4, 5$.

ε	$k \in \mathbb{X}$	$\pi_{\varepsilon,k}$	$\pi_{0,k}$	$\lvert\pi_{\varepsilon,k} - \pi_{0,\bar{d},k}\rvert$	Upper bound
0.15	1	0.09646	0.07576	0.02071	0.08738
0.15	2	0.23564	0.24242	0.00678	0.07510
0.15	3,4,5	0.22263	0.22727	0.00464	0.07283
0.1	1	0.08966	0.07576	0.01390	0.05825
0.1	2	0.23788	0.24242	0.00455	0.05007
0.1	3,4,5	0.22415	0.22727	0.00312	0.04855
0.05	1	0.08276	0.07576	0.00700	0.02913
0.05	2	0.24014	0.24242	0.00228	0.02503
0.05	3,4,5	0.22570	0.22727	0.00157	0.02427

Table 2.1. *Upper bounds for the rate of convergence*

Table 2.1 give values $\pi_{\varepsilon,k}$, $\pi_{0,\bar{d},k}$ and $|\pi_{\varepsilon,k} - \pi_{0,\bar{d},k}|$ computed with the use of rounded solutions of the system of linear equations [2.3] and the upper bounds given by inequality [2.21], for the values $\varepsilon = 0.15, 0.1, 0.05$.

Taking into account the specific forms of the damping matrix $\mathbf{D}$ and matrices $\mathbf{\Pi}_2$, we get that, in this case, coefficients $\pi^{(1)}_{0,\bar{d},k}[2] = 0, k = 1, \ldots, 5$, and, the coefficients of the eigenvalue representation given in relation [2.11] take the following forms, for $n \geq 1$:

$$p_{0,\bar{d},k}(n) = \begin{cases} \frac{5}{66} + \left(-\frac{223\sqrt{34}}{22440} + \frac{41}{660}\right)\left(-\frac{1}{15} - \frac{\sqrt{34}}{30}\right)^n \\ \quad + \left(-\frac{223\sqrt{34}}{22440} - \frac{41}{660}\right)\left(-\frac{1}{15} + \frac{\sqrt{34}}{30}\right)^n, \text{ for } k = 1, \\ \frac{8}{33} - \left(\frac{13\sqrt{34}}{5610} - \frac{7}{330}\right)\left(-\frac{1}{15} - \frac{\sqrt{34}}{30}\right)^n \\ \quad + \left(\frac{13\sqrt{34}}{5610} + \frac{7}{330}\right)\left(-\frac{1}{15} + \frac{\sqrt{34}}{30}\right)^n, \quad \text{for } k = 2, \\ \frac{5}{22} + \left(\frac{211\sqrt{34}}{36720} - \frac{37}{1080}\right)\left(-\frac{1}{15} - \frac{\sqrt{34}}{30}\right)^n \\ \quad + \left(\frac{19\sqrt{34}}{7480} + \frac{3}{220}\right)\left(-\frac{1}{15} + \frac{\sqrt{34}}{30}\right)^n, \quad \text{for } k = 3, \\ \frac{5}{22} + \left(\frac{211\sqrt{34}}{36720} - \frac{37}{1080}\right)\left(-\frac{1}{15} - \frac{\sqrt{34}}{30}\right)^n \\ \quad + \left(\frac{19\sqrt{34}}{7480} + \frac{3}{220}\right)\left(-\frac{1}{15} + \frac{\sqrt{34}}{30}\right)^n, \quad \text{for } k = 4, \\ \frac{5}{22} + \left(\frac{211\sqrt{34}}{36720} - \frac{37}{1080}\right)\left(-\frac{1}{15} - \frac{\sqrt{34}}{30}\right)^n \\ \quad + \left(\frac{19\sqrt{34}}{7480} + \frac{3}{220}\right)\left(-\frac{1}{15} + \frac{\sqrt{34}}{30}\right)^n, \quad \text{for } k = 5. \end{cases} \quad [2.22]$$

Finally, the coefficients in the asymptotic expansion [2.13] given by relation [2.12] take the following form, for $k \in \mathbb{X}$,

$$\tilde{\pi}^{(1)}_{0,\bar{d},1}[n] = \begin{cases} \frac{307}{2178}, & \text{for } n = 1, \\ \left(\frac{307}{4356} - \frac{853}{74052}\sqrt{34}\right)\left(\frac{1}{33} + \frac{\sqrt{34}}{33}\right)^{n-1} & \\ \quad + \left(\frac{307}{4356} + \frac{853}{74052}\sqrt{34}\right)\left(\frac{1}{33} - \frac{\sqrt{34}}{33}\right)^{n-1}, & \text{for } n > 1. \end{cases} \quad [2.23]$$

$$\tilde{\pi}^{(1)}_{0,\bar{d},2}[n] = \begin{cases} -\frac{50}{1089}, & \text{for } n = 1, \\ \left(-\frac{25}{1089} + \frac{107}{37026}\sqrt{34}\right)\left(\frac{1}{33} + \frac{\sqrt{34}}{33}\right)^{n-1} & \\ \quad - \left(\frac{25}{1089} + \frac{107}{37026}\sqrt{34}\right)\left(\frac{1}{33} - \frac{\sqrt{34}}{33}\right)^{n-1}, & \text{for } n > 1. \end{cases} \quad [2.24]$$

$$\tilde{\pi}^{(1)}_{0,\bar{d},3}[n] = \tilde{\pi}^{(1)}_{0,\bar{d},4}[n] = \tilde{\pi}^{(1)}_{0,\bar{d},5}[n]$$

$$= \begin{cases} -\frac{23}{726}, & \text{for } n = 1, \\ \left(-\frac{23}{1452} + \frac{71}{24684}\sqrt{34}\right)\left(\frac{1}{33} + \frac{\sqrt{34}}{33}\right)^{n-1} & \\ \quad - \left(\frac{23}{1452} + \frac{71}{24684}\sqrt{34}\right)\left(\frac{1}{33} - \frac{\sqrt{34}}{33}\right)^{n-1}, & \text{for } n > 1. \end{cases} \quad [2.25]$$

For example, the second-order expansions [2.13], for stationary probabilities $\pi_{\varepsilon,k}, k = 1, \ldots, 5$, take the forms given in relation [2.26]. The terms of the expansions, except the stationary probabilities (first terms) which are exact values, are computed correct to five decimal digits.

$$\pi_{\varepsilon,k} \approx \begin{cases} \frac{5}{66} + 0.14096\varepsilon - 0.01946\varepsilon^2 + O(\varepsilon^3) \text{ for } k = 1, \\ \frac{8}{33} - 0.04591\varepsilon + 0.00456\varepsilon^2 + O(\varepsilon^3) \text{ for } k = 2, \\ \frac{5}{22} - 0.03168\varepsilon + 0.00497\varepsilon^2 + O(\varepsilon^3) \text{ for } k = 3, \\ \frac{5}{22} - 0.03168\varepsilon + 0.00497\varepsilon^2 + O(\varepsilon^3) \text{ for } k = 4, \\ \frac{5}{22} - 0.03168\varepsilon + 0.00497\varepsilon^2 + O(\varepsilon^3) \text{ for } k = 5. \end{cases} \quad [2.26]$$

It can be observed from relation [2.26] that contributions of second terms in the above expansions are insignificant and, therefore, can be neglected even for a moderate value of $\varepsilon = 0.2$. For example, the first term 0.14096ε in the above asymptotic expansion for the stationary probability $\pi_{0,\bar{d},1} = \frac{5}{66} \approx 0.07576$ takes the value 0.02820 that is about 37.22% of the corresponding stationary probability. However, the second term $0.01946\varepsilon^2$ in the above asymptotic expansion takes the value 0.00078 that is only about 2.76% of the first term and about 1.03% of the corresponding stationary probability.

Let us also mention that the equalities given in relation [2.14] hold for coefficients given in relations [2.23]–[2.25] as well as, approximately up to the corresponding rounding corrections, for coefficients given in relation [2.26].

Let us now present an example where the second term in the corresponding asymptotic expansion for stationary probability cannot be ignored. Here, we consider a regularly perturbed Markov chain with matrices $\mathbf{P}_0$ and $\mathbf{D}$ given in relation [2.27], and the link graph shown in Figure 2.2.

$$\mathbf{P}_0 = \left\| \begin{matrix} 0 & 1 & 0 & 0 \\ \frac{1}{3} & 0 & \frac{1}{3} & \frac{1}{3} \\ 0 & \frac{1}{2} & 0 & \frac{1}{2} \\ 0 & \frac{1}{2} & \frac{1}{2} & 0 \end{matrix} \right\|, \ \mathbf{D} = \left\| \begin{matrix} \frac{1}{4} & \frac{1}{4} & \frac{1}{4} & \frac{1}{4} \\ \frac{1}{4} & \frac{1}{4} & \frac{1}{4} & \frac{1}{4} \\ \frac{1}{4} & \frac{1}{4} & \frac{1}{4} & \frac{1}{4} \\ \frac{1}{4} & \frac{1}{4} & \frac{1}{4} & \frac{1}{4} \end{matrix} \right\|. \quad [2.27]$$

In this case, condition $\mathbf{A}_{1,1}$ also holds.

The eigenvalues $\rho_k = \rho_{1,k}, k = 1, \ldots, 4$ of the matrix $\mathbf{P}_0$ are

$$\rho_1 = 1, \ \rho_2 = -\frac{1}{4} - \frac{\sqrt{33}}{12}, \ \rho_3 = -\frac{1}{2}, \ \rho_4 = -\frac{1}{4} + \frac{\sqrt{33}}{12}.$$

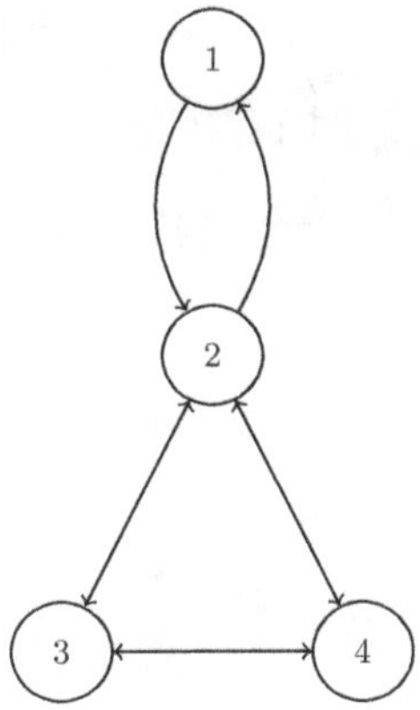

Figure 2.2. *A simple information network*

Computations (analogous to those performed above in the first example) give the following values for the coefficients of the eigenvalue representation, given in relation [2.11], for $n \geq 1$,

$$p_{0,\bar{d},k}(n) = \begin{cases} \frac{1}{8} + \left(\frac{1}{16} + \frac{\sqrt{33}}{528}\right)\left(-\frac{1}{4} - \frac{\sqrt{33}}{12}\right)^n \\ \quad + \left(\frac{1}{16} - \frac{\sqrt{33}}{528}\right)\left(-\frac{1}{4} + \frac{\sqrt{33}}{12}\right)^n, & \text{for } k = 1, \\ \frac{3}{8} - \left(\frac{1}{16} + \frac{3\sqrt{33}}{176}\right)\left(-\frac{1}{4} - \frac{\sqrt{33}}{12}\right)^n \\ \quad + \left(-\frac{1}{16} + \frac{3\sqrt{33}}{176}\right)\left(-\frac{1}{4} + \frac{\sqrt{33}}{12}\right)^n, & \text{for } k = 2, \\ \frac{1}{4} + \frac{\sqrt{33}}{132}\left(-\frac{1}{4} - \frac{\sqrt{33}}{12}\right)^n \\ \quad - \frac{\sqrt{33}}{132}\left(-\frac{1}{4} + \frac{\sqrt{33}}{12}\right)^n, & \text{for } k = 3, 4. \end{cases} \quad [2.28]$$

Finally, the coefficients in the asymptotic expansion [2.13] given by relation [2.12] take the following form, for $k \in \mathbb{X}$,

$$\tilde{\pi}^{(1)}_{0,\bar{d},1}[n] = \begin{cases} \frac{7}{64}, & \text{for } n = 1, \\ \frac{(-1)^{2n+1}}{16}\left(\frac{3+\sqrt{33}}{15+\sqrt{33}}\right)^n - \left(\frac{11+5\sqrt{33}}{176}\right)\left(\frac{-3+\sqrt{33}}{-15+\sqrt{33}}\right)^n \\ \quad + (-1)^{2n}\frac{5\sqrt{33}}{176}\left(\frac{3+\sqrt{33}}{15+\sqrt{33}}\right)^n, & \text{for } n > 1. \end{cases} \quad [2.29]$$

$$\tilde{\pi}^{(1)}_{0,\bar{d},2}[n] = \begin{cases} -\frac{3}{64}, & \text{for } n = 1, \\ (-1)^{1+2n}\left(\frac{33+\sqrt{33}}{176}\right)\left(\frac{3+\sqrt{33}}{15+\sqrt{33}}\right)^n & \\ \quad + \left(\frac{-33+\sqrt{33}}{176}\right)\left(\frac{-3+\sqrt{33}}{-15+\sqrt{33}}\right)^n, & \text{for } n > 1. \end{cases} \quad [2.30]$$

$$\tilde{\pi}^{(1)}_{0,\bar{d},3}[n] = \tilde{\pi}^{(1)}_{0,\bar{d},4}[n]$$
$$= \begin{cases} -\frac{1}{32}, & \text{for } n = 1, \\ \frac{\sqrt{33}}{11(3+\sqrt{33})}(-1)^n\left(\frac{-3-\sqrt{33}}{15+\sqrt{33}}\right)^n & \\ \quad + \frac{\sqrt{33}}{11(-3+\sqrt{33})}(-1)^n\left(\frac{-3+\sqrt{33}}{15-\sqrt{33}}\right)^n, & \text{for } n > 1. \end{cases}$$
[2.31]

Consequently, the second-order asymptotic expansions [2.13], for stationary probabilities $\pi_{\varepsilon,k}, k = 1, \ldots, 4$, take the forms given in relation [2.33]. The terms of the expansions, except the stationary probabilities (first terms) which are exact values, are computed correct to five decimal digits.

$$\pi_{\varepsilon,k} = \begin{cases} \frac{1}{8} + \frac{7}{64}\varepsilon - \frac{1}{512}\varepsilon^2 + O(\varepsilon^3), \text{ for } k = 1, \\ \frac{3}{8} - \frac{3}{64}\varepsilon - \frac{27}{512}\varepsilon^2 + O(\varepsilon^3), \text{ for } k = 2, \\ \frac{1}{4} - \frac{1}{32}\varepsilon + \frac{7}{256}\varepsilon^2 + O(\varepsilon^3), \text{ for } k = 3, \\ \frac{1}{4} - \frac{1}{32}\varepsilon + \frac{7}{256}\varepsilon^2 + O(\varepsilon^3), \text{ for } k = 4. \end{cases} \quad [2.32]$$

$$\approx \begin{cases} \frac{1}{8} + 0.10937\varepsilon - 0.00195\varepsilon^2 + O(\varepsilon^3), \text{ for } k = 1, \\ \frac{3}{8} - 0.04687\varepsilon - 0.05273\varepsilon^2 + O(\varepsilon^3), \text{ for } k = 2, \\ \frac{1}{4} - 0.03125\varepsilon + 0.02734\varepsilon^2 + O(\varepsilon^3), \text{ for } k = 3, \\ \frac{1}{4} - 0.03125\varepsilon + 0.02734\varepsilon^2 + O(\varepsilon^3), \text{ for } k = 4. \end{cases} \quad [2.33]$$

Let us mention again that the equalities given in relation [2.14] hold for coefficients given in relations [2.29]–[2.31] and [2.32], as well as (approximately up to the corresponding rounding corrections) for coefficients given in relation [2.33].

We see that in the third ($k = 3$) and fourth ($k = 4$) expansions in relation [2.33] for $\varepsilon = 0.2$, the term $\frac{1}{32}\varepsilon$ is about 2.5% of the limiting stationary probability 0.25, while the term $\frac{7}{256}\varepsilon^2$ is about 0.4% of 0.25, i.e. it improves the accuracy of approximation for the stationary probability by about 16%.

2.4. Coupling and ergodic theorems for perturbed Markov chains with damping component

In this section, we present coupling algorithms, and effective coupling-type upper bounds for the rate of convergence in ergodic theorems, for regularly and singularly perturbed Markov chains with damping component. We also formulate ergodic theorems for perturbed Markov chains with damping component in triangular array mode. Finally, we present the results of some numerical experiments supporting and illustrating the above results.

2.4.1. *Coupling for regularly perturbed Markov chains with damping component*

Let $\mathbf{A} = \|a_{ij}\|$ be an $m \times m$ a matrix with real-valued elements. Let us introduce functional,

$$Q(\mathbf{A}) = \min_{1 \leq i,j \leq m} \sum_{k=1}^{m} a_{ik} \wedge a_{jk}. \quad [2.34]$$

The following lemma presents some basic properties of functional $Q(\mathbf{A})$.

LEMMA 2.2.– *Functional $Q(\mathbf{A})$ possesses the following properties: (a) $Q(a\mathbf{A}) = aQ(\mathbf{A})$, for any $a \geq 0$; (b) $Q(\mathbf{A}) \geq a_1 Q(\mathbf{A}_1) + \cdots + a_n Q(\mathbf{A}_n)$, for any $m \times m$ matrices $\mathbf{A}_1, \ldots, \mathbf{A}_n$ with real-valued elements, numbers $a_1, \ldots, a_n \geq 0, a_1 \cdots + a_n = 1$ and matrix $\mathbf{A} = a_1\mathbf{A}_1 + \cdots + a_n\mathbf{A}_n$, for $n \geq 2$; (c) $Q(\mathbf{A}) \in [0, 1]$, for any stochastic matrix $\mathbf{A}$; (d) $Q(\mathbf{A}) = 1$, for any $m \times m$ stochastic damping-type matrix $\mathbf{A} = \|a_{ij}\|$, with elements $a_{ij} = a_j \geq 0, i, j = 1, \ldots, m$.*

The following useful proposition is a corollary of Lemma 2.2.

LEMMA 2.3.– *The following inequality takes place, for $\varepsilon \in (0, 1]$ and $N \geq 1$,*

$$1 - Q(\mathbf{P}_\varepsilon^N) \leq (1 - Q(\mathbf{P}_0^N))(1 - \varepsilon)^N. \quad [2.35]$$

Let us introduce, for $N \geq 1$, the coefficient of ergodicity,

$$\Delta_N(\mathbf{P}_0) = (1 - Q(\mathbf{P}_0^N))^{1/N}. \quad [2.36]$$

Also let $\bar{p}' = \langle p'_1, \ldots, p'_m \rangle, \bar{p}'' = \langle p''_1, \ldots, p''_m \rangle$ be two probability distributions, and,

$$Q(\bar{p}', \bar{p}'') = \sum_{i=1}^{m} \min(p'_i, p''_i). \quad [2.37]$$

In Theorems 2.5 and 2.6, we give effective coupling-type upper bounds for the rate of convergence in the individual ergodic theorem for Markov chains with damping component. These theorems are based on the corresponding general coupling results for Markov chains given in Griffeath (1975), Pitman (1979) and Lindvall (2002) and specify and detail coupling upper bounds for the rate of convergence for Markov chains with damping component.

Note that condition $\mathbf{A}_{h,1}$ is not required in Theorem 2.5, as formulated below.

THEOREM 2.5.– *The following relation takes place for every* $\varepsilon \in (0, 1]$ *and* $\bar{p} \in \mathcal{P}_m$, $j \in \mathbb{X}$, $n \geq 0$,

$$|p_{\varepsilon,\bar{p},j}(n) - \pi_{\varepsilon,j}| \leq (1 - Q(\bar{p}, \bar{\pi}_\varepsilon))\Delta_N(\mathbf{P}_0)^{[n/N]N}(1 - \varepsilon)^{[n/N]N}. \qquad [2.38]$$

Note that we count $\Delta_N(\mathbf{P}_0)^0 = 1$, if $\Delta_N(\mathbf{P}_0) = 0$.

The proof of this theorem is obtained with the use of the coupling method and the accurate computing of the maximal N-step coupling probability for the corresponding coupling Markov chain with damping component.

It is worth noting that the weaker upper bound $(1 - \varepsilon)^n$ on the right-hand side of inequality [2.38] has been given for Markov chains with a general phase and damping component, in the recent paper (Avrachenkov *et al.* 2018).

In the case where condition $\mathbf{A}_{1,1}$ holds (i.e. the phase space $\mathbb{X}$ is one aperiodic class of communicative states for the Markov chain $X_{0,n}$), $1 - Q(\mathbf{P}_0^N) \to 0$ as $N \to \infty$, and thus, the following condition holds for N large enough:

$\mathbf{B}_{N,1}$: $\Delta_N(\mathbf{P}_0) < 1$.

By passing $\varepsilon \to 0$ in inequality [2.38], we get the following relation holding for $\bar{p} \in \mathcal{P}_m, j \in \mathbb{X}, n \geq 0$,

$$|p_{0,\bar{p},j}(n) - \pi_{0,j}| \leq (1 - Q(\bar{p}, \bar{\pi}_0)))\Delta_N(\mathbf{P}_0)^{[n/N]N}, \qquad [2.39]$$

where $\bar{\pi}_0 = \langle \pi_{0,j}, j \in \mathbb{X} \rangle$ is the stationary distribution for the Markov chain $X_{0,n}$.

Some components of this stationary distribution can be equal to 0. In this case, a set $\mathbb{X}_1 = \{j \in \mathbb{X} : \pi_{0,j} > 0\}$ is a closed, aperiodic class of communicative states, while set $\mathbb{X}_0 = \{j \in \mathbb{X} : \pi_{0,j} = 0\}$ is the class of transient states, for the Markov chain $X_{0,n}$. If the stationary distribution $\bar{\pi}_0 = \langle \pi_{0,j}, j \in \mathbb{X} \rangle$ is positive, then $\mathbb{X}_0 = \emptyset$. In this case, condition $\mathbf{B}_{N,1}$ is sufficient for the holding of condition $\mathbf{A}_{1,1}$.

In the case where condition $\mathbf{A}_{1,1}$ or equivalently $\mathbf{A}_{1,3}$ holds, the eigenvalue decomposition representation [2.8] implies that, under some minimal technical assumption given, for example, in Abola *et al.* (2019), the following relation holds:

$$\Delta_N(\mathbf{P}_0) \to |\rho_{1,2}| \text{ as } N \to \infty. \qquad [2.40]$$

Relation [2.40] shows that, in this case, the coupling upper bounds for rate of convergence in individual ergodic relations, given in Theorem 2.5, are usually asymptotically equivalent with analogous upper bounds, which can be obtained with the use of eigenvalue decomposition representation for transition probabilities. At the same time, computing coefficients of ergodicity $\Delta_N(\mathbf{P}_0)$ do not require solving of the polynomial equation, $\det(\rho\mathbf{I} - \mathbf{P}_0) = 0$, that is required for finding eigenvalues.

It is useful also to note that inequality [2.39] can also be transformed to the form of inequality [2.6]. Indeed, inequality [2.39] implies that, for $n \geq 1$,

$$\max_{i,j\in\mathbb{X}} |p_{0,ij}(n) - \pi_{0,j}| \leq \Delta_N(\mathbf{P}_0)^{(n/N-1)N}$$
$$= (1 - Q(\mathbf{P}_0^N))^{-1}\Delta_N(\mathbf{P}_0)^n. \quad [2.41]$$

Let us assume that condition $\mathbf{A}_{1,1}$ holds. In this case, inequality [2.6] holds with parameters, which, in this case, depend on $N \geq 1$,

$$\lambda_{1,N} = \Delta_N(\mathbf{P}_0),\ C_{1,N} = (1 - Q(\mathbf{P}_0^N))^{-1},\ F_{1,N} = \frac{C_{1,N}\lambda_{1,N}}{1 - \lambda_{1,N}}. \quad [2.42]$$

2.4.2. *Coupling for singularly perturbed Markov chains with damping component*

Let us assume that the following condition holds for some $N \geq 1$ and $h > 1$:

$\mathbf{B}_{N,h}$: The phase space $\mathbb{X} = \cup_{g=1}^h \mathbb{X}^{(g)}$, where: **(a)** $\mathbb{X}^{(g)}$, $g = 1, \ldots, h$ are non-intersecting subsets of $\mathbb{X}$, **(b)** $\mathbb{X}^{(g)}, g = 1, \ldots, h$ are non-empty, closed classes of states for the Markov chain $X_{0,n}$ and **(c)** $\Delta_N(\mathbf{P}_{0,g}) < 1$, for $g = 1, \ldots, h$.

Let us introduce the discrete distributions $\bar{f}_{\bar{p}} = \langle f_{\bar{p}}^{(g)}, g = 1, \ldots, h\rangle$, where $f_{\bar{p}}^{(g)} = \sum_{k\in\mathbb{X}^{(g)}} p_k, j = 1, \ldots, h$ and, also, define, for $g = 1, \ldots, h$, the conditional distributions $\bar{p}^{(g)} = \langle p_k^{(g)}, k \in \mathbb{X}^{(g)}\rangle$ using relation $p_k = p_k^{(g)} f_{\bar{p}}^{(g)}, k \in \mathbb{X}^{(g)}$. Obviously, $p_k^{(g)} = p_k / f_{\bar{p}}^{(g)}, k \in \mathbb{X}^{(g)}$, for $1 \leq g \leq h$ such that $f_{\bar{p}}^{(g)} > 0$. As usual, we can also define the above conditional distributions in some standard way, for example, as $p_k^{(g)} = \frac{1}{m_g}, k \in \mathbb{X}^{(g)}$, if $f_{\bar{p}}^{(g)} = 0$.

We also use the above notations for this case, where distribution $\bar{p}$ coincides with distributions $\bar{d} = \langle d_k, k \in \mathbb{X}\rangle$ and $\bar{\pi}_\varepsilon = \langle \pi_{\varepsilon,k}, k \in \mathbb{X}\rangle$.

Let $\varepsilon \in (0, 1]$ and $\bar{\pi}_\varepsilon = \langle \pi_{\varepsilon,k}, k \in \mathbb{X}\rangle$ be the stationary distribution for the Markov chain $X_{\varepsilon,n}$. Theorem 2.2 and Lemma 2.1 imply that representation

$\pi_{\varepsilon,k} = \varepsilon \sum_{l=0}^{\infty} p_{0,\bar{d},k}(l)(1-\varepsilon)^l = f_{\bar{d}}^{(g)} \pi_{\varepsilon,k}^{(g)}$ takes place for $k \in \mathbb{X}^{(g)}, g = 1, \ldots, h$, where $\pi_{\varepsilon,k}^{(g)} = \varepsilon \sum_{l=0}^{\infty} p_{0,\bar{d}^{(g)},k}(l)(1-\varepsilon)^l$, for $k \in \mathbb{X}^{(g)}, g = 1, \ldots, h$.

It is readily seen that $\bar{\pi}_{\varepsilon}^{(g)} = \langle \pi_{\varepsilon,k}^{(g)}, k \in \mathbb{X} \rangle$ is, for every $g = 1, \ldots, h$, the stationary distribution for the Markov chain $X_{\varepsilon,n}^{(g)}$, with the phase space $\mathbb{X}^{(g)}$ and the matrix of transition probabilities $\mathbf{P}_{\varepsilon,g} = (1-\varepsilon)\mathbf{P}_{0,g} + \varepsilon \mathbf{D}_g$, where $\mathbf{P}_{0,g} = \| p_{0,rk} \|_{r,k \in \mathbb{X}^{(g)}}$ is, according to condition $\mathbf{B}_{N,h}$, a stochastic matrix, while $\mathbf{D}_g = \| d_{rk}^{(g)} \|_{r,k \in \mathbb{X}^{(g)}}$ is the stochastic damping matrix with elements $d_{rk}^{(g)} = d_k^{(g)} = d_k / f_{\bar{d}}^{(g)}, k, r \in \mathbb{X}^{(g)}$. Note that the above relations also imply that, $f_{\bar{\pi}_{\varepsilon}}^{(g)} = \sum_{k \in \mathbb{X}^{(g)}} \pi_{\varepsilon,k} = \sum_{k \in \mathbb{X}^{(g)}} f_{\bar{d}}^{(g)} \pi_{\varepsilon,k}^{(g)} = f_{\bar{d}}^{(g)}$, for $g = 1, \ldots, h$.

Condition $\mathbf{B}_{N,h}$ implies that the Markov chain $X_{0,n}^{(g)}$ is exponentially ergodic, for every $g = 1, \ldots, h$. Let $\bar{\pi}_0^{(g)} = \langle \pi_{0,k}^{(g)}, k \in \mathbb{X}^{(g)} \rangle$ be the corresponding stationary distribution. According to the above remarks and Theorem 2.2, $\pi_{\varepsilon,k}^{(g)} \to \pi_{0,k}^{(g)}$ as $\varepsilon \to 0$, for $k \in \mathbb{X}^{(g)}, g = 1, \ldots, h$.

Below, we count $\Delta_N(\mathbf{P}_{0,g})^0 = 1$, if $\Delta_N(\mathbf{P}_{0,g}) = 0$.

THEOREM 2.6.– *Let condition $\mathbf{B}_{N,h}$ hold. Then, the following relation takes place, for $\bar{p} \in \mathcal{P}_m$, $k \in \mathbb{X}^{(g)}$, $g = 1, \ldots, h$, $n \geq 0$ and $\varepsilon \in (0, 1]$,*

$$
\begin{aligned}
|p_{\varepsilon,\bar{p},k}(n) - \pi_{\varepsilon,k}| \leq & \, ((f_{\bar{d}}^{(g)}(1 - Q(\bar{\pi}_{\varepsilon}^{(g)}, \bar{\pi}_0^{(g)})) \\
& + f_{\bar{p}}^{(g)}(1 - Q(\bar{p}^{(g)}, \bar{\pi}_0^{(g)})))\Delta_N(\mathbf{P}_{0,g})^{[n/N]N} \\
& + |f_{\bar{p}}^{(g)} - f_{\bar{d}}^{(g)}| \pi_{0,k}^{(g)})(1-\varepsilon)^n.
\end{aligned} \quad [2.43]
$$

The proof of this theorem is based on the above relations for probabilities $p_{\varepsilon,\bar{p},k}(n)$, stationary probabilities $\pi_{\varepsilon,k}$ and application of coupling inequalities given in Theorem 2.5 and relation [2.39] to Markov chains $X_{\varepsilon,n}^{(g)}, g = 1, \ldots, h$ and $X_{0,n}^{(g)}, g = 1, \ldots, h$.

2.4.3. *Ergodic theorems for perturbed Markov chains with damping component in the triangular array mode*

The following theorem takes place.

THEOREM 2.7.– *Let condition $\mathbf{B}_{N,1}$ hold for some $N \geq 1$. Then, for $\bar{p} \in \mathcal{P}_m$, $k \in \mathbb{X}$ and any $n_\varepsilon \to \infty$ as $\varepsilon \to 0$,*

$$
p_{\varepsilon,\bar{p},k}(n_\varepsilon) \to \pi_{0,k} = \pi_{0,\bar{d},k} \text{ as } \varepsilon \to 0. \quad [2.44]
$$

Ergodic theorems for singularly perturbed Markov chains with damping component in the triangular array mode take much more complex forms.

THEOREM 2.8.– *Let condition $\mathbf{B}_{N,h}$ hold for some $N \geq 1$ and $h > 1$. Then, the following ergodic relations take place, for $\bar{p} \in \mathcal{P}_m$, $k \in \mathbb{X}^{(g)}$, $g = 0, \ldots, h$:*

(i) If $n_\varepsilon \to \infty$ and $\varepsilon n_\varepsilon \to \infty$ as $\varepsilon \to 0$, then,

$$p_{\varepsilon,\bar{p},k}(n_\varepsilon) \to \pi_{0,\bar{p},k}(\infty) = \pi_{0,\bar{d},k} \text{ as } \varepsilon \to 0. \quad [2.45]$$

(ii) If $n_\varepsilon \to \infty$ and $\varepsilon n_\varepsilon \to t \in (0, \infty)$ as $\varepsilon \to 0$, then,

$$p_{\varepsilon,\bar{p},k}(n_\varepsilon) \to \pi_{0,\bar{p},k}(t) = \pi_{0,\bar{p},k}e^{-t} + \pi_{0,\bar{d},k}(1 - e^{-t}) \text{ as } \varepsilon \to 0. \quad [2.46]$$

(iii) If $n_\varepsilon \to \infty$ and $\varepsilon n_\varepsilon \to 0$ as $\varepsilon \to 0$, then,

$$p_{\varepsilon,\bar{p},k}(n_\varepsilon) \to \pi_{0,\bar{p},k}(0) = \pi_{0,\bar{p},k} \text{ as } \varepsilon \to 0. \quad [2.47]$$

The proofs of theorems 2.7 and 2.8 are based on the accurate application of renewal-type equation [2.2] that can be derived for n-step transition probabilities of the Markov chain with damping component $X_{\varepsilon,n}$, the series representation for its stationary distribution given in theorem 2.1 and the coupling inequalities given in theorems 2.5 and 2.6.

We would also like to mention the work of: Silvestrov (1978, 1979, 2018); Englund and Silvestrov (1997); Gyllenberg and Silvestrov (2000, 2008); Englund (2001); Petersson (2016); and Silvestrov and Petersson (2014); these contain results related to ergodic theorems in triangular array mode and to so-called quasi-stationary ergodic theorems for perturbed regenerative processes, Markov chains and semi-Markov processes.

2.4.4. *Numerical examples*

In this section, we present some numerical examples that support and illustrate upper bounds for the rate of convergence in ergodic theorems, given in theorems 2.5 and 2.6, as well as ergodic theorems in triangular array mode for Markov chains with damping component, given in theorem 2.8.

Let us consider again the Markov chain with damping component introduced in the first example presented in section 2.3.4, in relation [2.15] and Figure 2.1.

Figure 2.3 displays the asymptotical behavior of the ergodicity coefficient $\Delta_N(\mathbf{P}_0)$ (computed with the use of relations [2.34] and [2.35]) as $N \to \infty$. In this case, coefficients of ergodicity $\Delta_N(\mathbf{P}_0) \to |\rho_2| = \frac{1}{3}$ as $N \to \infty$ are consistent with relation [2.40].

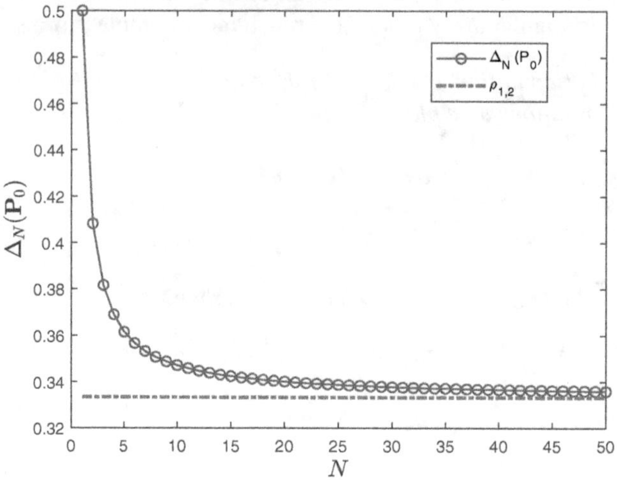

Figure 2.3. *Ergodicity coefficient $\Delta_N(\mathbf{P}_0)$. For a color version of this figure, see www.iste.co.uk/dimotikalis/analysis1.zip*

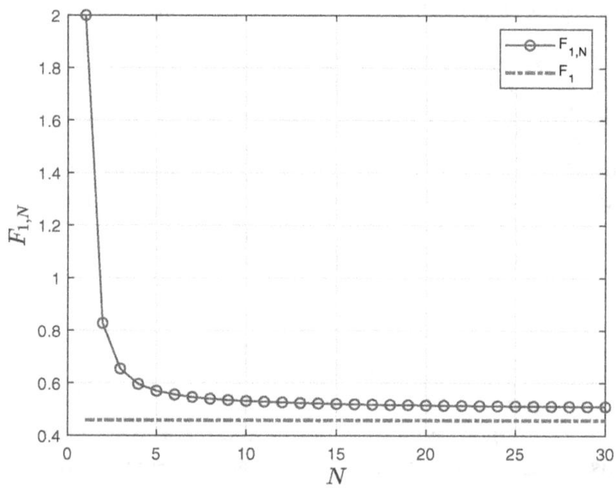

Figure 2.4. *Constants appearing in inequality [2.7]. For a color version of this figure, see www.iste.co.uk/dimotikalis/analysis1.zip*

Figure 2.4 shows the values of constants $F_1 = C_1\rho_{1,2}/(1 - \rho_{1,2})$, computed using relations [2.7] and [2.10], and constants $F_{1,N} = C_{1,N}\lambda_{1,N}/(1 - \lambda_{1,N}) = \Delta_N(P_0)/(1 - Q(P_0^N))(1 - \Delta_N(P_0))$, computed using relations [2.7] and [2.41], for several initial values of N.

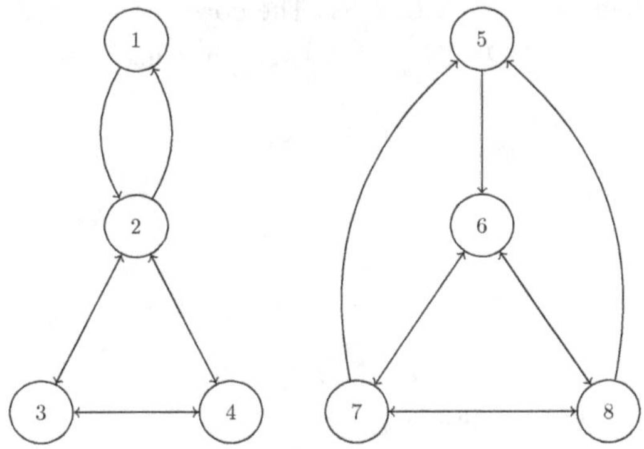

Figure 2.5. *A simple two-disjoint information network*

Figure 2.4 shows that the upper bounds in inequality [2.7], obtained with the use of the coupling method, are asymptotically equivalent with the upper bounds in inequality [2.7], obtained with the use of the eigenvalue decomposition representation. At the same time, the coupling upper bounds have a computationally much simpler form.

Let us now present an information network that captures results supporting and illustrating ergodic theorems for perturbed Markov chains with damping component in the triangular array mode. In this example, we consider an ergodic singularly perturbed Markov chain with damping component as stated in theorem 2.8. We use a two-disjoint information network associated with a singularly perturbed Markov chain with the matrices $\mathbf{P}_0$ and $\mathbf{D}$ given in relation [2.48] and the link graph shown in Figure 2.5.

$$\mathbf{P}_0 = \left\| \begin{matrix} 0 & 1 & 0 & 0 & 0 & 0 & 0 & 0 \\ \frac{1}{3} & 0 & \frac{1}{3} & \frac{1}{3} & 0 & 0 & 0 & 0 \\ 0 & \frac{1}{2} & 0 & \frac{1}{2} & 0 & 0 & 0 & 0 \\ 0 & \frac{1}{2} & \frac{1}{2} & 0 & 0 & 0 & 0 & 0 \\ 0 & 0 & 0 & 0 & 0 & 1 & 0 & 0 \\ 0 & 0 & 0 & 0 & 0 & 0 & \frac{1}{2} & \frac{1}{2} \\ 0 & 0 & 0 & 0 & \frac{1}{3} & \frac{1}{3} & 0 & \frac{1}{3} \\ 0 & 0 & 0 & 0 & \frac{1}{3} & \frac{1}{3} & \frac{1}{3} & 0 \end{matrix} \right\|, \mathbf{D} = \left\| \begin{matrix} \frac{1}{8} & \frac{1}{8} & \frac{1}{8} & \frac{1}{8} & \frac{1}{8} & \frac{1}{8} & \frac{1}{8} & \frac{1}{8} \\ \frac{1}{8} & \frac{1}{8} & \frac{1}{8} & \frac{1}{8} & \frac{1}{8} & \frac{1}{8} & \frac{1}{8} & \frac{1}{8} \\ \frac{1}{8} & \frac{1}{8} & \frac{1}{8} & \frac{1}{8} & \frac{1}{8} & \frac{1}{8} & \frac{1}{8} & \frac{1}{8} \\ \frac{1}{8} & \frac{1}{8} & \frac{1}{8} & \frac{1}{8} & \frac{1}{8} & \frac{1}{8} & \frac{1}{8} & \frac{1}{8} \\ \frac{1}{8} & \frac{1}{8} & \frac{1}{8} & \frac{1}{8} & \frac{1}{8} & \frac{1}{8} & \frac{1}{8} & \frac{1}{8} \\ \frac{1}{8} & \frac{1}{8} & \frac{1}{8} & \frac{1}{8} & \frac{1}{8} & \frac{1}{8} & \frac{1}{8} & \frac{1}{8} \\ \frac{1}{8} & \frac{1}{8} & \frac{1}{8} & \frac{1}{8} & \frac{1}{8} & \frac{1}{8} & \frac{1}{8} & \frac{1}{8} \\ \frac{1}{8} & \frac{1}{8} & \frac{1}{8} & \frac{1}{8} & \frac{1}{8} & \frac{1}{8} & \frac{1}{8} & \frac{1}{8} \end{matrix} \right\| \qquad [2.48]$$

The phase space $\mathbb{X}$ of the Markov chain $X_{0,n}$ with matrix of transition probabilities $\mathbf{P}_0$ split in two closed classes of communicative states $\mathbb{X} = \mathbb{X}^{(1)} \cup \mathbb{X}^{(2)}$, where

$\mathbb{X}^{(1)} = \{1, 2, 3, 4\}$ and $\mathbb{X}^{(2)} = \{5, 6, 7, 8\}$. The corresponding matrices of transition probabilities for these classes, respectively, $\mathbf{P}_{0,1}$ and $\mathbf{P}_{0,2}$ take the following forms:

$$\mathbf{P}_{0,1} = \left\| \begin{matrix} 0 & 1 & 0 & 0 \\ \frac{1}{3} & 0 & \frac{1}{3} & \frac{1}{3} \\ 0 & \frac{1}{2} & 0 & \frac{1}{2} \\ 0 & \frac{1}{2} & \frac{1}{2} & 0 \end{matrix} \right\|, \ \mathbf{P}_{0,2} = \left\| \begin{matrix} 0 & 1 & 0 & 0 \\ 0 & 0 & \frac{1}{2} & \frac{1}{2} \\ \frac{1}{3} & \frac{1}{3} & 0 & \frac{1}{3} \\ \frac{1}{3} & \frac{1}{3} & \frac{1}{3} & 0 \end{matrix} \right\|.$$

In this case, condition $\mathbf{A}_{2,1}$ holds.

We are going to get the second-order asymptotic expansion for stationary probabilities $\pi_{\varepsilon,k}, k \in \mathbb{X}$ given by relation [2.13]. In fact, we can use Theorem 2.4 to give the corresponding expansions for stationary probabilities $\pi^{(1)}_{\varepsilon,k}, k \in \mathbb{X}^{(1)}$ and $\pi^{(2)}_{\varepsilon,k}, k \in \mathbb{X}^{(2)}$. Then, these expansions can be transformed into the corresponding expansions for $\pi_{\varepsilon,k}, k \in \mathbb{X}$, using relations $\pi_{\varepsilon,k} = f^{(g)}_{\bar{d}} \pi^{(g)}_{\varepsilon,k}, k \in \mathbb{X}^{(g)}, g = 1, 2$ and taking into account that, in this case, probabilities $f^{(g)}_{\bar{d}} = \sum_{k \in \mathbb{X}^{(g)}} d_k = \frac{1}{2}, g = 1, 2$.

We can easily note that matrix $\mathbf{P}_{0,1}$ coincides with matrix $\mathbf{P}_0$ given in relation [2.27]. Thus, the second-order asymptotic expansions for stationary probabilities $\pi^{(1)}_{\varepsilon,k}, k \in \mathbb{X}^{(1)}$ take the forms given in relation [2.33], and, thus, the corresponding asymptotic expansions for stationary probabilities $\pi_{\varepsilon,k} = \frac{1}{2}\pi^{(1)}_{\varepsilon,k}, k \in \mathbb{X}^{(1)}$ take the following forms:

$$\pi_{\varepsilon,k} = \begin{cases} \frac{1}{16} + \frac{7}{128}\varepsilon - \frac{1}{1024}\varepsilon^2 + O(\varepsilon^3), & \text{for } k = 1, \\ \frac{3}{16} - \frac{3}{128}\varepsilon - \frac{27}{1024}\varepsilon^2 + O(\varepsilon^3), & \text{for } k = 2, \\ \frac{1}{8} - \frac{1}{64}\varepsilon + \frac{7}{512}\varepsilon^2 + O(\varepsilon^3), & \text{for } k = 3, \\ \frac{1}{8} - \frac{1}{64}\varepsilon + \frac{7}{512}\varepsilon^2 + O(\varepsilon^3), & \text{for } k = 4. \end{cases} \quad [2.49]$$

Similarly, the asymptotic expansion $\pi^{(2)}_{\varepsilon,k}$ for phase space $\mathbb{X}^{(2)}$ can be computed. The eigenvalues corresponding to $\mathbf{P}_{0,2}$ are

$$\rho_{2,1} = 1, \ \rho_{2,2} = -\frac{1}{3} - \frac{\sqrt{2}}{3}\mathbf{i}, \ \rho_{2,3} = -\frac{1}{3} + \frac{\sqrt{2}}{3}\mathbf{i}, \ \rho_{2,4} = -\frac{1}{3},$$

where $\mathbf{i}^2 = -1$. Hence, from relation [2.11], we obtain eigenvalue decomposition of $\mathbf{P}_{0,2}$, for $k \in \mathbb{X}^{(2)}, n \geq 1$, as

$$p_{0,\bar{d},k}(n) = \begin{cases} \frac{1}{6} + \left(\frac{\sqrt{2}\mathrm{i}}{16(1+\sqrt{2}\mathrm{i})}\right)\left(-\frac{1}{3} - \frac{\sqrt{2}}{3}\mathbf{i}\right)^n \\ \quad + \left(\frac{(-1+2\sqrt{2}\mathrm{i})}{24(-2+\sqrt{2}\mathrm{i})}\right)\left(-\frac{1}{3} + \frac{\sqrt{2}}{3}\mathbf{i}\right)^n, \text{ for } k = 5, \\ \frac{1}{3} - \left(\frac{1}{8+\sqrt{128}\mathrm{i}}\right)\left(-\frac{1}{3} - \frac{\sqrt{2}}{3}\mathbf{i}\right)^n \\ \quad - \left(\frac{-4-\sqrt{2}\mathrm{i}}{24(-2+\sqrt{2}\mathrm{i})}\right)\left(-\frac{1}{3} + \frac{\sqrt{2}}{3}\mathbf{i}\right)^n, \text{ for } k = 6, \\ \frac{1}{4} - \frac{\sqrt{2}\mathrm{i}}{32}\left(-\frac{1}{3} - \frac{\sqrt{2}}{3}\mathbf{i}\right)^n \\ \quad + \frac{\sqrt{2}\mathrm{i}}{32}\left(-\frac{1}{3} + \frac{\sqrt{2}}{3}\mathbf{i}\right)^n, \qquad \text{for } k = 7, 8. \end{cases} \quad [2.50]$$

Next, the coefficients $\tilde{\pi}^{(2)}_{0,\bar{d},k}[n], k \in \mathbb{X}^{(2)}$ (computed from relation [2.12]), for $n \geq 1$, take the form

$$\tilde{\pi}^{(2)}_{0,\bar{d},5}[n] = \begin{cases} \frac{5}{72}, & \text{for } n = 1, \\ \left(\frac{5}{144} + \frac{\sqrt{2}\mathrm{i}}{144}\right)\left(\frac{1}{3} + \frac{\sqrt{2}\mathrm{i}}{6}\right)^{n-1} \\ \quad + \left(\frac{5}{144} - \frac{\sqrt{2}\mathrm{i}}{144}\right)\left(\frac{1}{3} - \frac{\sqrt{2}\mathrm{i}}{6}\right)^{n-1}, & \text{for } n > 1. \end{cases} \quad [2.51]$$

$$\tilde{\pi}^{(2)}_{0,\bar{d},6}[n] = \begin{cases} -\frac{1}{36}, & \text{for } n = 1, \\ \left(-\frac{1}{72} + \frac{5\sqrt{2}\mathrm{i}}{144}\right)\left(\frac{1}{3} + \frac{\sqrt{2}\mathrm{i}}{6}\right)^{n-1} \\ \quad - \left(\frac{1}{72} + \frac{5\sqrt{2}\mathrm{i}}{144}\right)\left(\frac{1}{3} - \frac{\sqrt{2}\mathrm{i}}{6}\right)^{n-1}, & \text{for } n > 1. \end{cases} \quad [2.52]$$

$$\tilde{\pi}^{(2)}_{0,\bar{d},7}[n] = \tilde{\pi}^{(2)}_{0,\bar{d},8}[n] = \begin{cases} -\frac{1}{48}, & \text{for } n = 1, \\ \left(-\frac{1}{96} - \frac{\sqrt{2}\mathrm{i}}{48}\right)\left(\frac{1}{3} + \frac{\sqrt{2}\mathrm{i}}{6}\right)^{n-1} \\ \quad + \left(-\frac{1}{96} + \frac{\sqrt{2}\mathrm{i}}{48}\right)\left(\frac{1}{3} - \frac{\sqrt{2}\mathrm{i}}{6}\right)^{n-1}, & \text{for } n > 1. \end{cases} \quad [2.53]$$

Note that, according to the remarks made in section 2.3.3, coefficients $\tilde{\pi}^{(2)}_{0,\bar{d},k}[n]$ are real numbers. They are given in relations [2.51]–[2.53] in the compact form, which can create a false impression that these coefficients may be complex numbers. However, we can see that the use of the Newton binomial formula, and following multiplication in the corresponding expressions, yields cancellation of all complex terms.

The asymptotic expansions for $\pi_{\varepsilon,k} = \frac{1}{2}\pi^{(2)}_{\varepsilon,k}$, $k \in \mathbb{X}^{(2)}$ take the following forms:

$$\pi_{\varepsilon,k} = \begin{cases} \frac{1}{12} + \frac{5}{144}\varepsilon + \frac{1}{108}\varepsilon^2 + O(\varepsilon^3), & \text{for } k = 5, \\ \frac{1}{6} - \frac{1}{72}\varepsilon - \frac{7}{432}\varepsilon^2 + O(\varepsilon^3), & \text{for } k = 6, \\ \frac{1}{8} - \frac{1}{96}\varepsilon + \frac{1}{288}\varepsilon^2 + O(\varepsilon^3), & \text{for } k = 7, \\ \frac{1}{8} - \frac{1}{96}\varepsilon + \frac{1}{288}\varepsilon^2 + O(\varepsilon^3), & \text{for } k = 8. \end{cases} \quad [2.54]$$

Hence, by combining relations [2.49] and [2.54], we obtain the asymptotic expansions for stationary probabilities $\pi_{\varepsilon,k}$, $k \in \mathbb{X} = \mathbb{X}^{(1)} \cup \mathbb{X}^{(2)}$,

$$\pi_{\varepsilon,k} = \begin{cases} \frac{1}{16} + \frac{7}{128}\varepsilon - \frac{1}{1024}\varepsilon^2 + O(\varepsilon^3), & \text{for } k = 1, \\ \frac{3}{16} - \frac{3}{128}\varepsilon - \frac{27}{1024}\varepsilon^2 + O(\varepsilon^3), & \text{for } k = 2, \\ \frac{1}{8} - \frac{1}{64}\varepsilon + \frac{7}{512}\varepsilon^2 + O(\varepsilon^3), & \text{for } k = 3, \\ \frac{1}{8} - \frac{1}{64}\varepsilon + \frac{7}{512}\varepsilon^2 + O(\varepsilon^3), & \text{for } k = 4. \\ \frac{1}{12} + \frac{5}{144}\varepsilon + \frac{1}{108}\varepsilon^2 + O(\varepsilon^3), & \text{for } k = 5, \\ \frac{1}{6} - \frac{1}{72}\varepsilon - \frac{7}{432}\varepsilon^2 + O(\varepsilon^3), & \text{for } k = 6, \\ \frac{1}{8} - \frac{1}{96}\varepsilon + \frac{1}{288}\varepsilon^2 + O(\varepsilon^3), & \text{for } k = 7, \\ \frac{1}{8} - \frac{1}{96}\varepsilon + \frac{1}{288}\varepsilon^2 + O(\varepsilon^3), & \text{for } k = 8. \end{cases} \quad [2.55]$$

Let us mention again that the equalities given in relation [2.14] hold for coefficients given in relations [2.51]–[2.53] and [2.55].

Expansion [2.55] lets us illustrate the results presented in theorem 2.8. What is most interesting in this theorem is the asymptotic relation [2.46].

In order to illustrate this relation, we choose the initial distribution $\bar{p} = \langle 1, 0, 0, 0, 0, 0, 0, 0 \rangle$, which is concentrated in one ergodic class $\mathbb{X}^{(1)}$. In this case, probabilities $f^{(1)}_{\bar{p}} = 1$, $f^{(2)}_{\bar{p}} = 0$. Note that probabilities $f^{(1)}_{\bar{d}} = f^{(2)}_{\bar{d}} = \frac{1}{2}$ for the initial distribution $\bar{p} = \bar{d}$. Probabilities $\pi_{0,\bar{p},k}$ and $\pi_{0,\bar{d},k}$ are computed using relation [2.4]. The asymptotic relation [2.46] takes, for $k = 1$, the form $p_{\varepsilon,\bar{p},1}(n_\varepsilon) \to \pi_{0,\bar{p},1}(t)$ for $n_\varepsilon \to \infty$, $\varepsilon n_\varepsilon \to t$ and $\varepsilon \to 0$. Further, we choose $\varepsilon = 0.1$. According to the asymptotic relation [2.46], the values $p_{\varepsilon,\bar{p},1}(n_\varepsilon)$, for n_ε such that $\varepsilon n_\varepsilon \approx t$ can be expected to take values close to $\pi_{0,\bar{p},1}(t)$, for $1 \leq t \leq 3$. This, indeed, can be seen in Figure 2.6.

In addition, the result shows that the relative absolute errors $|\pi_{0,\bar{p},1}(t) - \pi_{0,\bar{d},1}|$ decreases dramatically from about 37 to 4% as $\varepsilon n_\varepsilon$ increases from 1 to 3, respectively.

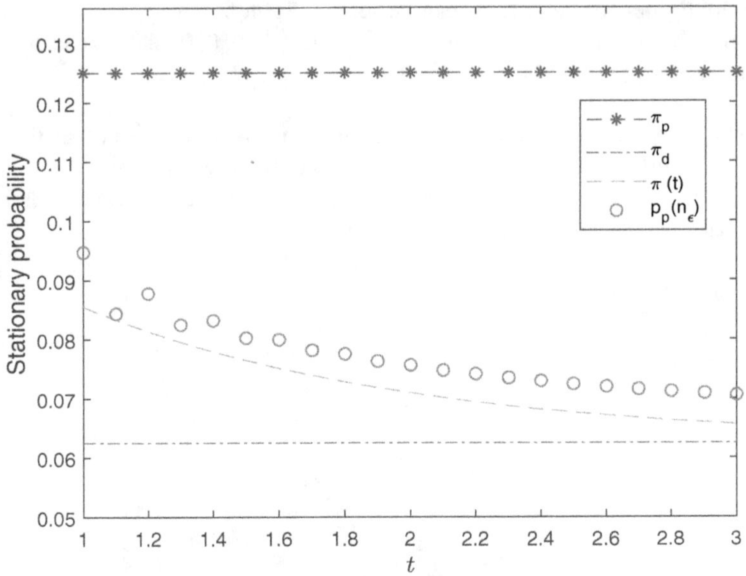

Figure 2.6. *Illustration of theorem 2.8 (ii) (π_p, $p_p(n_\varepsilon)$, $\pi(t)$ and π_d represent, respectively, $\pi_{0,\bar{p},1}$, $p_{\varepsilon,\bar{p},1}(n_\varepsilon)$, $\pi_{0,\bar{p},1}(t)$ and $p_{\varepsilon,\bar{d},1}$). For a color version of this figure, see www.iste.co.uk/dimotikalis/analysis1.zip*

ε	$k \in \mathbb{X}$	(n, N)	$p_{\varepsilon,\bar{p},k}(n)$	$\pi_{\varepsilon,k}$	$\lvert p_{\varepsilon,\bar{p},k}(n) - \pi_{\varepsilon,k}\rvert$	Upper bound
0.15	1	(2,2)	0.26490	0.07068	0.19422	0.47733
0.15	5	(2,2)	0.02938	0.08875	0.05937	0.06314
0.15	1	(5,5)	0.06793	0.07070	0.00277	0.14545
0.15	5	(5,5)	0.05183	0.08875	0.03692	0.03730
0.15	1	(30,5)	0.07118	0.07070	0.00048	0.00048
0.15	5	(30,5)	0.08811	0.08875	0.00064	0.00064

Table 2.2. *Upper bounds for the rate of convergence*

In Table 2.2, values for $p_{\varepsilon,\bar{p},k}(n)$, $\pi_{\varepsilon,k}$ and $|p_{\varepsilon,\bar{p},k}(n) - \pi_{\varepsilon,k}|$ are given, where $p_{\varepsilon,\bar{p},k}(n)$ and $\pi_{\varepsilon,k}$ are computed with the use of rounded solutions of relation [2.48] and [2.55], respectively. The upper bounds given by inequality [2.43], for $\varepsilon = 0.15$, $\bar{p} = (1, 0, 0, 0, 0, 0, 0, 0)$ and the quantities $Q(\bar{p}^{(g)}, \bar{\pi}_0^{(g)})$ and $Q(\bar{\pi}_\varepsilon^{(g)}, \bar{\pi}_0^{(g)})$, both computed by relation [2.37], are also given. The ergodicity coefficient $\Delta_N(\mathbf{P}_{0,g})$, for $g = 1, 2$, is obtained with the use of relations [2.34]–[2.36]. The results indicate that we can always obtain effective upper bounds for singularly perturbed Markov chains with damping component, without the use of the second largest eigenvalue. Moreover, the bounds are tighter for large n, as can be seen in Table 2.2. Figure 2.7 displays the overall picture for probability $p_{\varepsilon,\bar{p},1}(n)$ (for $\bar{p} = (1,0,0,0,0,0,0,0)$) as a function of bivariate parameter (ε, n) in the domain $[0.02, 0.2] \times [1, 2, \dots, 50]$.

The surface presented in Figure 2.7 smoothly approximates values of the above function in rectangle $[0.02, 0.2] \times [1, 50]$. The dotted curve represents the function, which takes the values $\pi_{0,\bar{p},1}(\varepsilon n)$ at points $(0.2, n)$ (where n is considered as a continuous parameter taking values in the interval $[1, 50]$).

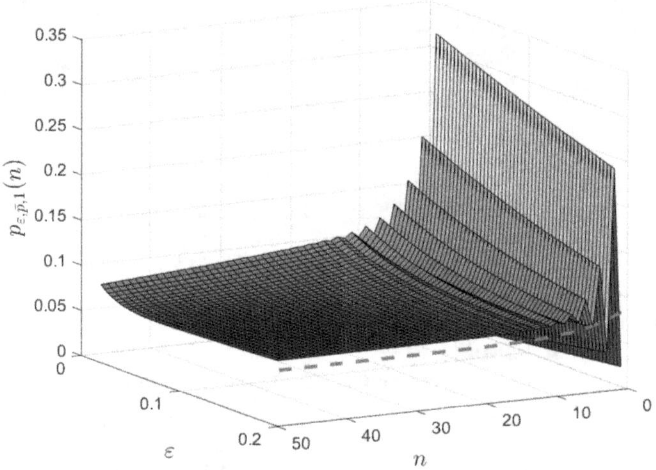

Figure 2.7. *Probability $p_{\varepsilon,\bar{p},1}(n)$. For a color version of this figure, see www.iste.co.uk/dimotikalis/analysis1.zip*

Figure 2.8 displays the overall pictures for functions $\pi_{0,\bar{p},1}(\varepsilon n)$ (for $\bar{p} = (1,0,0,0,0,0,0,0)$) at points of rectangle $[0.02, 0.2] \times [1, 50]$ (again, n is considered as a continuous parameter).

Figures 2.6–2.8 support and illustrate the asymptotic relation [2.46], given in theorem 2.8, and let us conclude that we should be careful when using probabilities $p_{\varepsilon,\bar{p},j}(n_\varepsilon)$ as approximations for stationary probabilities for the Markov chain $X_{0,n}$.

The computational examples presented in this chapter illustrate how the results of asymptotic perturbation analysis, presented in theorems 2.1–2.8, can be used in experimental studies of Markov chains with damping components associated with information networks.

We hope to present the results of more comprehensive experimental studies in future publications.

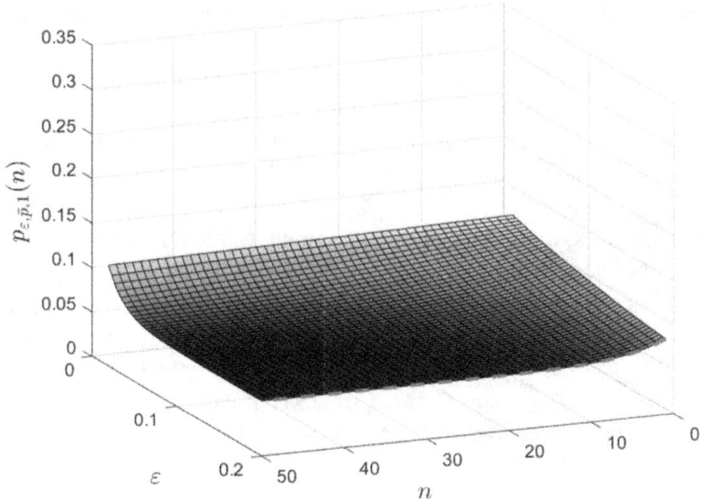

Figure 2.8. *Function* $\pi_{0,\bar{p},1}(\varepsilon n), (\varepsilon, n) \in [0.02, 0.2] \times [1, 50]$. *For a color version of this figure, see www.iste.co.uk/dimotikalis/analysis1.zip*

2.5. Acknowledgments

This research was supported by the Swedish International Development Cooperation Agency (Sida), the International Science Programme (ISP) (namely the International Programme in the Mathematical Sciences (IPMS)) and Sida Bilateral Research Programmes for research and education capacity development in Mathematics in Uganda and Tanzania. The authors are also grateful to the research environment Mathematics and Applied Mathematics (MAM), Division of Applied Mathematics, Mälardalen University, for providing an excellent and inspiring environment for research in mathematics.

2.6. References

Abola, B., Biganda, P.S., Engström, C., Mango, J.M., Kakuba, G., Silvestrov, S. (2018). PageRank in evolving tree graphs. In *Stochastic Processes and Applications*, Silvestrov, S., Rančić, M., Malyarenko, A. (eds). Springer, Cham.

Abola, B., Biganda, P.S., Silvestrov, D.S., Silvestrov, S., Engström, C., Mango, J.M., Kakuba, G. (2019). Perturbed Markov chains and information networks [Online]. Available at: arXiv:1901.11483v3[math.PR], 59 pp.

Abola, B., Biganda, P.S., Engström, C., Mango, J.M., Kakuba, G., Silvestrov, S. (2020). Updating of PageRank in evolving tree graphs. In *Data Analysis and Applications 3: Computational, Classification, Financial, Statistical and Stochastic Methods*, Makrides, A., Karagrigoriou, A., Skiadas, C.H. (eds). ISTE Ltd, London and Wiley, New York.

Andersson, F.K. and Silvestrov, S.D. (2008). The mathematics of internet search engines. *Acta Appl. Math.*, 104(2), 211–242.

Avrachenkov, K.E., Filar, J.A., Howlett, P.G. (2013). *Analytic Perturbation Theory and Its Applications*. SIAM, Philadelphia.

Avrachenkov, K., Piunovskiy, A., Zhang, Y. (2018). Hitting times in Markov chains with restart and their application to network centrality. *Methodol. Comput. Appl. Probab.*, 20(4), 1173–1188.

Battiston, S., Puliga, M., Kaushik, R., Tasca, P., Caldarelli, G. (2012). Debtrank: Too central to fail? Financial networks, the FED and systemic risk. *Scientific Reports*, 2(541).

Biganda, P.S., Abola, B., Engström, C., Silvestrov, S. (2017). PageRank, connecting a line of nodes with multiple complete graphs. In *Proceedings of the 17th Applied Stochastic Models and Data Analysis International Conference with the 6th Demographics Workshop*. London.

Biganda, P.S., Abola, B., Engström, C., Magero, J.M., Kakuba, G., Silvestrov, S. (2018). Traditional and lazy PageRanks for a line of nodes connected with complete graphs. In *Stochastic Processes and Applications*, Silvestrov, S., Rančić, M., Malyarenko, A. (eds). Springer, Cham.

Biganda, P.S., Abola, B., Engström, C., Mango, J.M., Kakuba, G., Silvestrov, S. (2020). Exploring the relationship between ordinary PageRank, lazy PageRank and random walk with backstep PageRank for different graph structures. In *Data Analysis and Applications 3: Computational, Classification, Financial, Statistical and Stochastic Methods*, Makrides, A., Karagrigoriou, A., Skiadas, C.H. (eds). ISTE Ltd, London and Wiley, New York.

Bini, D.A., Latouche, G., Meini, B. (2005). *Numerical Methods for Structured Markov Chains. Numerical Mathematics and Scientific Computation*. Oxford University Press, New York.

Brin, S. and Page, L. (1998). The anatomy of a large-scale hypertextual web search engine. *Comp. Networks, ISDN Syst.*, 30(1–7), 107–117.

Englund, E. (2001). Nonlinearly perturbed renewal equations with applications. Doctoral Dissertation, Umeå University, Sweden.

Englund, E. and Silvestrov, D.S. (1997). Mixed large deviation and ergodic theorems for regenerative processes with discrete time. In *Proceedings of the Second Scandinavian–Ukrainian Conference in Mathematical Statistics*. Umeå, Sweden.

Engström, C. (2016). PageRank in evolving networks and applications of graphs in natural language processing and biology. Doctoral Dissertation. Mälardalen University, Västerås.

Engström, C. and Silvestrov, S. (2014). Generalisation of the damping factor in PageRank for weighted networks. In *Modern Problems in Insurance Mathematics*, Silvestrov, D.S., Martin-Löf, A. (eds). Springer, Cham.

Engström, C. and Silvestrov, S. (2016a). PageRank, a look at small changes in a line of nodes and the complete graph. In *Engineering Mathematics II. Algebraic, Stochastic and Analysis Structures for Networks, Data Classification and Optimization*, Silvestrov, S., Rančić, M. (eds). Springer, Cham.

Engström, C. and Silvestrov, S. (2016b). PageRank, connecting a line of nodes with a complete graph. In *Engineering Mathematics II. Algebraic, Stochastic and Analysis Structures for Networks, Data Classification and Optimization*, Silvestrov, S. and Rančić, M. (eds). Springer, Cham.

Engström, C. and Silvestrov, S. (2017). PageRank for networks, graphs, and Markov chains. *Teor. Ĭmovirn. Mat. Stat.*, 96, 61–83.

Feller, W. (1968). *An Introduction to Probability Theory and Its Applications*, 3rd edition. Wiley, New York.

Gambini, A., Krzyanowski, V., Pokarowski, P. (2008). Aggregation algorithms for perturbed Markov chains with applications to networks modeling. *SIAM J. Sci. Comput.*, 31(1), 45–73.

Gleich, D.F. (2015). PageRank beyond the Web. *SIAM Rev.*, 57(3), 321–363.

Griffeath, D. (1975). A maximal coupling for Markov chains. *Z. Wahrsch. verw. Gebiete*, 31, 95–106.

Gyllenberg, M. and Silvestrov, D.S. (2000). Nonlinearly perturbed regenerative processes and pseudo-stationary phenomena for stochastic systems. *Stoch. Process. Appl.*, 86, 1–27.

Gyllenberg, M. and Silvestrov, D.S. (2008). *Quasi-Stationary Phenomena in Nonlinearly Perturbed Stochastic Systems*. Walter de Gruyter, Berlin.

Hartfiel, D.J. and Meyer, C.D. (1998). On the structure of stochastic matrices with a subdominant eigenvalue near 1. *Linear Algebra Appl.*, 272(1–3), 193–203.

Haveliwala, T. and Kamvar, S. (2003). The second eigenvalue of the Google matrix. Technical Report 2003-20, Stanford University, USA.

Kalashnikov, V.V. (1989). Coupling method, development and applications. Supplement to the Russian edition of the book by Nummelin, E. (1984). *General Irredusible Markov Chains and Non-negative Operators*. Cambridge University Press (Russian edition, MIR, Moscow, 1989, 176–190).

Kamvar, S.D., Schlosser, M.T., Garcia-Molina, H. (2003). The eigentrust algorithm for reputation management in p2p networks. *Proceedings of the 12th International Conference on World Wide Web*. New York, USA.

Konstantinov, M., Gu, D.W., Mehrmann, V., Petkov, P. (2003). *Perturbation Theory for Matrix Equations. Studies in Computational Mathematics*. North-Holland, Amsterdam.

Korolyuk, V.S. and Limnios, N. (2005). *Stochastic Systems in Merging Phase Space*. World Scientific, Singapore.

Korolyuk, V.S. and Korolyuk, V.V. (1999). *Stochastic Models of Systems. Mathematics and Its Applications*. Kluwer, Dordrecht.

Langville, A.N. and Meyer, C.D. (2012). *Google's PageRank and Beyond: The Science of Search Engine Rankings*. Princeton University Press, Princeton.

Lindvall, T. (2002). *Lectures on the Coupling Method*. Wiley, New York.

Mitrophanov, A.Y. (2005). Sensitivity and convergence of uniformly ergodic Markov chains. *J. Appl. Prob.*, 42, 1003–1014.

Ni, Y. (2011). Nonlinearly perturbed renewal equations: Asymptotic results and applications. Doctoral Dissertation. Mälardalen University, Västerås.

Ni, Y., Silvestrov, D.S., Malyarenko, A. (2008). Exponential asymptotics for nonlinearly perturbed renewal equation with non-polynomial perturbations. *J. Numer. Appl. Math.*, 1(96), 173–197.

Petersson, M. (2016). Perturbed discrete time stochastic models. Doctoral Dissertation, Stockholm University, Sweden.

Pitman, J.W. (1979). On coupling of Markov chains. *Z. Wahrsch. verw. Gebiete*, 35, 315–322.

Silvestrov, D.S. (1978). The renewal theorem in a series scheme 1. *Teor. Veroyatn. Mat. Stat.*, 18, 144–161 (English translation in *Theory Probab. Math. Statist.*, 18, 155–172).

Silvestrov, D.S. (1979). The renewal theorem in a series scheme 2. *Teor. Veroyatn. Mat. Stat.*, 20, 97–116 (English translation in *Theory Probab. Math. Statist.* 20, 113–130).

Silvestrov, D.S. (1983). Method of a single probability space in ergodic theorems for regenerative processes 1. *Math. Operat. Statist., Ser. Optim.*, 14, 285–299.

Silvestrov, D.S. (1984a). Method of a single probability space in ergodic theorems for regenerative processes 2. *Math. Operat. Statist., Ser. Optim.*, 15, 601–612.

Silvestrov, D.S. (1984b). Method of a single probability space in ergodic theorems for regenerative processes 3. *Math. Operat. Statist., Ser. Optim.*, 15, 613–622.

Silvestrov, D.S. (1994). Coupling for Markov renewal processes and the rate of convergence in ergodic theorems for processes with semi-Markov switchings. *Acta Appl. Math.*, 34, 109–124.

Silvestrov, D.S. (2018). Individual ergodic theorems for perturbed alternating regenerative processes. In *Stochastic Processes and Applications*, Silvestrov, S., Rančić, M., Malyarenko, A. (eds). Springer, Cham.

Silvestrov, D.S. and Petersson, M. (2014). Exponential expansions for perturbed discrete time renewal equations. In *Applied Reliability Engineering and Risk Analysis*, Karagrigoriou, A., Lisnianski, A., Kleyner, A., Frenkel, I. (eds). Wiley, New York.

Silvestrov, D.S. and Pezinska, G. (1985). On maximally coinciding random variables. *Theor. Veroyatn. Mat. Stat.*, 32, 102–105 (English translation in *Theory Probab. Math. Statist.*, 32, 113–115).

Silvestrov, D.S. and Silvestrov, S. (2016). Asymptotic expansions for stationary distributions of perturbed semi-Markov processes. In *Engineering Mathematics II. Algebraic, Stochastic and Analysis Structures for Networks, Data Classification and Optimization*, Silvestrov, S. and Rančić, M. (eds). Springer, Cham.

Silvestrov, D.S. and Silvestrov, S. (2017a). *Nonlinearly Perturbed Semi-Markov Processes*. Springer, Cham.

Silvestrov, D.S. and Silvestrov, S. (2017b). Asymptotic expansions for stationary distributions of nonlinearly perturbed semi-Markov processes 1. *Methodol. Comput. Appl. Probab.*, 21(3), 945–964 (2019).

Silvestrov, D.S. and Silvestrov, S. (2017c). Asymptotic expansions for stationary distributions of nonlinearly perturbed semi-Markov processes 2. *Methodol. Comput. Appl. Probab.*, 21(3), 965–984 (2019).

Silvestrov, D.S. and Silvestrov, S. (2017d). Asymptotic expansions for power-exponential moments of hitting times for nonlinearly perturbed semi-Markov processes. *Teor. Ĭmovirn. Mat. Stat.*, 97, 171–187 (Also, in *Theory Probab. Math. Statist.*, 97, 183–200).

Silvestrov, D.S., Petersson, M., Hössjer, O. (2018). Nonlinearly perturbed birth-death-type models. In *Stochastic Processes and Applications*, Silvestrov, S., Rančić, M., Malyarenko, A. (eds). Springer, Cham.

Stewart, W.J. (1994). *Introduction to the Numerical Solution of Markov Chains*. Princeton University Press, Princeton.

Stewart, G.W. (1998). *Matrix Algorithms: Volume 1: Basic Decompositions*. SIAM, Philadelphia.

Stewart, G.W. (2001). *Matrix Algorithms: Volume 2: Eigensystems*. SIAM, Philadelphia.

Sun, Y. and Han, J. (2013). Mining heterogeneous information networks: A structural analysis approach. *ACM SIGKDD Explor. Newslet.*, 14(2), 20–28.

Yin, G.G. and Zhang, Q. (2005). *Discrete-Time Markov Chains. Two-Time-Scale Methods and Applications*. Springer, New York.

Yin, G.G. and Zhang, Q. (2013). Continuous-time Markov chains and applications. A two-time-scale approach. *Stochastic Modelling and Applied Probability*, 2nd edition, Springer, New York.

3

PageRank and Perturbed Markov Chains

PageRank is a widely used hyperlink-based algorithm for estimating the relative importance of nodes in networks. Since many real-world networks are large sparse networks, efficient calculation of PageRank is complicated. Moreover, we need to overcome dangling effects in some cases as well as slow convergence of the transition matrix. Primitivity adjustment with a damping (perturbation) parameter is one of the essential procedures known to ensure convergence of the transition matrix. If the perturbation parameter is not small enough, the transition matrix loses information due to the shift of information to the teleportation matrix. We formulate the PageRank problem as a first- and second-order Markov chains perturbation problem. Using numerical experiments, we compare convergence rates for different values of perturbation parameter on different graph structures and investigate the difference in ranks for the two problems.

3.1. Introduction

PageRank is a link-based criterion that captures the importance of web pages and provides a ranking of the pages in Google, subject to a user's search query (Brin and Page 1998; Langville and Meyer 2012; Avrachenkov *et al.* 2013; Engström and Silvestrov 2014, 2016a,b, 2017; Engström 2016; Biganda *et al.* 2017). The theory behind PageRank is built from the Perron–Frobenius theory (Berman and Plemmons 1994) and Markov chains (Norris 1998), whose matrix of transition probabilities, as defined in the literature, such as Haveliwala and Kamvar (2003), Andersson and Silvestrov (2008) and Langville and Meyer (2012), takes the form $\mathbf{G} = c\mathbf{P}+(1-c)\mathbf{D}$, where $\mathbf{P}$ is a row stochastic matrix that represents a real link structure of the web, $\mathbf{D}$ is a rank-one stochastic matrix (also known as damping matrix) that guarantees $\mathbf{G}$ to be

Chapter written by Pitos Seleka BIGANDA, Benard ABOLA, Christopher ENGSTRÖM, Sergei SILVESTROV, Godwin KAKUBA and John Magero MANGO.

primitive and $c \in (0,1)$ is a damping factor, which ensures convergence of the power method applied on $\mathbf{G}$ to yield a stationary distribution $\bar{\pi}$, which is termed here as the PageRank vector (Langville and Meyer 2012).

PageRank is a widely used hyperlink-based algorithm for estimating the relative importance of nodes in networks (Brin and Page 1998), and since many real-world networks are large sparse networks, efficient calculation of PageRank is complicated. Moreover, we need to overcome dangling effects in some cases as well as slow convergence of the transition matrix. Primitivity adjustment with a damping (perturbation) parameter $\varepsilon \in (0, \varepsilon_0]$ (for fixed $\varepsilon_0 \simeq 0.15$) is one of the essential procedures that is known to ensure convergence of the transition matrix (Langville and Meyer 2012). If ε is not small enough, the transition matrix loses information due to the shift of information to the teleportation matrix (Silvestrov and Silvestrov 2017).

Fundamental for the concept of PageRank is to use the stationary distribution $\bar{\pi}$ of the Markov chain on the web graph to rank web pages. However, the mathematics of PageRank is general and applies to any network in different domains, for instance, in bibliometrics, social and information network analysis; link prediction and recommendation and many others (Gleich 2015). In addition, there exist in the literature algorithms that are similar to PageRank. For instance, the EigenTrust algorithm applied in peer-to-peer networks (Kamvar *et al.* 2003), the DeptRank algorithm in financial networks (Battiston *et al.* 2012) and the PageRank-based vulnerable lines screening algorithm applied in electrical power transmission networks (Ma *et al.* 2019).

By considering how well PageRank and related algorithms can perform, especially in terms of rate of convergence and ranking behavior, variants of PageRank, with respect to some types of networks, have been studied (Biganda *et al.* 2018, 2020). Futhermore, for updating PageRank subject to changes in a network structure, the interested reader may refer to Biganda *et al.* (2017) and Abola *et al.* (2018, 2020).

Computation of PageRank depends solely on damping parameter c. The value of c must be correctly chosen to avoid convergence and stability problems (Andersson and Silvestrov 2008). Langville and Meyer (2012) and Avrachenkov *et al.* (2013) point out that the damping parameter plays an important role in regulating the level of uniform noise introduced to the system. If $c = 1$, there are several absorbing states for a random walk defined by $\mathbf{P}$. But with $0 < c < 1$, the Markov chain induced by $\mathbf{G}$ is ergodic (Avrachenkov *et al.* 2013), and as a result, the convergence of the PageRank algorithm is guaranteed (Langville and Meyer 2012). In this chapter, we formulate the PageRank problem as a first- and second-order Markov chains perturbation problem. Using numerical experiments, we compare convergence rates for the two problems for different values of ε on different graph structures and investigate the difference in ranks for the two problems.

In fact, PageRank is the stationary distribution of a singularly perturbed Markov chain $\mathbf{P}_\varepsilon = (1-\varepsilon)\mathbf{P} + \varepsilon\mathbf{D}$ with perturbation parameter $\varepsilon = 1 - c$. As will be shown in the next section, the matrix polynomial $\mathbf{P}_\varepsilon$ is the first-order linear perturbation problem of the form $\mathbf{P}_\varepsilon = \mathbf{P} + \varepsilon\mathbf{P}_1$. In this chapter, we formulate the matrix polynomial of the form $\mathbf{P}_\varepsilon = \frac{1-\varepsilon}{1+\varepsilon^2}\mathbf{P} + \frac{\varepsilon}{1+\varepsilon^2}\mathbf{D}_1 + \frac{\varepsilon^2}{1+\varepsilon^2}\mathbf{D}_2$, which we name as a second-order perturbed matrix or second-order perturbed Markov chain. For convenience, we will denote $\mathbf{P}_{\varepsilon,1}$ and $\mathbf{P}_{\varepsilon,2}$ for first- and second-order perturbed Markov chains (PMCs), respectively. Following a compromised value of $c = 0.85$ by Brin and Page (1998), we will use $\varepsilon \in (0, 0.15]$ to compare the convergence rates and ranking behavior between the two PageRank perturbation problems.

It is worth mentioning here that our formulation can be viewed as a way of understanding the effect of promoting (for instance, giving favor to) some nodes within the network or between inter-networks. However, one big challenge here is how best the matrices $\mathbf{P}_1$ and $\mathbf{P}_2$ should be defined in order to maintain the second-order perturbed matrix $\mathbf{P}_\varepsilon$ stochastic and irreducible.

The structure of this chapter is as follows. In section 3.2, we present the PageRank problem as a first-order PMC, followed by the second-order perturbation problem of the PageRank in section 3.3. Section 3.4 discusses convergences of the two PageRank problems, and finally, the overall conclusion is given in section 3.5.

3.2. PageRank of the first-order perturbed Markov chain

As mentioned earlier, PageRank is the stationary distribution of the singularly perturbed Markov chain with perturbation parameter $\varepsilon = 1 - c$, $c \in (0,1)$ and matrix of transition probabilities $\mathbf{P}_{\varepsilon,1}$ (Abola *et al.* 2019), which takes the form

$$\mathbf{P}_{\varepsilon,1} = (1-\varepsilon)\mathbf{P} + \varepsilon\mathbf{D}, \qquad [3.1]$$

where $\mathbf{P}$ is the $n \times n$ hyperlink matrix and $\mathbf{D} = \bar{e}\bar{u}^\top$ is a rank-one stochastic matrix, where $\bar{e}$ is a column n-vector of ones, while $\bar{u}$ is the uniform column vector. Usually $\bar{u} = (1/n)\bar{e}$, where n is the size of the network (total number of nodes). In addition, $\mathbf{P} = \mathbf{A} + \bar{g}\bar{u}^\top$, where $\mathbf{A}$ is the weighted adjacency matrix such that an element $a_{ij} = 0$ means there is no outlink from node i to j, otherwise $a_{ij} = \frac{1}{d_i}$, where d_i is the number of outlinks from node i. The n-vector $\bar{g}$ has elements equal to 1 for dangling nodes in the network and zero otherwise.

It follows that relation [3.1] may be formulated as a first-order linear perturbation problem $\mathbf{P}_{\varepsilon,1} = \mathbf{P} + \varepsilon\mathbf{P}_1$, where $\mathbf{P}_1 = \mathbf{D} - \mathbf{P}$, and according to Avrachenkov *et al.* (2013), its corresponding stationary distribution is the PageRank vector $\bar{\pi}_{\varepsilon,1} = \bar{\pi} + \varepsilon\bar{\pi}_1 + O(\varepsilon^2)$. Clearly, $\lim_{\varepsilon\to 0}\bar{\pi}_{\varepsilon,1} = \bar{\pi}$, where $\bar{\pi}$ is the stationary distribution (or PageRank) of the unperturbed Markov chain $\mathbf{P}$. Note that computational procedures on how to find the asymptotic expansion $\bar{\pi}_\varepsilon = \bar{\pi} + \varepsilon\bar{\pi}_1 + \varepsilon^2\bar{\pi}_2 + \cdots$ for any perturbed Markov chain $\mathbf{P}_\varepsilon$ may be obtained in the literature, for instance, in Avrachenkov *et al.* (2013) and Abola *et al.* (2019).

In the next section, we formulate the PageRank as a second-order perturbation problem. But first we give the following definition, useful for comparison purposes in the following sections.

DEFINITION 3.1.– *The normalized PageRank vector $\bar{\pi}_{\varepsilon,1}$ of the first-order perturbed Markov chain is the eigenvector corresponding to the dominant eigenvalue equal to 1 of matrix $\mathbf{P}_{\varepsilon,1} = (1-\varepsilon)\left(\mathbf{A} + \bar{g}\bar{u}^\top\right) + \varepsilon\bar{e}\bar{u}^\top$, where $\varepsilon \in (0,1)$, but usually $\varepsilon = 0.15$. Its corresponding non-normalized PageRank, denoted by $\bar{r}_{\varepsilon,1}$, is given by*

$$\bar{r}_{\varepsilon,1} = n\bar{u}^\top\left(\mathbf{I} - (1-\varepsilon)\mathbf{A}\right)^{-1}, \qquad [3.2]$$

where $\mathbf{I}$ is the identity matrix of size n (total number of nodes), and all other variables are as defined in [3.1]. *In addition, $\|\bar{\pi}_{\varepsilon,1}\|_1 = 1$, $\|\bar{r}_{\varepsilon,1}\|_1 \neq 1$, $\bar{\pi}_{\varepsilon,1} = \frac{\bar{r}_{\varepsilon,1}}{\|\bar{r}_{\varepsilon,1}\|_1}$ and $\bar{r}_{\varepsilon,1} \propto \bar{\pi}_{\varepsilon,1}$ (Langville and Meyer 2012; Engström and Silvestrov 2016a), where $\|\bar{x}\|_1 = \sum_i |x_i|$.*

3.3. PageRank of the second-order perturbed Markov chain

We define a second-order perturbation of the matrix $\mathbf{P}$ as

$$\mathbf{P}_{\varepsilon,2} = \frac{1-\varepsilon}{1+\varepsilon^2}\mathbf{P} + \frac{\varepsilon}{1+\varepsilon^2}\mathbf{D}_1 + \frac{\varepsilon^2}{1+\varepsilon^2}\mathbf{D}_2, \qquad [3.3]$$

where $\mathbf{P}$ is the hyperlink matrix as defined in relation [3.1]. The question is how best can we define matrices $\mathbf{D}_1$ and $\mathbf{D}_2$. We will consider a number of cases as described below.

Case 1. $\mathbf{D}_2 = \mathbf{D}_1^2$, where $\mathbf{D}_1 = \bar{e}\bar{u}^\top$. Following the definition of vectors $\bar{u}$ and $\bar{e}$, in this case, $\mathbf{D}_2 = \mathbf{D}_1$. Hence, relation [3.3] can be rewritten as

$$\mathbf{P}_{\varepsilon,2} = \frac{1-\varepsilon}{1+\varepsilon^2}\left(\mathbf{A} + \bar{g}\bar{u}^\top\right) + \frac{\varepsilon+\varepsilon^2}{1+\varepsilon^2}\bar{e}\bar{u}^\top. \qquad [3.4]$$

Recall that the PageRank is the normalized eigenvector associated with the dominant eigenvalue 1 of a modified link matrix and normalization equation $\bar{\pi}_\varepsilon\bar{e} = 1$ (Langville and Meyer 2012). Thus, we have

$$\bar{\pi}_{\varepsilon,2} = \bar{\pi}_{\varepsilon,2}\mathbf{P}_{\varepsilon,2} = \frac{1-\varepsilon}{1+\varepsilon^2}\bar{\pi}_{\varepsilon,2}\left(\mathbf{A} + \bar{g}\bar{u}^\top\right) + \frac{\varepsilon+\varepsilon^2}{1+\varepsilon^2}\bar{\pi}_{\varepsilon,2}\bar{e}\bar{u}^\top,$$

which reduces to

$$\bar{\pi}_{\varepsilon,2} = \frac{1-\varepsilon}{1+\varepsilon^2}\bar{\pi}_{\varepsilon,2}\mathbf{A} + \left(\frac{(1-\varepsilon)k_\varepsilon + \varepsilon + \varepsilon^2}{1+\varepsilon^2}\right)\bar{u}^\top, \qquad [3.5]$$

where $k_\varepsilon = \bar{\pi}_{\varepsilon,2}\bar{g}$.

Let $\bar{r}_{\varepsilon,2}$ be a non-normalized PageRank corresponding to the stationary distribution $\bar{\pi}_{\varepsilon,2}$. According to Engström and Silvestrov (2016a), $\bar{r}_{\varepsilon,2}$ is proportional to $\bar{\pi}_{\varepsilon,2}$. In particular, if $\bar{\pi}_{\varepsilon,2}$ itself is taken as the PageRank, then [3.5] yields

$$\bar{\pi}_{\varepsilon,2} = \zeta_\varepsilon \bar{u}^\top \left(\mathbf{I} - \frac{1-\varepsilon}{1+\varepsilon^2}\mathbf{A}\right)^{-1}, \qquad [3.6]$$

where $\zeta_\varepsilon = \frac{(1-\varepsilon)k_\varepsilon + \varepsilon + \varepsilon^2}{1+\varepsilon^2}$ is a scalar factor for every fixed ε. A useful non-normalized version of the PageRank is obtained when the scalar factor ζ_ε is replaced by a quantity independent of ε, that is, the total number of nodes n:

$$\bar{r}_{\varepsilon,2} = n\bar{u}^\top \left(\mathbf{I} - \frac{1-\varepsilon}{1+\varepsilon^2}\mathbf{A}\right)^{-1}. \qquad [3.7]$$

EXAMPLE 3.1.– *In this example, we illustrate the PageRank of the second-order perturbed Markov chain and compare it with the corresponding PageRank of the first-order discussed in the literature. We do this by using a five-node simple line graph that is linked by two outside nodes to one of the nodes in the line, making the whole network to comprise seven nodes, as shown in Figure 3.1. The matrices* $\mathbf{P}$ *and* $\mathbf{A}$ *for this network take the form*

$$\mathbf{P} = \begin{pmatrix} \frac{1}{7} & \frac{1}{7} & \frac{1}{7} & \frac{1}{7} & \frac{1}{7} & \frac{1}{7} & \frac{1}{7} \\ 1 & 0 & 0 & 0 & 0 & 0 & 0 \\ 0 & 1 & 0 & 0 & 0 & 0 & 0 \\ 0 & 0 & 1 & 0 & 0 & 0 & 0 \\ 0 & 0 & 0 & 1 & 0 & 0 & 0 \\ 0 & 0 & 1 & 0 & 0 & 0 & 0 \\ 0 & 0 & 1 & 0 & 0 & 0 & 0 \end{pmatrix}, \quad \mathbf{A} = \begin{pmatrix} 0 & 0 & 0 & 0 & 0 & 0 & 0 \\ 1 & 0 & 0 & 0 & 0 & 0 & 0 \\ 0 & 1 & 0 & 0 & 0 & 0 & 0 \\ 0 & 0 & 1 & 0 & 0 & 0 & 0 \\ 0 & 0 & 0 & 1 & 0 & 0 & 0 \\ 0 & 0 & 1 & 0 & 0 & 0 & 0 \\ 0 & 0 & 1 & 0 & 0 & 0 & 0 \end{pmatrix}. \qquad [3.8]$$

Using relation [3.2] *with matrix* $\mathbf{A}$ *as given in* [3.8], $\bar{r}_{\varepsilon,1} = (7 - 16\varepsilon + 16\varepsilon^2 - 7\varepsilon^3 + \varepsilon^4, 6 - 10\varepsilon + 6\varepsilon^2 - \varepsilon^3, 5 - 5\varepsilon + \varepsilon^2, 2 - \varepsilon, 1, 1, 1)$ *for nodes* $n_1, \ldots, n_5, \nu_1, \nu_2$, *in that order. This gives a normalization constant* $\|\bar{r}_{\varepsilon,1}\|_1 = 23 - 32\varepsilon + 23\varepsilon^2 - 8\varepsilon^3 + \varepsilon^4$. *For instance, the normalized PageRank for node* n_3 *is given by*

$$\pi_{\varepsilon,1,n_3} = \frac{5 - 5\varepsilon + \varepsilon^2}{23 - 32\varepsilon + 23\varepsilon^2 - 8\varepsilon^3 + \varepsilon^4} \qquad [3.9]$$

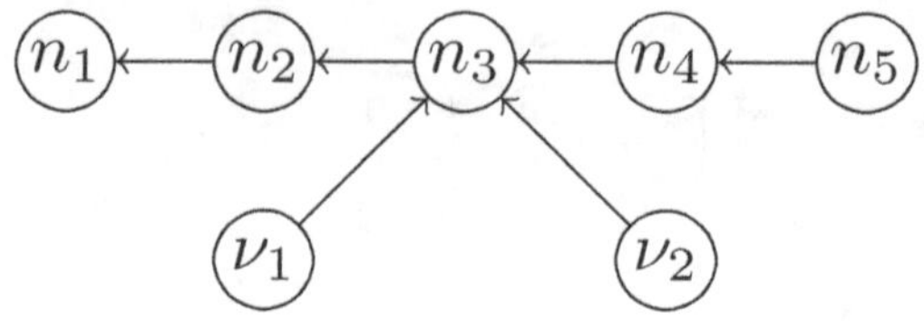

Figure 3.1. *A simple line network*

for $\varepsilon \in (0,1)$. *In a similar way, normalized PageRank values for all other nodes can be computed.*

For the second-order Markov chain perturbation problem, the non-normalized PageRank $\bar{r}_{\varepsilon,2}$ *can be computed by using relation* [3.7] *and normalized in a similar way to that described for the first-order case.*

We have presented, in Figures 3.2 and 3.3, normalized PageRank values for some specific nodes. Nodes that are not presented were found to have PageRank functional values similar to those presented here. Figure 3.2 presents the PageRank values for nodes n_3 *and* ν_1, *while Figure 3.3 gives the PageRank values for nodes* n_1 *and* n_5, *for each value of perturbation parameter* $\varepsilon \in (0, 0.3]$. *We see from the two figures that, except for node* n_1, *which is a dangling node, the PageRank values corresponding to the second-order perturbed Markov chain are higher than those corresponding to the first-order one for* $\varepsilon \in (0.05, 0.3]$. *However, for very small values of perturbation parameters, say* $0 < \varepsilon < 0.05$, *the two Markov chains give the same PageRank values for all given nodes in Figures 3.2 and 3.3.*

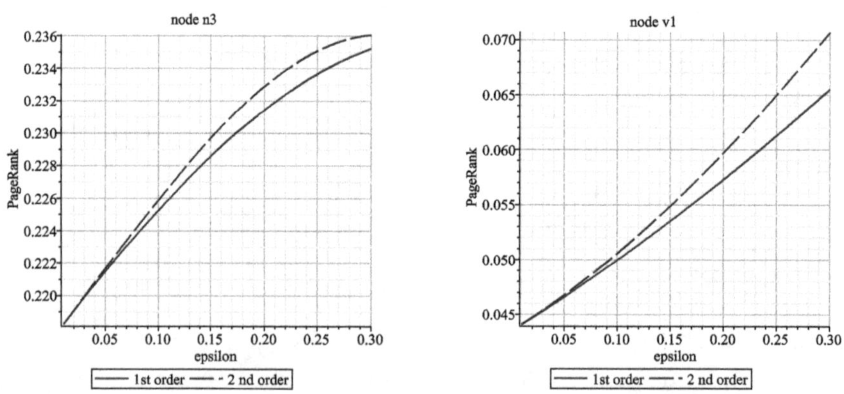

Figure 3.2. *Normalized PageRank as a function of perturbation parameter* ε *for first- and second-order perturbed Markov chains. For a color version of this figure, see www.iste.co.uk/dimotikalis/analysis1.zip*

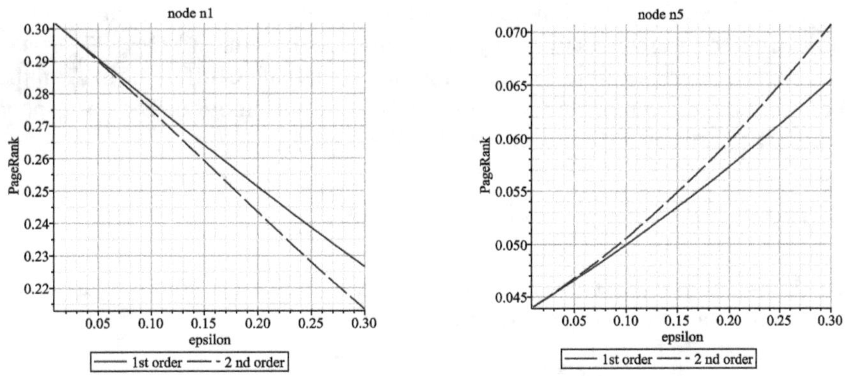

Figure 3.3. *Normalized PageRank as a function of perturbation parameter ε for first- and second-order perturbed Markov chains. For a color version of this figure, see www.iste.co.uk/dimotikalis/analysis1.zip*

Node label	n_1	n_2	n_3	n_4	n_5	ν_1	ν_2
1st-order MC ($\varepsilon = 0.15$)	0.26413	0.24780	0.22859	0.09898	0.05350	0.05350	0.05350
2nd-order MC ($\varepsilon = 0.1346$)	0.26413	0.24780	0.22859	0.09898	0.05350	0.05350	0.05350

Table 3.1. *Normalized PageRank values of first- and second-order perturbed Markov chains computed from relations* [3.2] *and* [3.7], *respectively*

Next, we would like to know the value of ε for which the second-order perturbed Markov chain gives equivalently the same PageRank values as the first-order perturbed Markov chain. Recall that traditional PageRank algorithms (here referred to as PageRank of the first-order perturbed Markov chain) use a parameter $c = 1-\varepsilon = 0.85$ (Brin and Page 1998). This is equivalent to $\varepsilon = 0.15$ in our current formulation of the first-order perturbed Markov chain. Now, setting $(1-\varepsilon)/(1+\varepsilon^2) = 0.85$, we obtain $\varepsilon \approx 0.1346$. This is the value of the perturbation parameter for the second-order perturbed Markov chain to give the same ranking values as the traditional PageRank. Table 3.1 outlines the rounded values of the normalized PageRank of first- and second-order perturbed Markov chains for the nodes in Figure 3.1. We see, from the table, that the normalized PageRank values of the first-order PMC at $\varepsilon = 0.15$ are the same as those of the second-order PMC at $\varepsilon = 0.1346$.

Case 2. Here, we consider the case where one node in the network is promoted by other nodes. By "promotion" we mean that some nodes in the network link to the particular node, including a loop on the node to be promoted. Under this setting, we wish to define a first-order perturbed Markov chain as

$$\mathbf{P}_{\varepsilon,1} = (1-\varepsilon)\tilde{\mathbf{P}} + \varepsilon\mathbf{D}, \qquad [3.10]$$

where $\tilde{\mathbf{P}} = \tilde{\mathbf{A}} + \bar{g}\bar{u}^\top$ is the link matrix of the resulting network with all promotions, with $\tilde{\mathbf{A}}$ being the weighted adjacency matrix after promotion and $\mathbf{D} = \bar{e}\bar{u}^\top$ is a rank-one stochastic matrix as in [3.1]. We emphasize here that $\tilde{\mathbf{A}}$ differs from $\mathbf{A}$ in that $\mathbf{A}$ is the weighted adjacency matrix of a network before promotion. The corresponding second-order perturbed Markov chain takes the form

$$\mathbf{P}_{\varepsilon,2} = \frac{1-\varepsilon}{1+\varepsilon^2}\mathbf{P} + \frac{\varepsilon}{1+\varepsilon^2}\mathbf{D}_1 + \frac{\varepsilon^2}{1+\varepsilon^2}\mathbf{D}_2, \qquad [3.11]$$

where $\mathbf{P} = \mathbf{A} + \bar{g}\bar{u}^\top$ is the hyperlink matrix and $\mathbf{D}_1$ is a row stochastic matrix representing promotional links and takes the form

$$\mathbf{D}_1 = \begin{matrix} \\ \\ i \\ \\ \\ \\ \end{matrix}\overset{\qquad\qquad i}{\begin{pmatrix} 0 & \dots & 1 & 0 & \dots & 0 \\ \vdots & \ddots & \vdots & 0 & \dots & 0 \\ 0 & \dots & 1 & \vdots & \ddots & \vdots \\ 0 & \dots & 1 & 0 & \dots & 0 \\ \vdots & \ddots & \vdots & \vdots & \ddots & \vdots \\ 0 & \dots & 1 & 0 & \dots & 0 \end{pmatrix}},$$

where i is the promoted node (or $\mathbf{D}_1 = \bar{e}\bar{v}_i^\top$, where $\bar{v}_i$ is a vector whose i-th entry is 1 and 0 elsewhere) and $\mathbf{D}_2 = \mathbf{D}$. Since $\mathbf{P}$, $\mathbf{D}_1$ and $\mathbf{D}_2$ are row stochastic matrices, $\mathbf{P}_{\varepsilon,2}$ given by [3.11] is row stochastic for all $\varepsilon \in [0,1]$. The relation [3.11], multiplied from the left by $\bar{\pi}_{\varepsilon,2}$ and regrouped, yields

$$\bar{\pi}_{\varepsilon,2} = \hat{\zeta}_\varepsilon \bar{u}^\top \left(\mathbf{I} - \frac{1}{1+\varepsilon^2}[(1-\varepsilon)\mathbf{A} + \varepsilon\mathbf{D}_1]\right)^{-1}, \qquad [3.12]$$

where $\hat{\zeta}_\varepsilon = \frac{(1-\varepsilon)k_\varepsilon + \varepsilon^2}{1+\varepsilon^2}$ is a scalar factor for every fixed ε. Note that expressions for the matrix and the scalar factor $\hat{\zeta}_\varepsilon$ in [3.12] are slightly different from those for the corresponding matrix and the scalar factor ζ_ε appearing on the right-hand side of [3.6]. Useful variants of the non-normalized PageRank are obtained when we replace the scalar factors $\hat{\zeta}_\varepsilon$ and ζ_ε in [3.12] and [3.6] by the total number of nodes n, which is independent of ε. Such a simplification and re-scaling of the right-hand side of [3.12] yields the following expression for the corresponding non-normalized PageRank:

$$\bar{r}_{\varepsilon,2} = n\bar{u}^\top \left(\mathbf{I} - \frac{1}{1+\varepsilon^2}[(1-\varepsilon)\mathbf{A} + \varepsilon\mathbf{D}_1]\right)^{-1}. \qquad [3.13]$$

Then, it is in our interest to compare the PageRanks of the three perturbed Markov chains: [3.1], [3.10] and [3.11]. In line to this, we have considered the network in Figure 3.1, this time, by allowing every node in the network to link to node ν_1, as shown in Figure 3.4 (see dotted edges). The resulting matrices $\tilde{\mathbf{P}}$ and $\mathbf{D}_1$ take the form [3.14], and matrices $\mathbf{P}$ and $\mathbf{D}_2$ are of the form [3.15].

$$\tilde{\mathbf{P}} = \begin{pmatrix} 0&0&0&0&0&1&0 \\ \frac{1}{2}&0&0&0&0&\frac{1}{2}&0 \\ 0&\frac{1}{2}&0&0&0&\frac{1}{2}&0 \\ 0&0&\frac{1}{2}&0&0&\frac{1}{2}&0 \\ 0&0&0&\frac{1}{2}&0&\frac{1}{2}&0 \\ 0&0&\frac{1}{2}&0&0&\frac{1}{2}&0 \\ 0&0&\frac{1}{2}&0&0&\frac{1}{2}&0 \end{pmatrix}, \quad \mathbf{D}_1 = \begin{pmatrix} 0&0&0&0&0&1&0 \\ 0&0&0&0&0&1&0 \\ 0&0&0&0&0&1&0 \\ 0&0&0&0&0&1&0 \\ 0&0&0&0&0&1&0 \\ 0&0&0&0&0&1&0 \\ 0&0&0&0&0&1&0 \end{pmatrix} \qquad [3.14]$$

$$\mathbf{P} = \begin{pmatrix} \frac{1}{7}&\frac{1}{7}&\frac{1}{7}&\frac{1}{7}&\frac{1}{7}&\frac{1}{7}&\frac{1}{7} \\ 1&0&0&0&0&0&0 \\ 0&1&0&0&0&0&0 \\ 0&0&1&0&0&0&0 \\ 0&0&0&1&0&0&0 \\ 0&0&1&0&0&0&0 \\ 0&0&1&0&0&0&0 \end{pmatrix}, \quad \mathbf{D}_2 = \begin{pmatrix} \frac{1}{7}&\frac{1}{7}&\frac{1}{7}&\frac{1}{7}&\frac{1}{7}&\frac{1}{7}&\frac{1}{7} \\ \frac{1}{7}&\frac{1}{7}&\frac{1}{7}&\frac{1}{7}&\frac{1}{7}&\frac{1}{7}&\frac{1}{7} \\ \frac{1}{7}&\frac{1}{7}&\frac{1}{7}&\frac{1}{7}&\frac{1}{7}&\frac{1}{7}&\frac{1}{7} \\ \frac{1}{7}&\frac{1}{7}&\frac{1}{7}&\frac{1}{7}&\frac{1}{7}&\frac{1}{7}&\frac{1}{7} \\ \frac{1}{7}&\frac{1}{7}&\frac{1}{7}&\frac{1}{7}&\frac{1}{7}&\frac{1}{7}&\frac{1}{7} \\ \frac{1}{7}&\frac{1}{7}&\frac{1}{7}&\frac{1}{7}&\frac{1}{7}&\frac{1}{7}&\frac{1}{7} \\ \frac{1}{7}&\frac{1}{7}&\frac{1}{7}&\frac{1}{7}&\frac{1}{7}&\frac{1}{7}&\frac{1}{7} \end{pmatrix} \qquad [3.15]$$

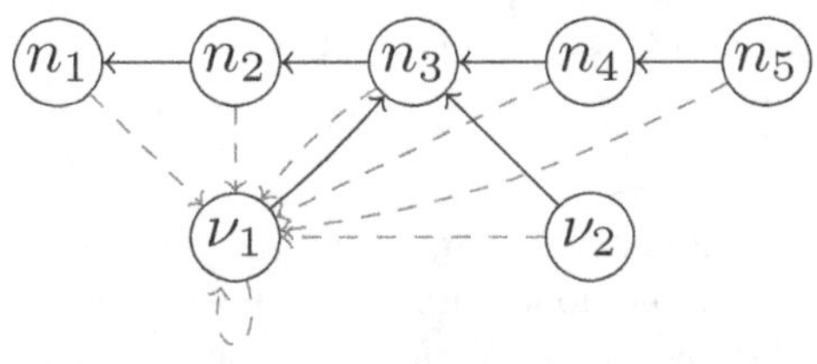

Figure 3.4. *A simple network with promotion of one node. The dotted edges represent added links due to promotion. For a color version of this figure, see www.iste.co.uk/dimotikalis/analysis1.zip*

In Figure 3.5, a comparison of rank values corresponding to the three perturbed Markov chains for nodes n_3 and ν_1 is presented.

Being the promoted node in the network, the results in Figure 3.5 show that node ν_1 has relatively higher PageRank of both first-order perturbed MC with promotion, and second-order perturbed MC, than the corresponding PageRank of first-order perturbed MC before promotion. Similar results can also be seen for node n_3, which is directly connected to ν_1. The only difference is that the PageRank of first-order perturbed MC with promotion decreases more sharply with an increase in the perturbation parameter value than for node ν_1. In addition, for all two nodes and at any value of ε, the PageRank of second-order perturbed MC is comparably higher than the corresponding PageRank of first-order perturbed MC before promotion. This complies with the reason why Brin and Page (1998) penalize spam web pages in their algorithms, to avoid these web pages and to have higher ranks in Google as spam web pages tend to link (or promote) between themselves.

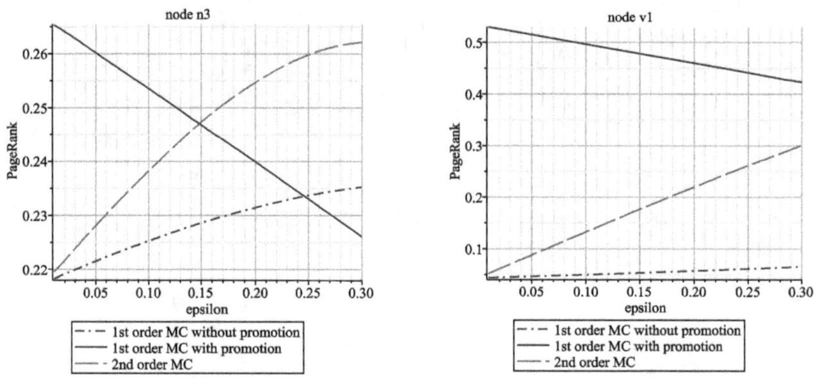

Figure 3.5. *Normalized PageRank as a function of perturbation parameter ε for first- and second-order perturbed Markov chains for nodes n_3 (left) and ν_1 (right). The first-order MC without promotion corresponds to relation* [3.2]*, the first-order MC with promotion corresponds to relation* [3.2] *but uses matrix $\tilde{\mathbf{P}}$ instead of $\mathbf{P}$, and the second-order MC corresponds to* [3.13]*. For a color version of this figure, see www.iste.co.uk/dimotikalis/analysis1.zip*

Case 3. Now we consider a network comprising some independent networks, where each of these networks has one node with the highest value of the PageRank of the first-order perturbed Markov chain. We identify such a node in each of the small networks and then allow the disjoint networks to interact by promoting the highest PageRank valued nodes. The second-order perturbed Markov chain is then formulated in a similar way to that in case 2 with matrix $\mathbf{D}_1$ built on the new added (dotted) links. Contrary to case 2, in this case, $\mathbf{D}_1$ has no uniform structure, but of course, it is a row stochastic matrix as in case 2. For instance, if we consider a network in Figure 3.6, which has two disjoint sub-networks $G_1 = \{1, 2\}$ and $G_2 = \{3, 4\}$, its corresponding matrix $\mathbf{D}_1$ takes the form given by relation [3.16].

$$\mathbf{D}_1 = \begin{pmatrix} 0\,0\,0\,1 \\ 0\,0\,0\,1 \\ 1\,0\,0\,0 \\ 1\,0\,0\,0 \end{pmatrix}. \qquad [3.16]$$

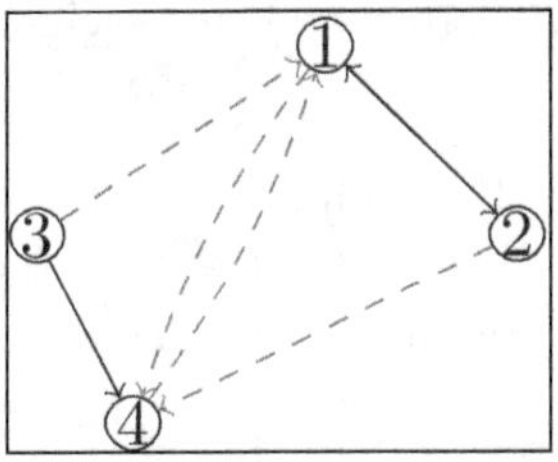

Figure 3.6. *An example of a network with two disjoint sub-networks. The dotted edges indicate new links between the two disjoint sub-networks due to promotion. For a color version of this figure, see www.iste.co.uk/dimotikalis/analysis1.zip*

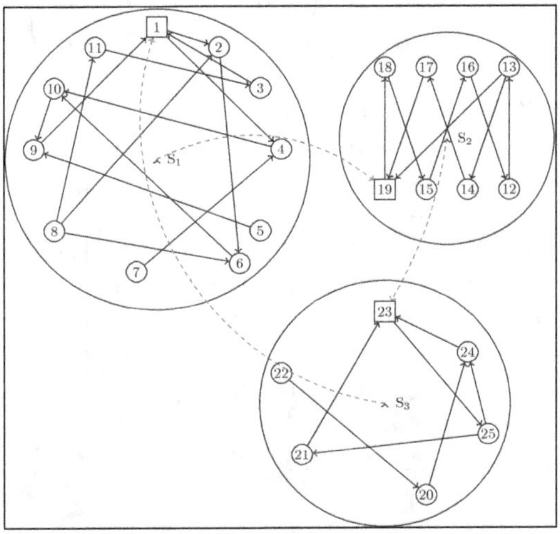

Figure 3.7. *A network with three independent sub-networks. A dotted edge from the center of one circle inscribing a sub-network to one node in the other sub-network means that all nodes in the particular sub-network link to the one node in the other sub-network as promotion case. For a color version of this figure, see www.iste.co.uk/dimotikalis/analysis1.zip.*

Now let us illustrate case 3 by using a network comprising three disjoint sub-networks, as shown in Figure 3.7. The weighted adjacency matrix $\mathbf{A}$ corresponding to this network takes the form

$$\mathbf{A} = \begin{pmatrix} \mathbf{A}_{11} & \mathbf{0} & \mathbf{0} \\ \mathbf{0} & \mathbf{A}_{22} & \mathbf{0} \\ \mathbf{0} & \mathbf{0} & \mathbf{A}_{33} \end{pmatrix}, \qquad [3.17]$$

where $\mathbf{A}_{11}$, $\mathbf{A}_{22}$ and $\mathbf{A}_{33}$ correspond to weighted adjacency matrices for sub-networks S_1, S_2 and S_3:

$$
\mathbf{A}_{11} = \begin{array}{c} \\ 1\\2\\3\\4\\5\\6\\7\\8\\9\\10\\11 \end{array}
\begin{array}{c}
\begin{array}{ccccccccccc} 1 & 2 & 3 & 4 & 5 & 6 & 7 & 8 & 9 & 10 & 11 \end{array}\\
\begin{pmatrix}
0 & \frac{1}{2} & 0 & \frac{1}{2} & 0 & 0 & 0 & 0 & 0 & 0 & 0\\
0 & 0 & 0 & 0 & 0 & 1 & 0 & 0 & 0 & 0 & 0\\
1 & 0 & 0 & 0 & 0 & 0 & 0 & 0 & 0 & 0 & 0\\
0 & 0 & 0 & 0 & 0 & 0 & 0 & 0 & 0 & 1 & 0\\
0 & 0 & 0 & 0 & 0 & 0 & 0 & 0 & 1 & 0 & 0\\
0 & 0 & 0 & 0 & 0 & 0 & 0 & 0 & 0 & 1 & 0\\
0 & 0 & 0 & 1 & 0 & 0 & 0 & 0 & 0 & 0 & 0\\
0 & \frac{1}{3} & 0 & 0 & 0 & \frac{1}{3} & 0 & 0 & 0 & 0 & \frac{1}{3}\\
1 & 0 & 0 & 0 & 0 & 0 & 0 & 0 & 0 & 0 & 0\\
0 & 0 & 0 & 0 & 0 & 0 & 0 & 0 & 1 & 0 & 0\\
0 & 0 & 1 & 0 & 0 & 0 & 0 & 0 & 0 & 0 & 0
\end{pmatrix}
\end{array}.
$$

$$
\mathbf{A}_{22} = \begin{array}{c} \\ 12\\13\\14\\15\\16\\17\\18\\19 \end{array}
\begin{array}{c}
\begin{array}{cccccccc} 12 & 13 & 14 & 15 & 16 & 17 & 18 & 19 \end{array}\\
\begin{pmatrix}
0 & 1 & 0 & 0 & 0 & 0 & 0 & 0\\
0 & 0 & \frac{1}{2} & 0 & 0 & 0 & 0 & \frac{1}{2}\\
0 & 0 & 0 & 0 & 0 & 1 & 0 & 0\\
0 & 0 & 0 & 0 & 1 & 0 & 0 & 0\\
1 & 0 & 0 & 0 & 0 & 0 & 0 & 0\\
0 & 0 & 0 & 0 & 0 & 0 & 0 & 1\\
0 & 0 & 0 & 1 & 0 & 0 & 0 & 0\\
0 & 0 & 0 & 0 & 0 & 0 & 1 & 0
\end{pmatrix}
\end{array}.
$$

$$
\mathbf{A}_{33} = \begin{array}{c} \\ 20\\21\\22\\23\\24\\25 \end{array}
\begin{array}{c}
\begin{array}{cccccc} 20 & 21 & 22 & 23 & 24 & 25 \end{array}\\
\begin{pmatrix}
0 & 0 & 0 & 0 & 1 & 0\\
0 & 0 & 0 & 1 & 0 & 0\\
1 & 0 & 0 & 0 & 0 & 0\\
0 & 0 & 0 & 0 & 0 & 1\\
0 & 0 & 0 & 1 & 0 & 0\\
0 & \frac{1}{2} & 0 & 0 & \frac{1}{2} & 0
\end{pmatrix}
\end{array}.
$$

The matrix $\mathbf{D}_1$, due to promotional links, is a 25×25 matrix whose first column $\mathbf{D}_1[:,1]$ has 1 in the row entries from 20 to 25 and 0 elsewhere, the 19th column $\mathbf{D}_1[:,19]$ has 1 in the row entries from 1 to 11 and 0 elsewhere and the 23rd column has 1 in the row entries from 12 to 19 and 0 elsewhere. All other remaining columns have 0 entries. Of course, this matrix is row stochastic, as described above. In addition, the columns 1, 19 and 23 correspond to nodes with high rank values in the sub-networks S_1, S_2 and S_3, as shown in Figure 3.7.

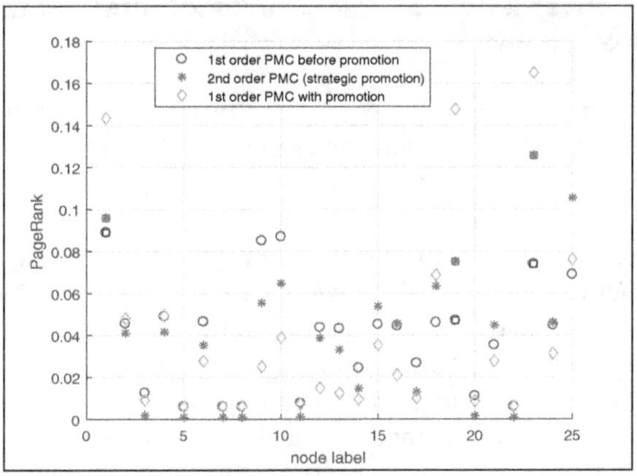

Figure 3.8. *PageRank values of first- and second-order perturbed Markov chains for all nodes in the network. The first-order MC before promotion corresponds to relation* [3.2], *the first-order MC with promotion corresponds to relation* [3.2] *but uses matrix* $\tilde{\mathbf{P}} = \tilde{\mathbf{A}} + \bar{g}\bar{u}^\top$ *instead of* $\mathbf{P} = \mathbf{A} + \bar{g}\bar{u}^\top$*, and the second-order MC corresponds to relation* [3.13]. *For a color version of this figure, see www.iste.co.uk/dimotikalis/analysis1.zip*

Furthermore, matrix $\tilde{\mathbf{A}}$ corresponding to the network in Figure 3.7, i.e. the weighted adjacency matrix after promotion, is defined in a similar way to that in case 2.

Then, Figure 3.8 displays the PageRank values of nodes in the two networks before and after promotion, as given in Figure 3.7. The following can be observed from Figure 3.8:

– the first-order perturbed MC built on the new network after some nodes having been promoted favors the promoted nodes only, i.e. these nodes have extremely higher PageRank values, while other nodes within sub-networks have either the same or fewer values than their corresponding PageRank values by the first-order perturbed Markov chain before promotion;

– the second-order perturbed MC gives relatively higher PageRank values for promoted nodes (1, 19 and 23, denoted by square markers) than the corresponding values by the first-order perturbed MC in the network;

– for sub-networks A and B, which are slightly dense, the difference between the PageRank values for the three PMCs is insignificant when compared to sub-network C.

Generally, the PageRank problem of the second-order perturbed Markov chains can be seen as a way to control over-scoring of some vertices when strategic promotion is made.

3.4. Rates of convergence of PageRanks of first- and second-order perturbed Markov chains

It is known from the literature (e.g. Haveliwala and Kamvar 2003; Kamvar and Haveliwala 2003; Langville and Meyer 2004, 2012; Andersson and Silvestrov 2008; Biganda *et al.* 2020) that the convergence of the power method algorithm for computation of PageRank depends on the second largest (in absolute value) eigenvalue $|\lambda_2|$ of $\mathbf{P}_\varepsilon$, where $|\lambda_2| \leq c$ such that the higher the value of $|\lambda_2|$, the slower the convergence. To our formulations of the two PageRank problems, this is equivalent to state that $|\lambda_2| \leq 1-\varepsilon$ and $|\lambda_2| \leq \frac{1}{1+\varepsilon^2}$ for the first- and second-order perturbed Markov chains, respectively. This shows that numerically the power method for computing PageRank of the second-order perturbed Markov chains converges slightly slower compared to the power method for PageRank of first-order perturbed Markov chains, since $\frac{1}{1+\varepsilon^2} \geq 1-\varepsilon$ for all $\varepsilon \in (0,1)$. However, an application of a power inner–outer iteration, introduced in Gu *et al.* (2015), improves the convergence better than the power method.

To apply the power inner–outer iteration, we re-define [3.11] as follows:

$$\mathbf{P}(c^*) = c^*\mathbf{P}^* + (1-c^*)\mathbf{D}_2, \qquad [3.18]$$

where $\mathbf{P}^* = (1-\varepsilon)\mathbf{P} + \varepsilon\mathbf{D}_1$ and $c^* = \frac{1}{1+\varepsilon^2}$. Pre-multiply both sides of [3.18] by $\bar{\pi}$ to obtain

$$\bar{\pi}\mathbf{P}(c^*) = c^*\bar{\pi}\mathbf{P}^* + (1-c^*)\bar{v}, \quad \bar{v} = \bar{\pi}\mathbf{D}_2 \qquad [3.19]$$

from which, at stationary, we get

$$\bar{\pi} = c^*\bar{\pi}\mathbf{P}^* + (1-c^*)\bar{v}$$

or

$$\bar{\pi}(\mathbf{I} - c^*\mathbf{P}^*) = (1-c^*)\bar{v}. \qquad [3.20]$$

We define a parameter $\beta \in (0, c^*)$. Then, by Gu *et al.* (2015), we have

$$\bar{\pi}(\mathbf{I} - \beta\mathbf{P}^*) = (c^* - \beta)\bar{\pi}\mathbf{P}^* + (1-c^*)\bar{v} \qquad [3.21]$$

and the iterative scheme takes the form

$$\bar{\pi}^{(k+1)} = \bar{\pi}^{(k)}\mathbf{M}_2^{-1}\mathbf{N}_2\mathbf{M}_1^{-1}\mathbf{N}_1 + \bar{b}\mathbf{M}_2^{-1}(\mathbf{I} + \mathbf{N}_2\mathbf{M}_1^{-1}), \qquad [3.22]$$

where

$$\mathbf{M}_1 = \mathbf{I}, \quad \mathbf{N}_1 = c^*\mathbf{P}^*,$$
$$\mathbf{M}_2 = \mathbf{I} - \beta\mathbf{P}^*, \quad \mathbf{N}_2 = (c^* - \beta)\mathbf{P}^*,$$
$$\bar{b} = (1-c^*)\bar{v}.$$

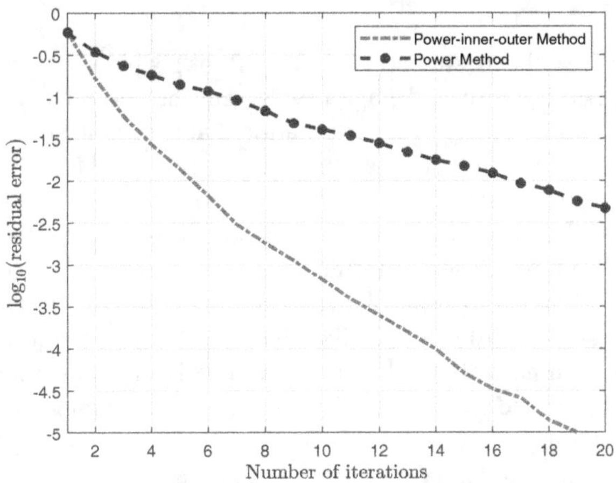

Figure 3.9. *Plot of rates of convergence of PageRanks of first- and second-order perturbed Markov chains for the network given in Figure 3.7. For a color version of this figure, see www.iste.co.uk/dimotikalis/analysis1.zip*

The convergence of the iterative scheme [3.22] depends on the spectral radius $\rho(\mathbf{M}_2^{-1}\mathbf{N}_2\mathbf{M}_1^{-1}\mathbf{N}_1)$, which if it is less than 1, the iterative sequence $\{\bar{\pi}^{(k)}\}$ converges to the unique solution $\bar{\pi}$ of [3.20] for all initial vectors $\bar{\pi}^{(0)}$. It follows from Gu *et al.* (2015, Theorem 2) that the iteration matrix $\mathbf{T}(c^*, \beta)$ of the power inner–outer method takes the form

$$\mathbf{T}(c^*, \beta) = c^*(c^* - \beta)(\mathbf{I} - \beta\mathbf{P}^*)^{-1}\mathbf{P}^{*2} \qquad [3.23]$$

and

$$\rho\left(\mathbf{T}(c^*, \beta)\right) \leq s < 1, \quad s = \frac{c^*(c^* - \beta)}{1 - \beta}, \qquad [3.24]$$

where β is always taken as either 0.5 or 0.6 Gleich *et al.* (2010).

Figure 3.9 illustrates the difference in rates of convergence between the two methods for computing PageRanks of second-order perturbed Markov chains (**case 3**). As can be seen, the power inner–outer method for computing the PageRank of the second-order perturbed Markov chains has an improved convergence speed when compared to the power method for computing the corresponding PageRank of first-order perturbed Markov chains. This computation involved the network given in Figure 3.7.

3.5. Conclusion

In this chapter, we have studied how PageRank, which was pioneered more than two decades ago by Brin and Page (1998), can be formulated as a first-order perturbed Markov chain. A theory on perturbation that is used to solve this PageRank formulation can be found in the literature, such as Avrachenkov *et al.* (2013); Silvestrov and Silvestrov (2017); Abola *et al.* (2019). The main contribution of this chapter is the formulation of the PageRank problem as a second-order perturbed Markov chain, where three different formulations or cases have been discussed. We have compared the two PageRank problems in terms of node-ranking in a network and found out that the second-order PMC give relatively higher PageRank values. This may make it suitable in network applications that involve identification of key nodes of the network and promote such nodes, for instance, in recommendation systems. In addition, by conducting numerical experiments on some randomly generated graph structures, we have compared rates of convergence for algorithms used for computing the two PageRank problems. We have found that the power method that is generally used to compute the PageRank problem of the first-order PMC converges slightly slower when applied to the second-order PMC PageRank problem. However, a variant of the power method known as "power inner–outer iteration" was found to give an improved convergence speed, better than the normal power method. Based on the aforementioned results, we conclude that the second-order PMC cannot be ignored since it is practically advantageous when strategic promotion is required. However, computationally it is challenging since the second eigenvalue of $\mathbf{P}_{\varepsilon,2}$ is very close to 1, leading to slow convergence. This calls for further numerical investigations of the problem.

3.6. Acknowledgments

This research was supported by the Swedish International Development Cooperation Agency (Sida), the International Science Programme (ISP) (namely the International Programme in the Mathematical Sciences (IPMS)) and Sida Bilateral Research Programmes for research and education capacity development in Mathematics in Tanzania and Uganda. The authors are also grateful to the research environment Mathematics and Applied Mathematics (MAM), Division of Applied Mathematics, Mälardalen University, for providing an excellent and inspiring environment for research in mathematics.

3.7. References

Abola, B., Biganda, P.S., Engström, C., Mango, J.M., Kakuba, G., Silvestrov, S. (2018). PageRank in evolving tree graphs. In *Stochastic Processes and Applications*, Silvestrov, S., Rančić, M., Malyarenko, A. (eds). Springer, Cham.

Abola, B., Biganda, P.S., Silvestrov, D.S, Silvestrov, S., Engström, C., Mango, J.M., Kakuba, G. (2019). Perturbed Markov chains and information networks [Online]. Available at: arXiv:1901.11483, 59 pp.

Abola, B., Biganda, P.S., Engström, C., Mango, J.M., Kakuba, G., Silvestrov, S. (2020). Updating of PageRank in evolving tree graphs. In *Data Analysis and Application 3: Computational, Classification, Financial, Statistical and Stochastic Methods*, Makrides, A., Karagrigoriou, A., Skiadas, C.H. (eds). ISTE Ltd, London, and Wiley, New York.

Andersson, F.K. and Silvestrov, S.D. (2008). The mathematics of internet search engines. *Acta Appl. Math.*, 104(2), 211–242.

Avrachenkov, K.E., Filar, J.A., Howlett, P.G. (2013). *Analytic Perturbation Theory and Its Applications*. SIAM, Philadelphia.

Battiston, S., Puliga, M., Kaushik, R., Tasca, P., Caldarelli, G. (2012). Debtrank: Too central to fail? Financial networks, the fed and systemic risk. *Scientific Reports*, 2(541).

Berman, A. and Plemmons, R.J. (1994). *Nonnegative Matrices in the Mathematical Sciences*. SIAM, Philadelphia.

Biganda, P.S., Abola, B., Engström, C., Silvestrov, S. (2017). PageRank, connecting a line of nodes with multiple complete graphs. In *Proceedings of 17th Applied Stochastic Models and Data Analysis International Conference with the 6th Demographics Workshop*. London, UK.

Biganda, P.S., Abola, B., Engström, C., Mango, J.M., Kakuba, G., Silvestrov, S. (2018). Traditional and lazy PageRanks for a line of nodes connected with complete graphs. In *Stochastic Processes and Applications*, Silvestrov, S., Rančić, M., Malyarenko, A. (eds). Springer, Cham.

Biganda, P.S., Abola, B., Engström, C., Mango, J.M., Kakuba, G., Silvestrov, S. (2020). Exploring the relationship between ordinary PageRank, lazy PageRank and random walk with backstep PageRank for different graph structures. In *Data Analysis and Applications 3: Computational, Classification, Financial, Statistical and Stochastic Methods*, Makrides, A., Karagrigoriou, A., Skiadas, C.H. (eds). ISTE Ltd, London, and Wiley, New York.

Brin, S. and Page, L. (1998). The anatomy of a large-scale hypertextual web search engine. *Comp. Networks, ISDN Syst.*, 30(1–7), 107–117.

Engström, C. (2016). PageRank in evolving networks and applications of graphs in natural language processing and biology. Doctoral Dissertation, Mälardalen University, Västerås.

Engström, C. and Silvestrov, S. (2014). Generalisation of the damping factor in PageRank for weighted networks. In *Modern Problems in Insurance Mathematics*, Silvestrov, D.S. and Martin-Löf, A. (eds). Springer, Cham.

Engström, C. and Silvestrov, S. (2016a). PageRank, a look at small changes in a line of nodes and the complete graph. In *Engineering Mathematics II. Algebraic, Stochastic and Analysis Structures for Networks, Data Classification and Optimization*, Silvestrov, S. and Rančić, M. (eds). Springer, Cham.

Engström, C. and Silvestrov, S. (2016b). PageRank, connecting a line of nodes with a complete graph. In *Engineering Mathematics II. Algebraic, Stochastic and Analysis Structures for Networks, Data Classification and Optimization*, Silvestrov, S. and Rančić, M. (eds). Springer, Cham.

Engström, C. and Silvestrov, S. (2017). PageRank for networks, graphs, and Markov chains. *Teor. Ĭmovirn. Mat. Stat.*, 96, 61–83 (reprinted in *Theor. Probab. Math. Statist.*, 96, 59–82, 2018).

Gleich, D.F. (2015). PageRank beyond the Web. *SIAM Rev.*, 57(3), 321–363.

Gleich, D.F., Gray, A.P., Greif, C., Lau, T. (2010). An inner-outer iteration for computing PageRank. *SIAM J. Sci. Comput.*, 32(1), 349–371.

Gu, C., Xie, F., Zhang, K. (2015). A two-step matrix splitting iteration for computing PageRank. *J. Comput. Appl. Math.*, 278, 19–28.

Haveliwala, T.H. and Kamvar, S.D. (2003). The second eigenvalue of the Google matrix. Technical report, Stanford InfoLab, Stanford.

Kamvar, S.D. and Haveliwala, T.H. (2003). The condition number of the PageRank problem. Technical report, Stanford InfoLab, Stanford.

Kamvar, S.D., Schlosser, M.T., Garcia-Molina, H. (2003). The eigentrust algorithm for reputation management in p2p networks. *Proceedings of the 12th International Conference on World Wide Web*. New York, USA.

Langville, A.N. and Meyer, C.D. (2004). Deeper inside PageRank. *Internet Math.*, 1(3), 335–380.

Langville, A.N. and Meyer, C.D. (2012). *Google's PageRank and Beyond: The Science of Search Engine Rankings*. Princeton University Press, Princeton.

Ma, Z., Shen, C., Liu, F., Mei, S. (2019). Fast screening of vulnerable transmission lines in power grids: A PageRank-based approach. *IEEE Trans. Smart Grid*, 10(2), 1982–1991.

Norris, J.R. (1998). *Markov Chains 2*. Cambridge University Press, Cambridge.

Silvestrov, D.S and Silvestrov, S. (2017). *Nonlinearly Perturbed Semi-Markov Processes*. Springer, Cham.

4

Doubly Robust Data-driven Distributionally Robust Optimization

Data-driven distributionally robust optimization (DD-DRO) via optimal transport has been shown to encompass a wide range of popular machine learning algorithms. The distributional uncertainty size is often shown to correspond to the regularization parameter. The type of regularization (e.g. the norm used to regularize) corresponds to the shape of the distributional uncertainty. We propose a data-driven robust optimization methodology to inform the transportation cost underlying the definition of the distributional uncertainty. Empirically, we show that this additional layer of robustification, which produces a method we call doubly robust data-driven distributionally robust optimization (DD-R-DRO), allows the generalization properties of regularized estimators to be enhanced while reducing testing error relative to state-of-the-art classifiers in a wide range of datasets.

4.1. Introduction

A wide class of popular machine learning estimators have recently been shown to be particular cases of data-driven distributionally robust optimization (DD-DRO) formulations with a distributional uncertainty set centered around the empirical distribution (Abadeh *et al.* 2015; Blanchet *et al.* 2016, 2017; Gao and Kleywegt 2016).

For example, regularized logistic regression (Lee *et al.* 2006), support vector machines and sqrt-Lasso (Belloni *et al.* 2011), among many other machine learning formulations can be represented as DD-DRO problems involving an uncertainty set composed of probability distributions, within a distance δ from the empirical distribution (Blanchet *et al.* 2016). The distance is measured in terms of a class of

Chapter written by Jose BLANCHET, Yang KANG, Fan ZHANG, Fei HE and Zhangyi HU.

suitably defined Wasserstein distances or, more generally, optimal transport distances between distributions (Blanchet *et al.* 2017).

This chapter aims to build an additional robustification layer on top of the DD-DRO formulation, which encompasses the machine learning algorithms mentioned earlier. Because of this second layer of robustification, we call our approach DD-R-DRO.

More specifically, we consider a parametric family of optimal transport distances and formulate a data-driven robust optimization (RO) problem for the selection of such a distance, which in turn is used to inform the distributional uncertainty region in the type of DD-DRO mentioned in the previous paragraph. In addition, we provide an iterative algorithm for solving such a RO problem.

In order to explain DD-R-DRO more precisely, let us discuss the different layers of robustness that are added in our various optimization formulations, and how these layers translate in terms of machine learning properties.

A DD-DRO problem takes the general form

$$\min_{\beta} \max_{P \in \mathcal{U}_\delta(P_n)} \mathbb{E}_P \left[l\left(X, Y, \beta\right)\right], \qquad [4.1]$$

where β is a decision variable, (X, Y) is a random element and $l\left(X, Y, \beta\right)$ is a loss incurred if the decision β is taken and (X, Y) is realized. The expectation $\mathbb{E}_P[\cdot]$ is taken under the probability model P. The set $\mathcal{U}_\delta\left(P_n\right)$ is called the distributional uncertainty set; it is centered around the empirical distribution P_n of the data, and it is indexed by the parameter $\delta > 0$, which measures the size of the distributional uncertainty.

The min–max problem in [4.1] can be interpreted as a game. We (the outer player) wish to learn a task using a class of machines indexed by β. An adversary (the inner player) is introduced to enhance out-of-sample performance. The adversary has a budget δ and can perturb the data, represented by P_n, in a certain way – this is important and we will return to this point. By introducing the artificial adversary and the distributional uncertainty, the DD-DRO formulation provides a direct mechanism to control the generalization properties of the learning procedure.

To further connect the DD-DRO representation [4.1] with more mainstream machine learning mechanisms, for the control of out-of-sample performance (such as regularization), we recall one of the explicit representations given in Blanchet *et al.* (2016).

In the context of generalized logistic regression (i.e. if the $l\left(x, y, \beta\right) = \log\left(1 + \exp\left(-y\beta^T x\right)\right)$, given an empirical sample $\mathcal{D}_n = \{(X_i, Y_i)\}_{i=1}^n$ with

$Y_i \in \{-1, 1\}$ and a judicious choice of the distributional uncertainty $\mathcal{U}_\delta(P_n)$, Blanchet *et al.* (2016) shows that

$$\min_{\beta} \max_{P \in \mathcal{U}_\delta(P_n)} \mathbb{E}_P[l(X, Y, \beta)] = \min_{\beta} \left(\mathbb{E}_{P_n}[l(X, Y, \beta)] + \delta \|\beta\|_p \right), \qquad [4.2]$$

where $\|\cdot\|_p$ is the l_p norm in $\mathbb{R}^d$ for $p \in [1, \infty)$ and $\mathbb{E}_{P_n}[l(X, Y, \beta)] = n^{-1} \sum_{i=1}^{n} l(X_i, Y_i, \beta)$.

The definition of $\mathcal{U}_\delta(P_n)$ turns out to be informed by the dual norm $\|\cdot\|_q$ with $1/p + 1/q = 1$. In simpler terms, the *shape* of the distributional uncertainty $\mathcal{U}_\delta(P_n)$ directly implies the *type of regularization* and the *size* of the distributional uncertainty, δ, dictates the regularization parameter.

The story behind the connection to sqrt-Lasso, support vector machines and other estimators is completely analogous to that given for [4.2]. A key point in most of the known representations, such as [4.2], is that they are only partially informed by data. Only the center, P_n, and the size, δ (via cross validation), are informed by data, but not the shape.

In recent work, Blanchet *et al.* (2017) proposes using metric learning procedures to inform the shape of the distributional uncertainty. But the procedure proposed in Blanchet *et al.* (2017), though data-driven, is not robustified.

One of the driving points of using robust optimization techniques in machine learning is that the introduction of an adversary can be seen as a tool used to control the testing error. While the data-driven procedure in Blanchet *et al.* (2017) is rich in the use of information, and is able to improve the generalization performance, the lack of robustification exposes the testing error to potentially high variability. Therefore, our contribution in this chapter is to design an RO procedure in order to choose the shape of $\mathcal{U}_\delta(P_n)$ using a suitable parametric family. In the context of logistic regression, for example, the parametric family that we consider includes the type of choice leading to [4.2] as a particular case. In turn, the choice of $\mathcal{U}_\delta(P_n)$ is applied to formulation [4.1] in order to obtain a *doubly robustified* estimator.

Figure 4.1 shows the various combinations of information and robustness, which have been studied in the literature so far. The figure shows four diagrams. Diagram (A) represents the standard empirical risk minimization (ERM), which uses the information fully, but often leads to high variability in testing error and, therefore, poor out-of-sample performance. Diagram (B) represents DD-DRO where only the center, P_n, and the size of the uncertainty, δ, are data-driven; this choice controls out-of-sample performance but does not use data to shape the type of perturbation, thus potentially resulting in testing error bounds, which might be pessimistic. Diagram (C) represents DD-DRO with data-driven shape information for perturbation type using metric learning techniques; this construction can reduce the testing error bounds at the expense of increase in the variability of the testing error estimates. Diagram (D) represents DD-R-DRO; the shape of the perturbation allowed for the adversary player is estimated using an RO procedure; this double robustification, as we will show in the numerical experiments, is able to control the variability present in the third diagram.

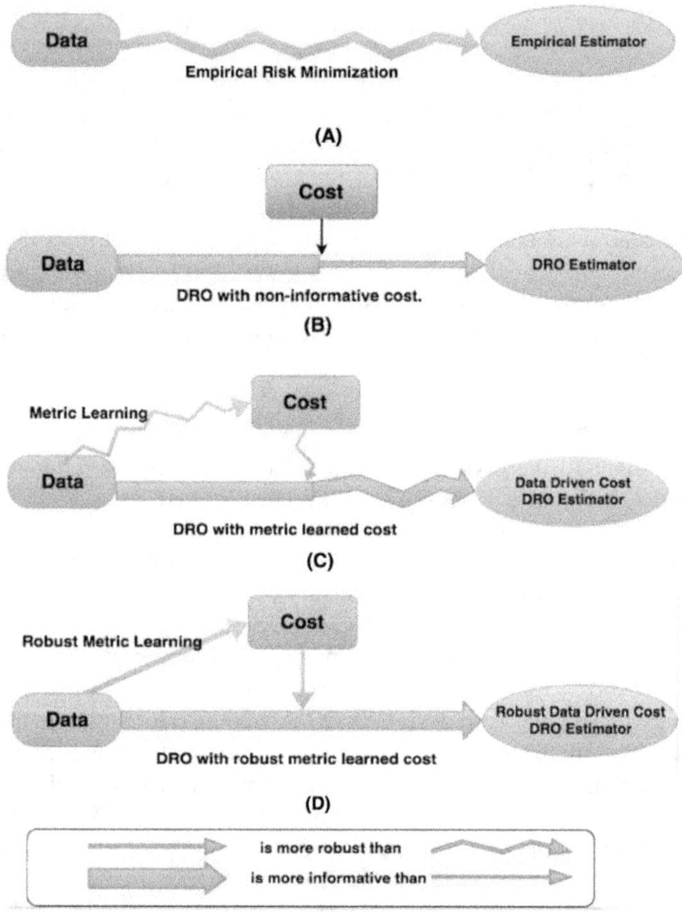

Figure 4.1. *Four diagrams illustrating information on robustness. For a color version of this figure, see www.iste.co.uk/dimotikalis/analysis1.zip*

In the diagrams, the straight arrows represent the use of a robustification procedure, a wide arrow represents the use of a high degree of information and a wiggly arrow indicates potentially noisy testing error estimates.

The contributions of this chapter can be stated, in order of importance, as follows:

1) the fourth diagram, DD-R-DRO, illustrates the main contribution of this chapter, namely, a double robustification approach, which reduces the generalization error, uses information efficiently and controls variability;

2) an explicit RO formulation for metric learning tasks;

3) iterative procedures for the solution of these RO problems.

4.2. DD-DRO, optimal transport and supervised machine learning

Let us consider a supervised machine learning classification problem, where we have a response $Y \in \{-1, 1\}$ and predictors $X \in \mathbb{R}^d$. Underlying, there is a general loss function $l(x, y, \beta)$ and a class of classifiers indexed by the parameter β. The goal for the supervised learning problem is to find the optimal decision parameter β minimizing the risk function under the true probability P_{true}, i.e.

$$\min_{\beta} \mathbb{E}_{P_{true}} \left[l\left(X, Y, \beta\right)\right]. \qquad [4.3]$$

Most of time, we do not have access to P_{true}, but we have a collection of observations as an independent sample of P_{true}. We can solve [4.3] by replacing P_{true} by the empirical measure P_n. **The empirical risk minimization would focus too much on the observed evidence, and the model faces the high risk of overfitting** (Esfahani and Kuhn 2015; Smith and Winkler 2006). As we discussed in the Introduction, our DD-DRO formalization is trying to address this issue. As defined in [4.1], we are trying to minimize the worst-case risk function rather than the ERM. We consider the worst case within a distributional uncertainty region, and it takes the form

$$\mathcal{U}_\delta\left(P_n\right) = \left\{P : D_c\left(P, P_n\right) \leq \delta\right\}, \qquad [4.4]$$

where $D_c\left(P, P_n\right)$ is a suitably defined notion of discrepancy between P and P_n, so that $D_c\left(P, P_n\right) = 0$ implies that $P = P_n$.

Other notions of discrepancy have been considered in the DRO literature, for example, the Kullback–Leibler divergence (or another divergence notion which depends on the likelihood ratio) is used (Hu and Hong 2013). Unfortunately, divergence criteria, which rely on the existence of the likelihood ratio between P and P_n, ultimately force P to share the same support as P_n, therefore potentially inducing undesirable out-of-sample performance.

Instead, we follow the approach in Abadeh *et al.* (2015), Esfahani and Kuhn (2015) and Blanchet *et al.* (2016), and define $D_c\left(P, P_n\right)$ as the optimal transport discrepancy between P and P_n.

4.2.1. *Optimal transport distances and discrepancies*

Assume that the cost function $c : \mathbb{R}^{d+1} \times \mathbb{R}^{d+1} \to [0, \infty]$ is lower semi-continuous. We also assume that $c(u, v) = 0$ if and only if $u = v$.

Given two distributions P and Q, with supports $\mathcal{S}_P$ and $\mathcal{S}_Q$, respectively, we define the optimal transport discrepancy, D_c, via

$$D_c(P, Q) = \inf\{E_\pi [c(U, V)] : \pi \in \mathcal{P}(\mathcal{S}_P \times \mathcal{S}_Q),\ \pi_U = P,\ \pi_V = Q\}, \quad [4.5]$$

where $\mathcal{P}(\mathcal{S}_P \times \mathcal{S}_Q)$ is the set of probability distributions π supported on $\mathcal{S}_P \times \mathcal{S}_Q$, and π_U and π_V denote the marginals of U and V under π, respectively. Because $c(\cdot)$ is non-negative, we have $D_c(P, Q) \geq 0$. Moreover, requiring that $c(u, v) = 0$ if and only if $u = v$ guarantees that $D_c(P, Q) = 0$ if and only if $P = Q$.

If, in addition, $c(\cdot)$ is symmetric (i.e. $c(u, v) = c(v, u)$), and there exists $\varrho \geq 1$ such that $c^{1/\varrho}(u, w) \leq c^{1/\varrho}(u, v) + c^{1/\varrho}(v, w)$ (i.e. $c^{1/\varrho}(\cdot)$ satisfies the triangle inequality), then it can be easily verified (see Villani (2008)) that $D_c^{1/\varrho}(P, Q)$ is a metric. For example, if $c(u, v) = \|u - v\|_q^\varrho$ for $q \geq 1$ (where $\|u - v\|_q$ denotes the l_q norm in $\mathbb{R}^{d+1}$), then $D_c(\cdot)$ is known at the Wasserstein distance of order ϱ.

Observe that [4.5] is obtained by solving a linear programming problem. For example, suppose that $Q = P_n$, so $Q \in \mathcal{P}(\mathcal{D}_n)$ and assume that the support $\mathcal{S}_P$ of P is finite. Then, using $U = (X, Y)$, $D_c(P, P_n)$ is obtained by computing

$$\min_\pi \sum_{u \in \mathcal{S}_P} \sum_{v \in \mathcal{D}_n} c(u, v)\, \pi(u, v) : \quad [4.6]$$

$$\text{s.t.} \sum_{u \in \mathcal{S}_P} \pi(u, v) = \frac{1}{n}\ \forall\, v \in \mathcal{D}_n$$

$$\sum_{v \in \mathcal{D}_N} \pi(u, v) = P(\{u\})\ \forall\, u \in \mathcal{X}_N,$$

$$\pi(u, v) \geq 0\ \forall\ (u, v) \in \mathcal{S}_P \times \mathcal{D}_n$$

A completely analogous linear program (LP), albeit an infinite-dimensional one, can be defined if $\mathcal{S}_P$ has infinitely many elements. This LP has been extensively studied in great generality in the context of optimal transport under the name of Kantorovich's problem (see Villani (2008)). Requiring $c(\cdot)$ to be lower semi-continuous guarantees the existence of an optimal solution to Kantorovich's problem.

Note that $D_c(P, P_n)$ can be interpreted as the minimal cost of rearranging (i.e. transporting the mass of) the distribution P_n into the distribution P. The rearrangement mechanism has a transportation cost $c(u,w) \geq 0$ for moving a unit of mass from location u in the support of P_n, to location w in the support of P. For instance, in the setting of [4.2], we have

$$c\left((x,y),(x',y')\right) = \|x-x'\|_q^2 I\left(y=y'\right) + \infty \cdot I\left(y \neq y'\right). \quad [4.7]$$

The infinite contribution in the definition of c (i.e. $\infty \cdot I\left(y \neq y'\right)$) indicates that the adversary player in the DRO formulation is not allowed to perturb the response variable.

4.3. Data-driven selection of optimal transport cost function

By suitably choosing $c(\cdot)$, we might further improve the generalization properties of the DD-DRO estimator based on [4.1]. To fix ideas, consider a suitably parameterized family of transportation costs as follows. Let Λ be a positive semi-definite matrix (denoted as $\Lambda \succeq 0$) and define $\|x\|_\Lambda^2 = x^T \Lambda x$. Inspired by [4.7], consider the cost function

$$c_\Lambda\left((x,y),(x',y')\right) = d_\Lambda^2\left(x,x'\right) I\left(y=y'\right) + \infty I\left(y \neq y'\right), \quad [4.8]$$

where $d_\Lambda^2\left(x,x'\right) = \|x-x'\|_\Lambda^2$. Then, Blanchet *et al.* (2017) shows that in the generalized logistic regression setting (i.e. $l(x,y,\beta) = \log(1+\exp\left(-y\beta^T x\right))$), if Λ is positive definite, we obtain

$$\min_\beta \max_{P:D_{c_\Lambda}(P,P_n)\leq\delta} \mathbb{E}\left[l(X,Y,\beta)\right] = \min_\beta \mathbb{E}_{P_n}\left[l(X,Y,\beta)\right] + \delta \|\beta\|_{\Lambda^{-1}}. \quad [4.9]$$

If the choice of Λ is data-driven in order to impose a penalty on transportation costs whose outcomes are highly impactful in terms of risk, then we would be able to control the risk bound induced by the DD-DRO formulation. This is the strategy studied in Blanchet *et al.* (2017), in which metric learning procedures have been precisely implemented in order to achieve such control. Our contribution, as we will explain in the next section, is the use of a robust optimization formulation to calibrate $c_\Lambda(\cdot)$. We emphasize that once $c_\Lambda(\cdot)$ is calibrated, it can be used for multiple learning tasks and arbitrary loss functions (not only logistic regression).

4.3.1. *Data-driven cost functions via metric learning procedures*

We quickly review the elements of standard metric learning procedures. In this chapter, we will concentrate on the binary response: our data is of the form $\mathcal{D}_n = \{(X_i,Y_i)\}_{i=1}^n$ and $Y_i \in \{-1,+1\}$, and the prediction variables are assumed to be standardized.

Motivated by applications such as social networks, in which there is a natural graph that can be used to connect instances in the data, we assume that one is given sets $\mathcal{M}$ and $\mathcal{N}$, where $\mathcal{M}$ is the set of the pairs that should be close (so that we can connect them) to each other, and $\mathcal{N}$, on the contrary, is characterizing the relations that the pairs should be far away (not connected). We define them as

$$\mathcal{M} = \{(X_i, X_j)\,|X_i \text{ and } X_j \text{ must connect}\},$$
$$\text{and } \mathcal{N} = \{(X_i, X_j)\,|X_i \text{ and } X_j \text{ should not connect}\}.$$

While it is typically assumed that $\mathcal{M}$ and $\mathcal{N}$ are given, we may always resort to the k-nearest neighbor (k-NN) method for the generation of these sets. This is the approach that we follow in our numerical experiments. But we emphasize that choosing any criterion for the definition of $\mathcal{M}$ and $\mathcal{N}$ should be influenced by the learning task in order to retain both interpretability and performance. In our experiments, we let (X_i, X_j) belong to $\mathcal{M}$ if, in addition to being sufficiently close (i.e. in the k-NN criterion), $Y_i = Y_j$. If $Y_i \neq Y_j$, then we have $(X_i, X_j) \in \mathcal{N}$.

In addition, we consider the relative constraint set $\mathcal{R}$ containing data triplets with relative relation defined as

$$\mathcal{R} = \{(i, j, k)\,|d_\Lambda(X_i, X_j) \text{ should be smaller than } d_\Lambda(X_i, X_k)\}.$$

Let us consider the following two formulations of metric learning, the so-called absolute metric learning formulation

$$\min_{\Lambda\succeq 0} \sum_{(i,j)\in\mathcal{M}} d_\Lambda^2(X_i, X_j) \quad \text{s.t.} \quad \sum_{(i,j)\in\mathcal{N}} d_A^2(X_i, X_j) \geq 1, \qquad [4.10]$$

and the relative metric learning formulation,

$$\min_{\Lambda\succeq 0} \sum_{(i,j,k)\in\mathcal{R}} \left(d_\Lambda^2\,(X_i, X_j) - d_\Lambda^2\,(X_i, X_k) + 1\right)_+. \qquad [4.11]$$

Both formulations have their merits: [4.10] exploits both the constraint sets $\mathcal{M}$ and $\mathcal{N}$, while [4.12] is only based on information in $\mathcal{R}$. Further intuition or motivation of these two formulations can be found in Xing *et al.* (2002) and Weinberger and Saul (2009), respectively. We will show how to formulate and solve the robust counterpart of the two representative examples by robustifying a single constraint set, or two sets simultaneously. For simplicity, we will only discuss these two formulations, but many metric learning algorithms are based on natural generalizations of these two forms, as mentioned in the survey (Bellet *et al.* 2013).

The assumption that the response variables arc binary is only evoked to construct the set $\mathcal{M}, \mathcal{N}$ and $\mathcal{R}$, and a similar idea can be generalized to the case where the response variable is continuous. For example, when $Y_i \in \mathbb{R}$, X_i and X_j should connect if and only if $|Y_i - Y_j|$ is small.

Once formulation [4.10] or [4.12] is considered and the matrix Λ has been calibrated, we may then consider the cost function $c_\Lambda(\cdot)$ in [4.8] and solve the problem [4.9]. This is the benchmark that we will consider in our numerical experiments. We will compare this approach with a method that chooses Λ using a robust optimization version of [4.10] or [4.12], which we will explain next.

4.4. Robust optimization for metric learning

In this section, we will review a robust optimization method to a metric learning optimization problem in order to learn a robust data-driven cost function. RO is a family of optimization techniques that deals with uncertainty or mis-specification in the objective function and constraints. RO was first proposed in Ben-Tal *et al.* (2009) and has attracted increasing attention in recent decades (El Ghaoui and Lebret 1997; Bertsimas *et al.* 2011). RO has been applied in machine learning to regularize statistical learning procedures, for example, in Xu *et al.* (2009a, 2009b), robust optimization was employed for SR-Lasso and support vector machines. We apply RO, as we will demonstrate, to reduce the variability in testing error when implementing DD-DRO.

4.4.1. *Robust optimization for relative metric learning*

The robust metric learning that we will use is based on the work of Huang *et al.* (2012). Consider the relative constraint set $\mathcal{R}$ containing data triplets with relative relation defined as

$$\mathcal{R} = \{(i,j,k)\,|d_\Lambda(X_i,X_j) \text{ should be smaller than } d_\Lambda(X_i,X_k)\}$$

and the relative metric learning formulation

$$\min_{\Lambda\succeq 0} \sum_{(i,j,k)\in\mathcal{R}} \left(d_\Lambda^2(X_i,X_j) - d_\Lambda^2(X_i,X_k) + 1\right)_+. \qquad [4.12]$$

Suppose we know that about $1-\alpha \in (0,1]$ of the constraints are noisy (the value of α is usually given by experience or can also be inferred by cross-validation), but we cannot exactly determine which part of them are noisy. Instead of optimizing over all subsets of constraints, we try to minimize the worst-case loss function over all possible $\alpha\,|\mathcal{R}|$ constraints (where $|\cdot|$ denotes the cardinality of a set) and obtain the following min–max formulation

$$\min_{\Lambda\succeq 0}\max_{q\in\mathcal{T}(\alpha)} \sum_{(i,j,k)\in\mathcal{R}} q_{i,j,k}\left(d_\Lambda^2(X_i,X_j) - d_\Lambda^2(X_i,X_k) + 1\right)_+, \qquad [4.13]$$

where $\mathcal{T}(\alpha)$ is a robust uncertainty set of the form

$$\mathcal{T}(\alpha) = \{q = \{q_{i,j,k} | (i,j,k) \in \mathcal{R}\} \, | 0 \leq q_{i,j,k} \leq 1, \sum_{(i,j,k)\in\mathcal{R}} q_{i,j,k} \leq \alpha \times |\mathcal{R}| \},$$

which is a convex and compact set.

First, we observe that the above minimax problem is equivalent to

$$\min_{\Lambda \succeq 0} \max_{q \in \mathcal{T}(\alpha)} \sum_{(i,j,k)\in\mathcal{R}} q_{i,j,k} \left(d_\Lambda^2 (X_i, X_j) - d_\Lambda^2 (X_i, X_k) + 1 \right), \qquad [4.14]$$

because $q_{i,j,k} = 0$ whenever $d_\Lambda^2(X_i, X_j) - d_\Lambda^2(X_i, X_k) + 1 < 0$. Therefore, our algorithm will be quite different from the original algorithm proposed by Huang *et al.* (2012) in the sense that our algorithm does not resort to any smooth technique; instead, we will use the primal-dual steepest descent algorithm.

Second, the objective function in [4.14] is convex in Λ and concave (linear) in q. Define

$$L(\Lambda, q) := \sum_{(i,j,k)\in\mathcal{R}} q_{i,j,k} \left(d_\Lambda^2 (X_i, X_j) - d_\Lambda^2 (X_i, X_k) + 1 \right) + \frac{\kappa}{2} \|\Lambda\|_F^2, \qquad [4.15]$$

where $\kappa > 0$ is the tuning parameter and $\|\cdot\|_F$ denotes the Frobenius norm. In addition, we define

$$f(\Lambda) := \max_{q\in\mathcal{T}(\alpha)} L(\Lambda, q), \quad g(q) := \min_{\Lambda\succeq 0} L(\Lambda, q) \qquad [4.16]$$

and since L is strongly convex in Λ, and linear in q, we also have

$$q(\Lambda) := \arg\max_{q\in\mathcal{T}(\alpha)} L(\Lambda, q), \quad \Lambda(q) := \arg\min_{\Lambda\succeq 0} L(\Lambda, q). \qquad [4.17]$$

Our iterative algorithm uses the fact that $f(\Lambda)$ and $q(\Lambda)$ can actually be obtained quickly by a simple sorting: for a fixed $\Lambda \succeq 0$, the inner maximization is linear in q, and the optimal q satisfies $q_{i,j,k} = 1$ whenever $(d_\Lambda(X_i, X_j) - d_\Lambda(X_i, X_k) + 1) \geq 0$ and ranks in the top $\alpha|\mathcal{R}|$ largest values and set $q_{i,j,k} = 0$ otherwise. If there are more than one optimal q's, then break the tie by choosing $q(\Lambda) := \arg\max_{q\in\mathcal{T}(\alpha)} L(\Lambda, q) - \kappa\|q - q^{(n)}\|_2^2$ at the $(n+1)$th iteration, where κ is a tuning parameter that is small enough, so that it ensures that $q(\Lambda)$ is chosen among all the optimal ones. On the other hand, since $L(\Lambda, q)$ is smooth in Λ, we can also obtain $g(q)$ and $\Lambda(q)$ quickly by gradient descent.

We summarize the primal-dual steepest descent algorithm in Algorithm 4.1.

Algorithm 4.1 Sequential coordinate-wise metric learning using relative relations

1: **Initialize** Set the iteration counter $n = 0$, the positive definite matrix $\Lambda^{(n)} = I_d$, and the tolerance $\epsilon = 10^{-3}$. Then randomly sample α proportion of elements from $\mathcal{R}$ to construct $q^{(n)}$. Compute $f(\Lambda^{(n)}), g(q^{(n)}), \Lambda(q^{(n)}), f(\Lambda(q^{(n)})), q(\Lambda^{(n)}), g(q(\Lambda^{(n)}))$ according to [4.16] and [4.17].

2: **While** $|\min\{f(\Lambda^{(n)}), f(\Lambda(q^{(n)}))\} - \max\{g(q^{(n)}), g(q(\Lambda^{(n)}))\}| > \epsilon$: **do**

3: (Line search) Generate the intermediate $\hat{\Lambda}^{(n+1)}, \hat{q}^{(n+1)}$ with perfect line search

$$\hat{\Lambda}^{(n+1)} = (1 - \gamma_n)\Lambda^{(n)} + \gamma_n \Lambda(q(\Lambda^{(n)})),$$

$$\hat{q}^{(n+1)} = (1 - \beta_n)q^{(n)} + \beta_n q(\Lambda(q^{(n)})).$$

where

$$\gamma_n = \arg\min_{\gamma \in [0,1]} f\big((1 - \gamma)\Lambda^{(n)} + \gamma \Lambda(q(\Lambda^{(n)}))\big),$$

$$\beta_n = \arg\max_{\beta \in [0,1]} g\big((1 - \beta)q^{(n)} + \beta q(\Lambda(q^{(n)}))\big).$$

4: (Update the iterates)

$$\Lambda^{(n+1)} = \arg\min_{\Lambda \in \{\hat{\Lambda}^{(n+1)}, \Lambda(q^{(n)})\}} f(\Lambda),$$

$$q^{(n+1)} = \arg\max_{q \in \{\hat{q}^{(n+1)}, q(\Lambda^{(n)})\}} g(q),$$

and then return to Step 2 with counter $n \leftarrow n + 1$.

5: **Output**

$$\bar{\Lambda} = \arg\min_{\Lambda \in \{\hat{\Lambda}^{(n+1)}, \Lambda(q^{(n)})\}} f(\Lambda),$$

$$\bar{q} = \arg\max_{q \in \{\hat{q}^{(n+1)}, q(\Lambda^{(n)})\}} g(q).$$

4.4.2. *Robust optimization for absolute metric learning*

The RO formulation of [4.10] that we present here appears to be novel in the literature. First, we write [4.10] in its Lagrangian form

$$\min_{\Lambda \succeq 0} \quad \max_{\lambda \geq 0} \quad \sum_{(i,j)\in\mathcal{M}} d_\Lambda^2(X_i, X_j) + \lambda\big(1 - \sum_{(i,j)\in\mathcal{N}} d_\Lambda^2(X_i, X_j)\big). \qquad [4.18]$$

Similar to $\mathcal{R}$, the side information sets $\mathcal{M}$ and $\mathcal{N}$ often suffer from noisiness or inaccuracy as well. Let us assume that about $1-\alpha$ proportion of the constraints in $\mathcal{M}$ and $\mathcal{N}$ are, respectively, inaccurate. We then construct robust uncertainty sets $\mathcal{W}(\alpha)$ and $\mathcal{V}(\alpha)$ from $\mathcal{M}$ and $\mathcal{N}$:

$$\mathcal{W}(\alpha) = \{\eta = \{\eta_{ij} : (i,j) \in \mathcal{M}\}\, |0 \leq \eta_{ij} \leq 1, \sum_{(i,j)\in\mathcal{M}} \eta_{ij} \leq \alpha \times |\mathcal{M}|\},$$

$$\mathcal{V}(\alpha) = \{\xi = \{\xi_{ij} : (i,j) \in \mathcal{N}\}\, |0 \leq \xi_{ij} \leq 1, \sum_{(i,j)\in\mathcal{N}} \xi_{ij} \geq \alpha \times |\mathcal{N}|\}.$$

Then, the RO formulation of [4.18] can be written as

$$\min_{\Lambda\succeq 0} \quad \max_{\lambda\geq 0} \quad \max_{\eta\in\mathcal{W}(\alpha),\xi\in\mathcal{V}(\alpha)} \sum_{(i,j)\in\mathcal{M}} \eta_{i,j} d_\Lambda^2(X_i, X_j) + \lambda\big(1 - \sum_{(i,j)\in\mathcal{N}} \xi_{i,j} d_\Lambda^2(X_i, X_j)\big) \qquad [4.19]$$

The switch of $\max_\lambda$ with $\max_{(\eta,\xi)}$ is valid in general. Also note that the Cartesian product $\mathcal{M}(\alpha) \times \mathcal{N}(\alpha)$ is a compact set, and the objective function is convex in Λ and concave (linear) in pair (η, ξ), so we can apply Sion's min–max theorem again (see Terkelsen (1973)) to switch the order of $\min_\Lambda$–$\max_{(\eta,\xi)}$. This leads to an iterative algorithm.

Define $q := (\eta, \xi)$ as the dual variable and

$$L(\Lambda, q) := \sum_{(i,j)\in\mathcal{M}} \eta_{i,j} d_\Lambda^2(X_i, X_j) + \lambda\big(1 - \sum_{(i,j)\in\mathcal{N}} \xi_{i,j} d_\Lambda^2(X_i, X_j)\big) + \frac{\kappa}{2}\|\Lambda\|_F^2, \qquad [4.20]$$

and

$$f(\Lambda) := \max_{q\in\mathcal{T}(\alpha)} L(\Lambda, q), \quad g(q) := \min_{\Lambda\succeq 0} L(\Lambda, q) \qquad [4.21]$$

and

$$q(\Lambda) := \arg\max_{q\in\mathcal{T}(\alpha)} L(\Lambda, q), \quad \Lambda(q) := \arg\min_{\Lambda\succeq 0} L(\Lambda, q). \qquad [4.22]$$

Similarly, $f(\Lambda)$ and $q(\Lambda)$ are easy to obtain. At the n-th step, given fixed $\Lambda^{(n-1)} \succeq 0$ and $\lambda > 0$ (it is easy to observe that optimal solution λ is positive, i.e. the constraint is active so we may safely assume $\lambda > 0$), the inner maximization problem looks like

$$\max_{\eta \in \mathcal{W}(\alpha)} \sum_{(i,j)\in\mathcal{M}} \eta_{i,j} d^2_{\Lambda^{(n-1)}}(X_i, X_j) + \lambda\big(1 - \min_{\xi\in\mathcal{V}(\alpha)} \sum_{(i,j)\in\mathcal{N}} \xi_{i,j} d^2_{\Lambda^{(n-1)}}(X_i, X_j)\big).$$

Analogous to the relative constraints case, the optimal η and ξ satisfy: $\eta_{i,j}$ is 1, if $d^2_{\Lambda^{(n-1)}}(X_i, X_j)$ ranks top α within $\mathcal{M}$ and equals 0 otherwise, while $\xi_{i,j} = 1$ if $d^2_{\Lambda^{(n-1)}}(X_i, X_j)$ ranks bottom α within $\mathcal{N}$ and equals 0 otherwise. So, we also define $\mathcal{M}_\alpha(\Lambda^{(n-1)})$ as a subset of $\mathcal{M}$, which contains the constraints with the largest α percent of $d_{\Lambda^{(n-1)}}(\cdot)$, and define $\mathcal{N}_\alpha(\Lambda^{(n-1)})$ as a subset of $\mathcal{N}$, which contains the constraints with the smallest α percent of $d_{\Lambda^{(n-1)}}(\cdot)$. Then, the optimal solution given, fixed $\Lambda^{(n-1)}$, can be reformulated as $\eta_{i,j} = 1$ if $(i,j) \in \mathcal{M}_\alpha(\Lambda^{(n-1)})$ and $\xi_{i,j} = 1$ if $(i,j) \in \mathcal{N}_\alpha(\Lambda^{(n-1)})$.

On the other hand, given fixed η and ξ, we can simplify the minimization problem $g(q)$ as

$$\min_{\Lambda\succeq 0} \sum_{(i,j)\in\mathcal{M}_\alpha(\Lambda^{(n-1)})} d^2_\Lambda(X_i, X_j) \quad \text{s.t.} \quad \sum_{(i,j)\in\mathcal{N}_\alpha(\Lambda^{(n-1)})} d^2_\Lambda(X_i, X_j) \geq 1.$$

This formulation of the minimization problem $g(q)$ takes the same form as [4.10], and it can thus be solved by the similar SDP algorithms presented in Xing *et al.* (2002). On the whole, we solve the minimax problem of [4.19] by using the same primal-dual algorithm presented in Algorithm 4.1.

Other robust methods have also been considered in the metric learning literature (see Zha *et al.* (2009) and Lim *et al.* (2013)), although the connections to RO are not fully exposed.

THEOREM 4.1.– *There exist saddle points* $(\bar{\Lambda}, \bar{q})$ *for the minimax problems* [4.15] *and* [4.20], *respectively. Algorithm 4.1 converges linearly to the common optimal value* $f(\bar{\Lambda}) = g(\bar{q})$ *in the sense that*

$$f(\Lambda^{(n+1)}) - f(\bar{\Lambda}) \leq \theta\big(f(\Lambda^{(n)}) - f(\bar{\Lambda})\big)$$
$$g(q^{(n+1)}) - g(\bar{q}) \leq \theta\big(g(q^{(n)}) - g(\bar{q})\big),$$

where $\theta \in (0, 1)$ *is a constant and the functions* f *and* g *are defined by* [4.16] *and* [4.21], *respectively.*

PROOF.– In both [4.15] and [4.20], the function L is strongly convex in Λ and q takes value in a bounded set, so it satisfies the condition of existence of saddle points (see Rockafeller (1970) and Zhu (1994)). Note that although L is not strongly concave in q, it is linear in q. At iteration $n+1$, if there is more than one optimal value of q, then we can use the proximal point algorithm to create a strongly convex–concave Lagrangian, i.e. we maximize $-\|q - q^{(n)}\|_2^2$ in order to break the tie. The linear convergence then follows from Theorem 3.3 in Zhu (1994). □

4.5. Numerical experiments

We proceed to numerical experiments to verify the performance of our DD-R-DRO method empirically, using six binary classification real datasets from the UCI machine learning database (Lichman 2013).

		BC	BN	QSAR	Magic	MB	SB
LR	Train	0 ± 0	$.008 \pm .003$	$.026 \pm .008$	$.213 \pm .153$	0 ± 0	0 ± 0
	Test	8.75 ± 4.75	2.80 ± 1.44	35.5 ± 12.8	17.8 ± 6.77	18.2 ± 10.0	14.5 ± 9.04
	Accur	$.762 \pm .061$	$.926 \pm .048$	$.701 \pm .040$	$.668 \pm .042$	$.678 \pm .059$	$.789 \pm .035$
LRL1	Train	$.185 \pm .123$	$.080 \pm .030$	$.614 \pm .038$	$.548 \pm .087$	$.401 \pm .167$	$.470 \pm .040$
	Test	$.428 \pm .338$	$.340 \pm .228$	$.755 \pm .019$	$.610 \pm .050$	$.910 \pm .131$	$.588 \pm .140$
	Accur	$.929 \pm .023$	$.930 \pm .042$	$.646 \pm .036$	$.665 \pm .045$	$.717 \pm .041$	$.811 \pm .034$
DD-DRO (absolute)	Train	$.022 \pm .019$	$.197 \pm .112$	$.402 \pm .039$	$.469 \pm .064$	$.294 \pm .046$	$.166 \pm .031$
	Test	$.126 \pm .034$	$.275 \pm .093$	$.557 \pm .023$	$.571 \pm .043$	$.613 \pm .053$	$.333 \pm .023$
	Accur	$.954 \pm .015$	$.919 \pm .050$	$.733 \pm .026$	$.727 \pm .039$	$.714 \pm .032$	$.887 \pm .011$
DD-R-DRO	Train	$.029 \pm .013$	$.078 \pm .031$	$.397 \pm .036$	$.420 \pm .063$	$.249 \pm .055$	$.194 \pm .031$
(absolute)	Test	$.126 \pm .023$	$.259 \pm .086$	$.554 \pm .019$	$.561 \pm .035$	$.609 \pm .044$	$.331 \pm .018$
$\alpha = 90\%$	Accur	$.954 \pm .012$	$.910 \pm .042$	$.736 \pm .025$	$.729 \pm .032$	$.709 \pm .025$	$.890 \pm .008$
DD-R-DRO	Train	$.040 \pm .055$	$.137 \pm .030$	$.448 \pm .032$	$.504 \pm .041$	$.351 \pm .048$	$.166 \pm .030$
(absolute)	Test	$.132 \pm .015$	$.288 \pm .059$	$.579 \pm .017$	$.590 \pm .029$	$.623 \pm .029$	$.337 \pm .013$
$\alpha = 50\%$	Accur	$.952 \pm .012$	$.918 \pm .037$	$.733 \pm .025$	$.710 \pm .033$	$.715 \pm .021$	$.888 \pm .008$
DD-DRO (relative)	Train	$.086 \pm .038$	$.436 \pm .138$	$.392 \pm .040$	$.457 \pm .071$	$.322 \pm .061$	$.181 \pm .036$
	Test	$.153 \pm .060$	$.329 \pm .124$	$.559 \pm .025$	$582 \pm .033$	$.613 \pm .031$	$.332 \pm .016$
	Accur	$.946 \pm .018$	$.916 \pm .075$	$.714 \pm .029$	$.710 \pm .027$	$.704 \pm .021$	$.890 \pm .008$
DD-R-DRO	Train	$.030 \pm .014$	$.244 \pm .121$	$.375 \pm .038$	$.452 \pm .067$	$.402 \pm .058$	$.234 \pm .032$
(relative)	Test	$.141 \pm .054$	$.300 \pm .108$	$.556 \pm .022$	$.577 \pm .032$	$.610 \pm .024$	$.332 \pm .011$
$\alpha = 90\%$	Accur	$.949 \pm .019$	$.921 \pm .070$	$.729 \pm .023$	$.717 \pm .025$	$.710 \pm .020$	$.892 \pm .007$
DD-R-DRO	Train	$.031 \pm .016$	$.232 \pm .094$	$.445 \pm .032$	$.544 \pm .057$	$.365 \pm .054$	$.288 \pm .029$
(relative)	Test	$.154 \pm .049$	$.319 \pm .078$	$.570 \pm .019$	$.594 \pm .018$	$.624 \pm .018$	$.357 \pm .008$
$\alpha = 50\%$	Accur	$.948 \pm .019$	$.918 \pm .081$	$.705 \pm .023$	$.699 \pm .028$	$.698 \pm .018$	$.881 \pm .005$
Num predictors		30	4	30	10	20	56
Train size		40	20	80	30	30	150
Test size		329	752	475	9990	125034	2951

Table 4.1. *Numerical results for real data sets*

We consider logistic regression (LR), regularized logistic regression (LRL1), DD-DRO with a cost function learned using absolute constraints (DD-DRO (absolute)) and its $\alpha = 50\%, 90\%$ level of doubly robust DRO (DD-R-DRO (absolute)); DD-DRO with cost function learned using relative constraints (DD-DRO (relative)) and its $\alpha = 50\%, 90\%$ level of doubly robust DRO (DD-R-DRO (relative)). For each data and each experiment, we randomly split the data into training and testing and fit models on the training set and evaluate on the testing set. **For all of the DRO models, we apply 5-fold cross-validation on the training set to select the optimal size of uncertainty δ.**

We reported the mean and standard deviation of training error, testing error and testing accuracy via 200 independent experiments for each dataset and summarized the detailed results and dataset information (including split setting) in Table 4.1.

For solving the DD-DRO and DD-R-DRO problem, we apply the smoothing approximation algorithm introduced in Blanchet *et al.* (2017), in order to solve the DRO problem.

We observe that the doubly robust DRO framework, in general, gets robust improvement compared to its non-robust counterpart with $\alpha = 90\%$. More importantly, the robust methods tend to enjoy the variance reduction property due to RO. Also, as the robust level increases, i.e. $\alpha = 50\%$, where we believe in higher noise in cost function learning, we can observe that the doubly robust-based approach seems to shrink towards LRL1 and benefits less from the data-driven cost structure.

4.6. Discussion and conclusion

We have proposed a novel methodology, DD-R-DRO, which calibrates a transportation cost function by using a data-driven approach based on RO. In turn, DD-R-DRO uses this cost function in the description of a DRO formulation based on an optimal transport uncertainty region. The overall methodology is doubly robust. On one hand, DD-DRO, which fully uses the training data to estimate the underlying transportation cost, enhances out-of-sample performance by allowing an adversary to perturb the data (represented by the empirical distribution), in order to obtain bounds on the testing risk which are tight. On the other hand, the tightness of bounds might come at the cost of potentially introducing noise in the testing error performance. The second layer of robustification, as shown in the numerical examples, precisely mitigates the presence of this noise.

4.7. References

Abadeh, S.S., Esfahani, P.M. and Kuhn, D. (2015). Distributionally robust logistic regression. *Advances in Neural Information Processing Systems*, pp. 1576–1584.

Bellet, A., Habrard, A. and Sebban, M. (2013). A survey on metric learning for feature vectors and structured data. arXiv preprint arXiv:1306.6709.

Belloni, A., Chernozhukov, V. and Wang, L. (2011). Square-root lasso: Pivotal recovery of sparse signals via conic programming. *Biometrika*, 98(4), 791–806.

Ben-Tal, A., El Ghaoui, L. and Nemirovski, A. (2009). *Robust Optimization*. Princeton University Press, Princeton.

Bertsimas, D., Brown, D.B. and Caramanis, C. (2011). Theory and applications of robust optimization. *SIAM Review*, 53(3), 464–501.

Blanchet, J., Kang, Y. and Murthy, K. (2016). Robust wasserstein profile inference and applications to machine learning. arXiv preprint arXiv:1610.05627.

Blanchet, J., Kang, Y., Zhang, F. and Murthy, K. (2017). Data-driven optimal cost selection for distributionally robust optimization. arXiv preprint arXiv:1705.07152.

El Ghaoui, L. and Lebret, H. (1997). Robust solutions to least-squares problems with uncertain data. *SIAM Journal on Matrix Analysis and Applications*, 18(4), 1035–1064.

Esfahani, P.M. and Kuhn, D. (2015). Data-driven distributionally robust optimization using the wasserstein metric: Performance guarantees and tractable reformulations. arXiv preprint arXiv:1505.05116.

Gao, R. and Kleywegt, A.J. (2016). Distributionally robust stochastic optimization with wasserstein distance. arXiv preprint arXiv:1604.02199.

Hu, Z. and Hong, L.J. (2013). Kullback-Leibler divergence constrained distributionally robust optimization [Online]. Available at: http://www.optimization-online.org/DB_FILE/2012/11/3677.pdf.

Huang, K., Jin, R., Xu, Z. and Liu, C.-L. (2012). Robust metric learning by smooth optimization. arXiv preprint arXiv:1203.3461.

Lee, S.-I., Lee, H., Abbeel, P. and Ng, A.Y. (2006). Efficient regularized logistic regression. *AAAI*, volume 6, pp. 401–408.

Lichman, M. (2013). UCI machine learning repository [Online]. Available at: https://archive.ics.uci.edu/ml.

Lim, D., McFee, B. and Lanckriet, G.R. (2013). Robust structural metric learning. *ICML-13*, pp. 615–623.

Rockafeller, R.T. (1970). *Convex Analysis*. Princeton University Press, Princeton.

Smith, J.E. and Winkler, R.L. (2006). The optimizer's curse: Skepticism and postdecision surprise in decision analysis. *Management Science*, 52(3), 311–322.

Terkelsen, F. (1973). Some minimax theorems. *Mathematica Scandinavica*, 31(2), 405–413.

Villani, C. (2008). *Optimal Transport: Old and New*, volume 338. Springer Science & Business Media, Cham.

Weinberger, K.Q. and Saul, L.K. (2009). Distance metric learning for large margin nearest neighbor classification. *Journal of Machine Learning Research*, 10(Feb), 207–244.

Xing, E.P., Ng, A.Y., Jordan, M.I. and Russell, S. (2002). Distance metric learning with application to clustering with side-information. *NIPS*, volume 15, p. 12.

Xu, H., Caramanis C. and Mannor S. (2009a). Robust regression and lasso. *NIPS-2009*, pp. 1801–1808.

Xu, H., Caramanis C. and Mannor S. (2009b). Constantine Caramanis, and Shie Mannor. Robustness and regularization of support vector machines. *JMLR*, 10(Jul), 1485–1510.

Zha, Z.-J., Mei, T., Wang, M., Wang, Z. and Hua, X.-S. (2009). Robust distance metric learning with auxiliary knowledge. *IJCAI*, pp. 1327–1332.

Zhu, C. (1994). Solving large-scale minimax problems with the primal-dual steepest descent algorithm. *Mathematical Programming*, 67(1–3), 53–76.

5

A Comparison of Graph Centrality Measures Based on Lazy Random Walks

When working with a network, it is often of interest to locate the "most important" nodes in the network. A common way to do this is by using some graph centrality measures. Since what constitutes as an important node varies from one network to another, or even in applications on the same network, there is a large number of different centrality measures proposed in the literature. Due to the large amount of different centrality measures proposed in different fields, there is also a large amount of very similar or equivalent centrality measures (in the sense that they give the same ranks). In this chapter, we focus on the centrality measures based on powers of the adjacency matrix and those based on random walk. In this case, we show how some of these centrality measures are related, as well as their lazy variants. We will perform some experiments to demonstrate the similarities between the centrality measures.

5.1. Introduction

In any network, it is important to identify key nodes within it. According to Roy *et al.* (2010), Das *et al.* (2018), Li and Zhang (2018), one of the most interesting tools used to establish the importance of a node from a complex network is its *centrality measure*. The concept of centralities is said to have originated between the 1940s and 1950s (Bavelas 1948, 1950; Leavitt 1951). To date, more than 100 different centrality measures have been proposed to describe several research fields in network analysis (Jalili *et al.* 2015).

A centrality measure captures the importance of a node in a network with respect to a specific metric or application. The appropriateness in the use of a particular centrality

Chapter written by Collins ANGUZU, Christopher ENGSTRÖM and Sergei SILVESTROV.

measure largely depends on the application involved (Bloch *et al.* 2019). For instance, Katz centrality is preferred in understanding the comparative importance or influence of individuals in a social network (Ballester *et al.* 2006), while eigenvector centrality is critical in determining whether a community aggregates information (Golub and Jackson 2010) in the right way.

In Engström (2016), a significant number of centralities are considered and categorized, mainly into categories of those based on the shortest path and the powers of the adjacency matrix. The latter category has attracted considerable attention among scientific communities due to its low computational cost and ability to enhance the understanding of large-scale tendencies across the network (Noel and Jajodia 2005). In this study, we consider the powers of the adjacency matrix category for directed graphs and focus on its properties associated with the different random walks, in particular, the traditional and lazy random walks.

By knowing the connections and similarities between different centralities, we can also learn how to use and generalize the results obtained from one centrality with others. One example is showing how results relating to the computation of PageRank (Biganda *et al.* 2018; Engström and Silvestrov 2019) can be applied to other similar centrality measures. The different centrality measures can be expressed in linear form or power series form. The latter embraces the notion of powers of the adjacency matrix. It is therefore important to understand the preliminaries on linear systems and Neumann series (Berman and Plemmons 1994; Benzi and Klymko 2015).

In Benzi and Klymko (2015), the authors analyzed the relationship between degree centrality, eigenvector centrality and various centrality measures based on the diagonal entries (for undirected graphs) and row sums of certain (analytic) functions of the adjacency matrix of the graph. Such measures included the Katz centrality, sub-graph centrality and total communicability among others. Their main result showed that degree and eigenvector centrality are limiting cases of the parameterized measures. Clearly, in Benzi and Klymko (2015), the focus was on centrality measures that can be expressed in terms of the adjacency matrix of the undirected network. It is natural to investigate the relationships between centrality measures, such as the alpha centrality, beta centrality and cumulative nomination centrality for directed networks, which in our opinion has not been explored, particularly for the lazy variants of the centrality measures.

Ghosh and Lerman (2011) introduced a normalized version of alpha centrality and used it to study network structure, to rank nodes and find the community structure of the network. In particular, they used the expected number of paths and extended the modularity-maximization method for community detection. To this end, they established the relationship between alpha, eigenvector, degree and Katz centralities. In such a formulation, the matrix structure is assumed to be symmetric and the convergence is guaranteed. We would like to remark that these assumptions may fail in some networks, for instance, in a cyclic network, which is known to have periodicity behavior (Pak and Żuk 2002). In such a case, we need to modify the network by adding self-loops to all vertices. This modification results in the notion of laziness in random walks, with a number of advantages. Among others, lazy random walk breaks periodicity in a network and it also takes care of the behavior of the random walker (Biganda *et al.* 2018), based on the network features. In the context of probability theory, a lazy random walk on a graph is understood as follows: assume $\mathbf{A}$ is a weighted adjacency matrix, in short, a transition matrix associated with a certain graph. Suppose a walker moves between any pair of nodes with probability β or decides to stay put in the current node with probability $1 - \beta$, this kind of random walk is referred to as a lazy random walk.

A lazy random walk has applications in science and engineering, for instance, Zhang and Hancock (2005) apply a lazy variant of random walks in image segmentation and noted that it outperformed the traditional random walk because it is able to detect weak boundaries and complex texture segmentation. In Zhou and Schölkopf (2004), a lazy random walk was applied in a semi-supervised learning algorithm. Based on this background, we have been motivated to extend the lazy random walk formulation to address network structures in which limiting behavior exhibits periodic nature.

In this chapter, we consider seven centrality measures based on the powers of the adjacency matrix, i.e. degree centrality, eigenvector centrality, Katz centrality, alpha centrality, beta centrality, cumulative nomination centrality and PageRank centrality. We describe the relationships between them and perform some experimental work to establish the relationships on heat maps. The above treatment applies to lazy variants of some selected centrality measures as well.

The rest of this chapter is organized as follows. In section 5.1, we give notations and abbreviations and introduce the concept of linear systems and the Neumann series that is applied throughout this chapter. In section 5.2, we review the different centrality measures and give a few derivations coupled with a summary of the power series expressions for the centrality measures. In section 5.3, we do a generalization, with an emphasis on lazy variants of alpha, Katz and cumulative nomination centrality measures, their convergence conditions, relations and a summary of their mathematical formulations. In section 5.4, we have implemented the experimental results by the use of heat maps. We discuss them in section 5.5, and a brief conclusion follows in section 5.6.

5.1.1. *Notations and abbreviations*

The words graph and network will be used interchangeably to mean a collection of vertices (also called states when it comes to Markov chains), with edges (links) connecting the vertices. "Centrality" and "centrality measure" will be used interchangeably to mean the same thing. We will refer to the adjacency matrix of a graph by $\mathbf{A}$ and λ refers to the dominant eigenvalue of $\mathbf{A}$. Let A_{ij} be the elements of $\mathbf{A}$ in the ith row and the jth column, then A_{ij} is equal to 1 if a link exists directly from nodes i and j and 0 otherwise. Matrix $\mathbf{A}^\top$ will mean the transposed form of $\mathbf{A}$, and $\vec{e}$ denotes the all one vector. The matrix $\mathbf{A}_w$ is the scaled or weighted adjacency matrix whose elements are $\frac{1}{d_i}$ if a link exists between node i and node j and 0 otherwise, where d_i is the out degree of i. We will use $\vec{c}_r^{(t)}(i)$ throughout this chapter to denote the centrality measure of the type r of node i, and the superscript (t) represents the time step of iteration. $_l\vec{c}$ will mean the lazy variant of the centrality measure $\vec{c}$.

5.1.2. *Linear systems and the Neumann series*

Given a linear system

$$\mathbf{A}\vec{x} = \vec{u}, \qquad [5.1]$$

where $\vec{x}$ is a vector of the ranks of the nodes and $\vec{u}$ is a vector of constants, the solution to the system [5.1] is

$$\vec{x} = \mathbf{A}^{-1}\vec{u}, \qquad [5.2]$$

for $\mathbf{A}$ invertible. Note that the matrix is not necessarily non-singular. By scaling and shifting the adjacency matrix $\mathbf{A}$, we get

$$(\lambda\mathbf{I} - \mathbf{A})\vec{x} = \vec{u}.$$

The solution exists if the constant λ is greater than the spectral radius of matrix $\mathbf{A}$; hence, $(\lambda\mathbf{I} - \mathbf{A})$ is invertible. Let us state, without proof, an important lemma that is sometimes called the Neumann lemma for non-negative matrices. For further discussion on this topic, refer to the book by Berman and Plemmons (1994).

LEMMA 5.1.– *The non-negative matrix $\mathbf{T} \in \mathbb{R}^{n\times n}$ is convergent, i.e. the spectral radius $\rho(\mathbf{T}) < 1$, if and only if $(\mathbf{I} - \mathbf{T})^{-1}$ exist and*

$$(\mathbf{I} - \mathbf{T})^{-1} = \sum_{k=0}^{\infty} \mathbf{T}^k, \qquad [5.3]$$

where $\mathbf{I} \in \mathbb{R}^{n\times n}$ is an identity matrix. Note that the class of matrix $(\mathbf{I} - \mathbf{T})$ is referred to as the M-matrix.

This class of matrix has many practical applications, for example, splitting matrices, in order to establish bounds on eigenvalues and convergence criteria for iterative schemes of linear systems. The generalization of the M-matrix for monotonic functions is well described in Berman and Plemmons (1994).

The following section presents an overview of the centrality measures based on the powers of the adjacency matrix. Each of them is presented with the formula for its computation, defining the terms therein.

5.2. Review on some centrality measures

5.2.1. *Degree centrality*

Degree centrality is the simplest form of graph centrality measures. It is defined in Zhang and Luo (2017) as the total number of direct connections a node has with other nodes. Realistically, degree centrality refers to the risk that a node has to immediately acquire whatever is flowing through the network. Degree centrality is categorized into in-degree centrality, $\vec{c}_d = \mathbf{A}^\top \vec{e}$, and out-degree centrality, $\vec{c}_{\bar{d}} = \mathbf{A}\vec{e}$.

5.2.2. *Katz status and β-centralities*

According to Katz (1953), the score $\vec{c}_z$ of a node in a graph is the sum of the weights, α^k, of all paths of length k leading to the node. The constant α is some *attenuation factor,* and k is the length of the path. This factor is chosen in such a way that all summations converge (Boldi and Vigna 2014). As a result of connection between the adjacency matrix $\mathbf{A}$ and the number of paths connecting the pairs of nodes, the Katz centrality measure takes the form

$$\vec{c}_z = \sum_{k=1}^{\infty} \alpha^k (\mathbf{A}^\top)^k \vec{e}, \qquad [5.4]$$

where $\alpha < \frac{1}{|\lambda|}$, $|\lambda| > 0$ and $\alpha > 0$.

The β-centrality, on the other hand, is a variant of the Bonacich centrality, which allows each node's centrality to contribute to the centrality of other nodes in the network (Bonacich 1972). The β-centrality of a node is obtained by

$$\vec{c}_\beta = \frac{\beta}{\alpha} \sum_{k=1}^{\infty} \alpha^k \mathbf{A}^k \vec{e}, \qquad [5.5]$$

where λ and α are defined as in formula [5.4]. The constant β is only a scaling factor and does not affect the ranking in any way; hence, Katz and β-centralities are equivalent, except for the fact that Katz uses the transposed adjacency matrix and β-centrality uses the non-transposed matrix. The equivalence here, however, holds the condition that $\mathbf{A}$ is symmetric. It is important to note that usually, when β-centrality is defined, α and β are swapped; we have used this definition so that α will be consistent between different centralities.

5.2.3. *Eigenvector and cumulative nomination centralities*

Eigenvector centrality considers the fact that the importance of a node depends on how important its neighbors are (Hagan *et al.* (2015) and Engström (2016)). The eigenvector centrality of any node j, denoted $\vec{c}_e(j)$, in a given graph is written as

$$\vec{c}_e(j) = \frac{1}{\lambda} \sum_{i=1}^{n} (\mathbf{A}^\top)_{ij} \vec{c}_e(i), \qquad [5.6]$$

In matrix form, the formula [5.6] can be written as

$$\mathbf{A}^\top \vec{c}_e = \lambda \vec{c}_e,$$

since at steady state, $\vec{c}_e(j)$ and $\vec{c}_e(i)$ attain a negligible difference. On the other hand, cumulative nomination centrality, denoted $\vec{c}_c$, is concerned with central individuals (nodes) being nominated more often in social networks (Poulin *et al.* 2000). The total value of the nomination of a node j, acquired after t iterations, is the value of j's nomination initially (which is always set as 1 (Lü *et al.* 2016) at the beginning) plus the sum of the nomination values of the neighbors of j up to $t-1$ iterations. Let $\vec{c}_c^{(t-1)}(j)$ be the accumulated number of nominations of node j up to $t-1$ iterations. Then, after t iterations, node j will have the following cumulative nomination:

$$\vec{c}_c^{(t)}(j) = \vec{c}_c^{(t-1)}(j) + \sum_i \mathbf{A}_{ji} \vec{c}_c^{(t-1)}(i)\,, \qquad [5.7]$$

In matrix form, equation [5.7] is written as

$$\vec{c}_c^{(t)} = \vec{c}_c^{(t-1)} + \mathbf{A}\vec{c}_c^{(t-1)}\,,$$

or simply

$$\vec{c}_c^{(t)} = (\mathbf{I} + \mathbf{A})\vec{c}_c^{(t-1)}\,. \qquad [5.8]$$

This is a power series iteration for matrix (I + A) with eigenvalue $(1+\lambda)$ and limit of convergence $\vec{c}_c$. And so, the normalized form of equation [5.8] is

$$\vec{c}_c^{(t)} = \frac{(\mathbf{I} + \mathbf{A})}{\|(\mathbf{I} + \mathbf{A})\vec{c}_c^{(t-1)}\|_1} \vec{c}_c^{(t-1)}\,,$$

where $\|\vec{z}\| = \sum_{i=1}^{n} |z_i|$, and it reduces to

$$\vec{c}_c = \frac{(\mathbf{I} + \mathbf{A})}{1+\lambda} \vec{c}_c\,,$$

after convergence. This further reduces to $\lambda\vec{c}_c = \mathbf{A}\vec{c}_c$. This means that the normalized cumulative nomination centrality reduces to an eigenvector centrality.

5.2.4. *Alpha centrality*

Alpha centrality (Bonacich and Lloyd 2004; Ghosh and Lerman 2011) measures the total number of paths from a node, exponentially attenuated by their length. The attenuation parameter in alpha centrality sets the length scale of interactions. The α-centrality hails from eigenvector centrality by allowing nodes to have external influence (global structure of network). The amount of influence received by each node i from all nodes j with $i \neq j$ is encoded in $e(i)$, and this process ends when

$$\vec{c}_\alpha(i) = \alpha(\mathbf{A}^\top)_{ij}\vec{c}_\alpha(j) + e(i),$$

where $\vec{c}_\alpha(j)$ refers to the alpha centrality of the nodes j that are one step away from node i in the graph. The $\vec{c}_\alpha(j)$ can be updated in parallel or sequentially, but more commonly, the parallel option is used. It is basically the difference between the Jacobi iteration and the Gauss–Seidel iteration for solving a linear system. Initially, a value is assigned to $\vec{c}_\alpha(j)$, which represents the initial rank values given to each j. At steady state, the alpha centrality $\vec{c}_\alpha$ of the entire graph is

$$\vec{c}_\alpha = (\mathbf{I} - \alpha\mathbf{A}^\top)^{-1}\vec{e}. \qquad [5.9]$$

By the Neumann series, the relation [5.9] becomes

$$\vec{c}_\alpha = ((\alpha\mathbf{A}^\top)^0 + (\alpha\mathbf{A}^\top)^1 + (\alpha\mathbf{A}^\top)^2 + \cdots)\vec{e},$$

$$= \sum_{k=0}^{\infty} \alpha^k (\mathbf{A}^\top)^k \vec{e} \qquad [5.10]$$

The α-centrality is similar to the Katz centrality, the only difference is that the summation is taken from $k = 0$ instead of $k = 1$ (equivalent to adding 1 to the final values); hence, they give the same ranking.

The beta centrality in equation [5.5] can be rewritten as

$$\vec{c}_\beta = \frac{\beta}{\alpha}\sum_{k=0}^{\infty} \alpha^k \mathbf{A}^k \vec{e} - \frac{\beta}{\alpha}\vec{e}. \qquad [5.11]$$

Comparing the relation [5.11] with that for alpha(out), $\vec{c}_{\bar{\alpha}}$, in Table 5.1, i.e.

$$\vec{c}_{\bar{\alpha}} = \sum_{k=0}^{\infty} \alpha^k \mathbf{A}^k \vec{e}, \qquad [5.12]$$

it is observed that

$$\vec{c}_\beta = \frac{\beta}{\alpha}\vec{c}_{\bar{\alpha}} - \frac{\beta}{\alpha}\vec{e}, \qquad [5.13]$$

where the constants β and α belong to the set $[0,1]$. Clearly, beta centrality $\vec{c}_\beta$ is a linear transformation of alpha centrality (out), $\vec{c}_{\bar{\alpha}}$. In this way, we are able to write β-centrality in terms of α-centrality. The two centralities should therefore give the same ranking since the term $\frac{\beta}{\alpha}\vec{e}$ is simply a translation. This relation is not explored in the literature.

5.2.5. *PageRank centrality*

DEFINITION 5.1.– *PageRank $\vec{r}$ is defined as the (right) eigenvector with eigenvalue one to the matrix:*

$$\mathbf{M} = c(\mathbf{A}_w + \vec{g}\vec{u}^\top)^\top + (1-c)\vec{u}\vec{e}^\top\,, \tag{5.14}$$

where $\vec{g}$ is a vector with zeros for nodes with outgoing edges and 1 for all nodes with no outgoing edges, $\vec{u}$ is a non-negative vector with norm $||\vec{u}||_1 = 1$ and $0 < c < 1$ is a scalar (Engström 2016).

Often, $\vec{r}$ is normalized such that the elements sum to one, PageRank can then be seen as the stationary distribution described by the transition matrix $\mathbf{M}$. Instead, we will use a non-normalized variation, which can be written using a power series and does not need the correction for dangling nodes.

From the eigenvalue equation, we get:

$$\vec{r} = \mathbf{M}\vec{r} = \left(c(\mathbf{A}_w + \vec{g}\vec{u}^\top)^\top + (1-c)\vec{u}\vec{e}^\top\right)\vec{r}\,,$$

$$\Leftrightarrow \left(\mathbf{I} - c\mathbf{A}_w^\top\right)\vec{r} = \left(c\vec{u}\vec{g}^\top + (1-c)\vec{u}\vec{e}^\top\right)\vec{r} = \vec{u}\left(c\vec{g}^\top + (1-c)\vec{e}^\top\right)\vec{r}\,.$$

Next, we use the fact that the right-hand side is proportional to $\vec{u}$ to get:

$$\left(\mathbf{I} - c\mathbf{A}_w^\top\right)\vec{r} \propto \vec{u} \Leftrightarrow \vec{r} \propto \left(\mathbf{I} - c\mathbf{A}_w^\top\right)^{-1}\vec{u}\,.$$

At this point, we can either normalize $\vec{r}$ or do as we have here and keep the non-normalized version of PageRank gained by solving the linear system

$$\left(\mathbf{I} - c\mathbf{A}_w^\top\right)\vec{r} = \vec{u}\,.$$

Rewriting this using the Neumann series for $\left(\mathbf{I} - c\mathbf{A}_w^\top\right)$, we get the power series formulation of PageRank we will use later:

$$\vec{r} = \left(\mathbf{I} - c\mathbf{A}_w^\top\right)^{-1}\vec{u} = \sum_{k=0}^{\infty} c^k(\mathbf{A}_w^\top)^k\vec{u}\,. \tag{5.15}$$

	Shifted power series	Power series	Steady state
A	β-centrality $\frac{\beta}{\alpha}\sum_{k=1}^{\infty}\alpha^k \mathbf{A}^k \vec{e}$	α-centrality (out) $\sum_{k=0}^{\infty}\alpha^k(\mathbf{A})^k\vec{e}$	Cumulative nomination $\lim_{n\to\infty} c^{(n+1)} = \frac{\vec{c}^{(n)} + \mathbf{A}\vec{c}^{(n)}}{\vert\vert\vec{c}^{(n)} + \mathbf{A}\vec{c}^{(n)}\vert\vert}$
$A^\top$	Katz $\sum_{k=1}^{\infty}\alpha^k(\mathbf{A}^\top)^k\vec{e}$	α-centrality $\sum_{k=0}^{\infty}\alpha^k(\mathbf{A}^\top)^k\vec{e}$	Eigenvector $\lim_{n\to\infty} \vec{c}^{(n+1)} = \frac{\mathbf{A}^\top\vec{c}^{(n)}}{\vert\vert\mathbf{A}^\top\vec{c}^{(n)}\vert\vert}$
$\mathbf{A}_w^\top$	Shifted PageRank $\sum_{k=1}^{\infty}c^k(\mathbf{A}_w^\top)^k\vec{u}$	PageRank $\sum_{k=0}^{\infty}c^k(\mathbf{A}_w^\top)^k\vec{u}$	Scaled eigenvector $\lim_{n\to\infty} \vec{c}^{(n+1)} = \frac{\mathbf{A}_w^\top\vec{c}^{(n)}}{\vert\vert\mathbf{A}_w^\top\vec{c}^{(n)}\vert\vert}$

Table 5.1. *Overview of centralities and their expression as a power series or power iteration. To guarantee convergence,* $0 < \alpha < 1/\rho(\mathbf{A})$, $0 < c < 1$, *additionally,* $\beta > 0$. *For the steady-state centralities, we require the matrix to be primitive for convergence, where* $\vec{c}$ *is a random positive vector. The steady-state variants converge to the same vectors as the corresponding ordinary variants*

5.2.6. *Summary of the centrality measures as steady state, shifted and power series*

Table 5.1 gives a summary of the formulas for the different centrality measures that have been described in section 5.2, together with the conditions necessary for convergence.

5.3. Generalizations of centrality measures

5.3.1. *Priors to centrality measures*

Centrality priors are visibly clear in random walk models. These models are finite and time-reversible Markov chains (Mao and Xiano 2018). Suppose there is a system with m states, and the initial probability distribution of these states is represented as $\pi_0 = \pi_{01}, \pi_{02}, \ldots, \pi_{0n}$. In this system, the states can be transited to each other. Specifically, if state i has different transition probabilities to other k states, the sum of these probabilities should be 1.0, i.e. $\sum_{j=1}^{n} p_{ij} = 1$, where p_{ij} is the transition probability from state i to state j. The transition probabilities of all state pairs can be represented as $\mathbf{P} = [p_{ij}]$ for $i, j = 1, 2, \ldots, n$. Thus, the random walk model can be clearly described by using the matrix representation (Liu and Yang 2008).

In some applications, when calculating some centrality measures such as the PageRank, it is important that nodes are not given equal initial ranks, for instance, by assigning the initial rank of one to all vertices. The importance of such nodes is recognized by assigning to them initial scores (or ranks) different from the rest. The initial vector $\vec{u}$ with this kind of diversity is referred to as a *prior*. For instance, in a social network, actors that are very popular or very influential are assigned high initial ranks and those that have little impact in the network are assigned very low or zero initial ranks. The same type of prior can be trivially applied to degree, Katz and α-centralities as well as variants based on the transposed adjacency matrix. For the centralities in the last column of Table 5.1, they are all defined by the steady-state vector; hence, a prior will not have any effect assuming that the corresponding matrix is primitive. However, we should note that priors are important because they affect the ranking of nodes in the case where the corresponding matrix is not primitive.

5.3.2. *Lazy variants of centrality measures*

Lazy variants of centrality measures help to break periodicity in the structure of networks and ensure convergence to the limiting probability distribution (Izaac *et al.* 2017). A lazy variant of a centrality measure considers the measure as a random walk. We consider lazy variants of a few of the centralities involved. In this case with some probability β, the walker stays at the same node, and with $(1-\beta)$, he uniformly moves to another node. Note that the definitions of lazy variants given here for either $\mathbf{A}$ or $\mathbf{A}^\top$ can be given in exactly the same way for the other centrality measures. For example, the results presented for cumulative nomination in section 5.3.5 are equally valid for eigenvector centrality.

5.3.3. *Lazy α-centrality*

In this section, we want to establish the relationship between the lazy α-centrality, ${}_l\vec{c}_\alpha$ and the normal α-centrality, $\vec{c}_\alpha$. The formula for α-centrality as per equation [5.9] is

$$\vec{c}_\alpha = (\mathbf{I} - \alpha\mathbf{A}^\top)^{-1}\vec{e}. \qquad [5.16]$$

When defining lazy α-centrality, we find two different cases. The first case, ${}_l\vec{c}_{\alpha,1}$ is the most natural generalization given by

$${}_l\vec{c}_{\alpha,1} = \left(\mathbf{I} - \alpha(\beta\mathbf{I} + (1-\beta)\mathbf{A}^\top)\right)^{-1}\vec{e}. \qquad [5.17]$$

The second case, ${}_l\vec{c}_{\alpha,2}$ is meant to ensure that parameter α has the same role in regard to convergence as for ordinary α-centrality. That is,

$${}_l\vec{c}_{\alpha,2} = \left(\mathbf{I} - \alpha(\lambda\beta\mathbf{I} + (1-\beta)\mathbf{A}^\top)\right)^{-1}\vec{e}. \qquad [5.18]$$

We now analyze the relationship between the parameters α, λ and β by looking at the two cases one at a time.

Case 1. Let us describe a relationship between the lazy α-centrality and the normal α-centrality given by relations [5.17] and [5.16], respectively. We then have the following lemma.

LEMMA 5.2.– *Consider the relation [5.17], where* $\mathbb{A}$ *is the adjacency matrix of a graph and* β *is the lazy factor. Let*

$$\alpha < \frac{1}{\lambda(1-\beta)+\beta}. \tag{5.19}$$

Then, the relation [5.17] has a unique solution, ${}_l\vec{c}_{\alpha,1}$*, which is the lazy* α*-centrality of the graph.*

PROOF.– The relation in [5.17] can be rewritten as

$$(\mathbf{I}-\alpha(\beta\mathbf{I}+(1-\beta)\mathbf{A}^\top))_l\vec{c}_{\alpha,1}=\vec{e}.$$

This expands to give

$$(1-\alpha\beta)_l\vec{c}_{\alpha,1}-\alpha(1-\beta)\mathbf{A}^\top{}_l\vec{c}_{\alpha,1}=\vec{e}.$$

Simplifying the terms on the left-hand side, we get

$$\left(\mathbf{I}-\frac{\alpha(1-\beta)}{1-\alpha\beta}\mathbf{A}^\top\right){}_l\vec{c}_{\alpha,1}=\frac{1}{1-\alpha\beta}\vec{e}.$$

This reduces further to

$${}_l\vec{c}_{\alpha,1}=\left(\mathbf{I}-\frac{\alpha(1-\beta)}{1-\alpha\beta}\mathbf{A}^\top\right)^{-1}\frac{1}{1-\alpha\beta}\vec{e}\,. \tag{5.20}$$

Comparing the result [5.20] to the relation [5.16], we can see that the lazy variant gives the same ranking as ordinary α-centrality, but for another parameter α (expression in front of $\mathbf{A}^\top$). For the convergence of the Neumann series of the corresponding matrix in relation [5.20],

$$\frac{\alpha(1-\beta)}{1-\alpha\beta}<\frac{1}{\lambda}.$$

Equivalently, we obtain

$$\alpha < \frac{1}{\lambda(1-\beta)+\beta}\,. \tag{5.21}$$

It is important to note here that α depends not only on λ, but also on β. From relation [5.21], if $\beta = 0$, then $\alpha < \frac{1}{\lambda}$, and if $\beta = 1$, $\alpha < 1$. □

Case 2. For case 2, we write the following lemma:

LEMMA 5.3.– *Suppose that the inequality* $\alpha < \frac{1}{\lambda}$ *holds. Then, relation* [5.18] *converges if*

$$\left|\frac{\alpha(1-\beta)}{1-\alpha\lambda\beta}\right| < \frac{1}{\lambda}. \qquad [5.22]$$

PROOF.– From the linear system [5.18]

$$_{l}\vec{c}_{\alpha,2} = (\mathbf{I} - \alpha(\lambda\beta\mathbf{I} + (1-\beta)\mathbf{A}^{\top}))^{-1}\vec{e}. \qquad [5.23]$$

A systematic simplification of the above relation yields

$$_{l}\vec{c}_{\alpha,2} = \left(\mathbf{I} - \frac{\alpha(1-\beta)}{1-\alpha\lambda\beta}\mathbf{A}^{\top}\right)^{-1} \frac{1}{1-\alpha\lambda\beta}\vec{e}. \qquad [5.24]$$

For convergence of the Neumann series of [5.24],

$$\left|\frac{\alpha(1-\beta)}{1-\alpha\lambda\beta}\right| < \frac{1}{\lambda}.$$

It is easy to remark here that the result in [5.22] has a big computational advantage. This follows the fact that if β is chosen such that $\frac{(1-\beta)}{1-\alpha\lambda\beta} < 1$, then the convergence is faster. From the same result above, we can also deduce that $\alpha < \frac{1}{\lambda}$. This implies that α only depends on λ, like in the normal case of α-centrality. As seen, both forms of lazy α-centrality can be reduced to ordinary Katz centrality with some scaling and a different damping parameter. The first one in equation [5.17] is more simply defined and calculated, but the convergence depends on both λ and β. The second one requires knowledge of the largest eigenvalue, but in return, α has the same role as it has for the ordinary Katz centrality in terms of convergence. □

5.3.4. *Lazy Katz centrality*

The ordinary Katz centrality is given by

$$\vec{c}_z = \sum_{k=1}^{\infty} \alpha^k (\mathbf{A}^{\top})^k \vec{e},$$

as given earlier in [5.4]. Taking the sum from $k = 0$, we get

$$\vec{c}_z = \sum_{k=0}^{\infty} \alpha^k (\mathbf{A}^{\top})^k \vec{e} - \vec{e}. \qquad [5.25]$$

Thus, a linear form of [5.25] becomes

$$\vec{c}_z = ((\mathbf{I} - \alpha\mathbf{A}^\top)^{-1} - \mathbf{I})\vec{e}\,. \qquad [5.26]$$

Now the lazy form of Katz centrality, ${}_l\vec{c}_z$, can be written as

$$\begin{aligned} {}_l\vec{c}_z &= (\mathbf{I} - \alpha(\beta\mathbf{I} + (1-\beta)\mathbf{A}^\top))^{-1}\vec{e} - \vec{e}\,, \\ ({}_l\vec{c}_z + \vec{e}) &= (\mathbf{I} - \alpha(\beta\mathbf{I} + (1-\beta)\mathbf{A}^\top))^{-1}\vec{e}\,. \end{aligned} \qquad [5.27]$$

The right-hand side of [5.27] gives the lazy α-centrality as in case 1 of section 5.3.3, which then reduces the equation to

$$({}_l\vec{c}_z + \vec{e}) = \left(\mathbf{I} - \frac{\alpha(1-\beta)}{1-\alpha\beta}\mathbf{A}^\top\right)^{-1} \frac{1}{1-\alpha\beta}\vec{e}\,,$$

and so,

$${}_l\vec{c}_z = \left(\mathbf{I} - \frac{\alpha(1-\beta)}{1-\alpha\beta}\mathbf{A}^\top\right)^{-1} \frac{1}{1-\alpha\beta}\vec{e} - \vec{e}\,. \qquad [5.28]$$

From [5.28] we deduce the following:

(i) the lazy Katz centrality is simply the lazy α-centrality minus the vector of all ones. As such, the ranking of the lazy Katz centrality and that of the lazy α centrality are the same, though the rank values may differ since

$${}_l\vec{c}_z = {}_l\vec{c}_\alpha - \vec{e}\,, \qquad [5.29]$$

where ${}_l\vec{c}_\alpha$ refers to lazy α-centrality;

(ii) the ranking of lazy Katz centrality is the same as the ranking of the normal Katz centrality, but the rank values differ on the basis of the parameters α and β in the coefficient $\frac{\alpha(1-\beta)}{1-\alpha\beta}$, of $\mathbf{A}^\top$.

Thus, the lazy Katz centrality, the normal Katz centrality and the lazy α-centrality are all equivalent in ranking with a possible variation in their rank values.

5.3.5. *Lazy cumulative nomination centrality*

Cumulative nomination centrality $\vec{c}_c(i)$ of a node i in time t is given by the previous value of nomination of node i plus the sum of the nomination values of the neighbors j in time $t-1$. That is,

$$\vec{c}_c^{(t)}(i) = \vec{c}_c^{(t-1)}(i) + \sum_{j \in A_i} \vec{c}_c^{(t-1)}(j)\,,$$

where A_i is a set of all nodes j, for which the digraph contains a directed edge (j, i) from node j to node i.

The normalized iterative form of the above equation can be written as

$$\vec{c}_c^{(t)} = \left(\frac{\mathbf{I}+\mathbf{A}}{1+\lambda}\right)\vec{c}_c^{(t-1)} . \qquad [5.30]$$

From relation [5.30], we write the lazy cumulative nomination centrality ${}_l\vec{c}_c^{(t)}$ as

$${}_l\vec{c}_c^{(t)} = \left(\frac{\beta\mathbf{I}+(1-\beta)\mathbf{A}}{\beta+(1-\beta)\lambda}\right){}_l\vec{c}_c^{(t-1)} , \qquad [5.31]$$

$$(\beta+(1-\beta)\lambda){}_l\vec{c}_c^{(t)} = (\beta\mathbf{I}+(1-\beta)\mathbf{A}){}_l\vec{c}_c^{(t-1)} .$$

If the limit exists as ${}_l\vec{c}_c$, then the stationary vector of relation [5.31] is approximated as

$$(\beta+(1-\beta)\lambda){}_l\vec{c}_c = (\beta\mathbf{I}+(1-\beta)\mathbf{A}){}_l\vec{c}_c ,$$

which simplifies to $\lambda{}_l\vec{c}_c = \mathbf{A}{}_l\vec{c}_c$. This means that regardless of the choice of $0 < \beta < 1$, the normalized cumulative nomination converges to the same vector, namely that of the eigenvector to the dominant eigenvalue of $\mathbf{A}$. The choice of β only affects the convergence speed of the algorithm.

We are not creating any lazy variants for degree centrality, since this would not change the ranking in any way.

In Table 5.2, we summarize the results of lazy variants of α-, β-, Katz, cumulative nomination and PageRank centralities.

5.4. Experimental results

According to Abdi (2007), Kendall tau compares two ordered sets by establishing the number (called symmetric difference distance) of distinct pairs between the two ordered sets. With $d_\Delta(p_1, p_2)$ denoting the symmetric difference distance between two sets of ordered pairs p_1 and p_2, the Kendall rank correlation coefficient is given by

$$\tau = \frac{\frac{1}{2}N(N-1) - d_\Delta(p_1,p_2)}{\frac{1}{2}N(N-1)} .$$

Here, we present our results with the data "email-Eu-core" of 1005 nodes, which is from the SNAP data repository https://snap.stanford.edu/data/. The source is a directed graph from "Networks with ground truth communities". Here, we use heat maps for pairwise Kendall correlation between the different centrality measures. We considered the following aspects of the implementation: a weighted Kendall correlation between the centrality measures, based on random walks or powers of the adjacency matrix, Kendall tau correlation for variation of the parameters β and α in β-centrality and Kendall tau correlation for the top 10% of nodes considered in the network.

Shifted power series	Power series	Steady state
Lazy β-centrality $\frac{\beta}{\alpha}\left(\sum_{k=1}^{\infty}\alpha^k(\beta\mathbf{I}+(1-\beta)\mathbf{A})^k\right)\vec{e}$	Lazy α-centrality (out) $\left(\sum_{k=0}^{\infty}\alpha^k(\beta\mathbf{I}+(1-\beta)\mathbf{A})^k\right)\vec{e}$	Lazy Cumulative Nomination ${}_l\vec{c}_c^{(n+1)}=\frac{\beta\vec{c}^{(n)}+(1-\beta)\mathbf{A}\vec{c}^{(n)}}{\lVert\beta\vec{c}^{(n)}+(1-\beta)\mathbf{A}\vec{c}^{(n)}\rVert}$
Lazy Katz $\left(\sum_{k=1}^{\infty}\alpha^k(\beta\mathbf{I}+(1-\beta)\mathbf{A}^\top)^k\right)\vec{e}$	Lazy α-centrality $\left(\sum_{k=0}^{\infty}\alpha^k(\beta\mathbf{I}+(1-\beta)\mathbf{A}^\top)^k\right)\vec{e}$	Lazy Eigenvector ${}_l\vec{c}_e^{(n+1)}=\frac{\beta\vec{c}^{(n)}+(1-\beta)\mathbf{A}^\top\vec{c}^{(n)}}{\lVert\beta\vec{c}^{(n)}+(1-\beta)\mathbf{A}^\top\vec{c}^{(n)}\rVert}$
Lazy shifted PageRank $\left(\sum_{k=1}^{\infty}\alpha^k(\beta\mathbf{I}+(1-\beta)\mathbf{A}_w^\top)^k\right)\vec{u}$	Lazy PageRank $\left(\sum_{k=0}^{\infty}\alpha^k(\beta\mathbf{I}+(1-\beta)\mathbf{A}_w^\top)^k\right)\vec{u}$	Lazy Scaled Eigenvector ${}_l\vec{c}_s^{(n+1)}=\frac{\beta\vec{c}^{(n)}+(1-\beta)\mathbf{A}_w^\top\vec{c}^{(n)}}{\lVert\beta\vec{c}^{(n)}+(1-\beta)\mathbf{A}_w^\top\vec{c}^{(n)}\rVert}$

Table 5.2. *Overview of lazy centralities and their expression as a power series or power iteration. The iteration formula in the last column is repeated until convergence*

DEFINITION 5.2.– *A weighted Kendall tau is defined as*

$$\tau_w = \frac{1}{\left[\sum_{i,j} w_{ij} - \sum_i w_{ii}\right] \sum_{i \neq j} w_{ij} sgn(i-j) sgn(R_i - R_j)},$$

where sgn(x)= -1, 0, 1 if x <, = or > 0, R_i, R_j *are the ranks of the independent and identically distributed (iid) random variables* x_i, x_j*, i,j = 1,2,...,n and* w_{ij} *is a weight function which is bounded and symmetric with* $w : (\mathbb{N})^2 \rightarrow \mathbb{R}$ *and*

$$w_{ij} = w_o exp(\alpha(i_1 + i_2 - 2n)) .$$

In this case, we have considered the absolute value of the rank difference. So, sgn(x)= 0, 1 if x = or > 0.

5.5. Discussion

The legend color band on the right-hand side of the heat maps indicate the Kendall tau correlation coefficients with the colors corresponding to them. In this case, we have only considered the absolute value of the coefficients. A high correlation between two variables signifies the identical rank value of their observations. The highest correlation is seen along the diagonals (see all figures), corresponding to the comparison of rank values of one centrality measure with itself on the given metric, and the rank values, which are identical. The axes labels are identically ordered on both axes on all the heat maps shown in this chapter.

In Figure 5.1, three main diagonal clusters are observed in the heat map. Correlation coefficients higher than 0.9 are seen for ranks due to associations between Katz or α-centralities and degree(in) or eigenvector centralities in the first cluster.

In the second cluster, we have PageRank(shift), or PageRank and deg(Scaled), or eigenvector (Scaled), while in the third class, we have β, or α(out) and degree(out), or cumulative nomination. It has been shown following equation [5.13] that beta centrality is a linear transformation of alpha centrality (out), and so, the two must produce the same ranking regardless of the different rank scores of the nodes. The rest of the associations show relatively low correlation coefficients, with the least being 0.55 for cumulative nomination and deg(Scaled) as one example of such pairs. In section 5.2.3 we argued that α- and Katz centralities are similar, with the only difference being that for the α-centrality, the summation starts at zero, while for the Katz centrality, it starts at 1. The experimental results in Figure 5.2 agree with our assertion that the two centralities give the same ranking for a particular network.

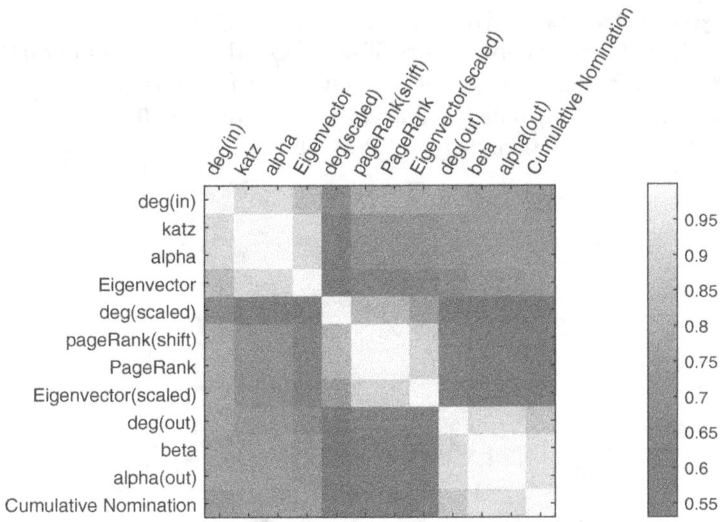

Figure 5.1. *Comparison of Kendall tau correlation coefficients between the centrality measures based on powers of the adjacency matrix. For a color version of this figure, see www.iste.co.uk/dimotikalis/analysis1.zip*

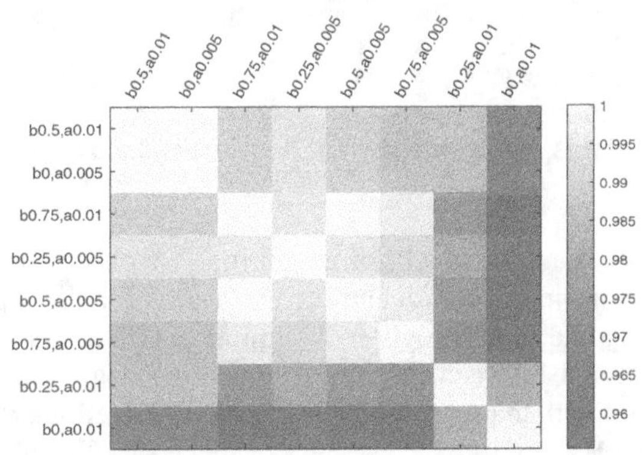

Figure 5.2. *Comparison of Kendall tau correlation coefficients between different pairs of values of β and α for β-centrality measure. The notations b0.xx,a0.yy mean β is equal to 0.xx and α is equal to 0.yy, respectively. For a color version of this figure, see www.iste.co.uk/dimotikalis/analysis1.zip*

Figure 5.2 compares the ranks of the nodes in the network for varying pairs of parameters α and β in the computation of β-centrality. The highest correlation coefficient is observed by comparing b0.5,a0.01 and b0,a0.005; b0.75,a0.01 and b0.5,a0.005, in addition to the self-comparison among the pairs of the parameters (along the diagonal). It is observed that generally, the β-centrality gives high values of correlation coefficient for the chosen pairs of α and β, with a minimum of 0.94. We conclude that when ranking the nodes of any network using the β-centrality, the best values to get identical ranking of the parameters α and β are b0.5,a0.01 with b0,a0.005 and b0.75,a0.01 with b0.5,a0.005.

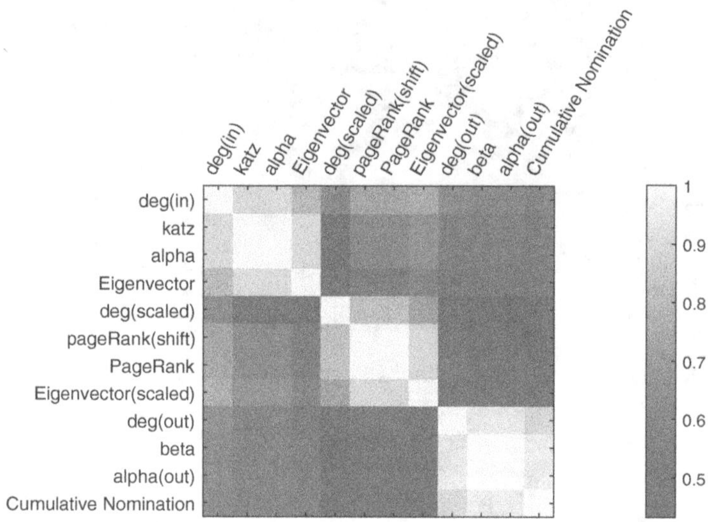

Figure 5.3. *Comparison of Kendall tau correlation coefficients between the centrality measures for the top 10% nodes in the network. For a color version of this figure, see www.iste.co.uk/dimotikalis/analysis1.zip*

In Figure 5.3, three diagonal blocks of length four are observed with coefficients of at least 0.8 between the different centrality measures for the top 10% nodes in ranking. The non-diagonal blocks register dissimilar ranking, with the least (= 0.5) being for the comparison between the block containing deg (out), β, α (out) and cumulative nomination with that containing deg (scaled), PageRank (shift), PageRank and eigenvector (scaled).

In Figure 5.4, we mainly focus on the lazy variants of the centrality measures. It is noted that a correlation coefficient higher than 0.85 is observed between the lazy alpha and each of the centralities deg(in), alpha and eigenvector. The implication of this is that the lazy alpha centrality gives almost the same ranking of nodes as the centralities mentioned. Lazy PageRank is seen to bear the same relationship with deg(scaled), PageRank and eigenvector(scaled). The above scenario does not leave out the set containing lazy alpha(out) with deg(out), alpha(out) and cumulative nomination centrality.

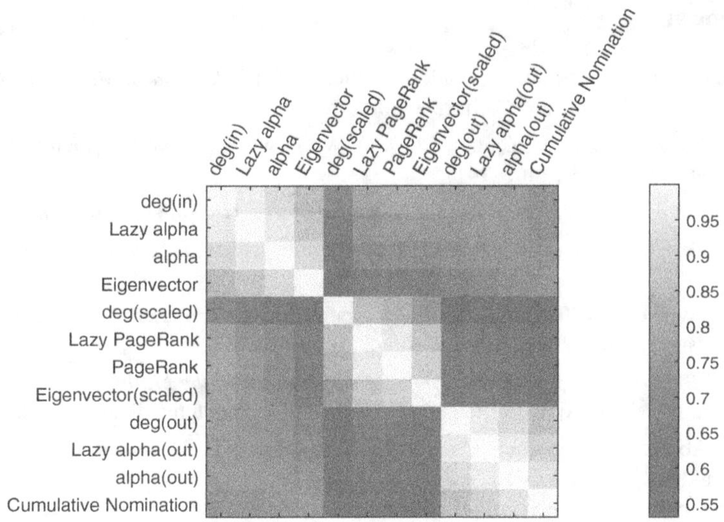

Figure 5.4. *Comparison of Kendall tau correlation coefficients between some centrality measures and the lazy variants. For a color version of this figure, see www.iste.co.uk/dimotikalis/analysis1.zip*

5.6. Conclusion

For decades, centralities have been used as a key measure of importance in the components of networks of varying types. In our exploration of the centralities based on the powers of the adjacency matrix or random walks, we derived the formulae for both ordinary and lazy variants of centrality measures and established relationships between them. We have reserved the numerical computations using our generalizations for the next chapter. With readily available data, we implemented the results using heat maps, which visibly agree with the relations in section 5.2.

5.7. Acknowledgments

This research was supported by the Swedish International Development Cooperation Agency (Sida), the International Science Programme (ISP) (namely the International Programme in the Mathematical Sciences (IPMS)) and Sida Bilateral Research Programmes for research and education capacity development in Mathematics in Uganda. The authors are also grateful to the Mathematics and Applied Mathematics (MAM), Division of Applied Mathematics, Mälardalen University, for providing an excellent and inspiring environment for research in mathematics.

5.8. References

Abdi, H. (2007). The Kendall rank correlation coefficient. *Encyclopedia of Measurement and Statistics*. Sage, Thousand Oaks, California, 508–510.

Ballester, C., Calvó-Armengol, A., Zenou, Y. (2006). Who's who in networks. Wanted: The key player. *Econometrica*, 74(5), 1403–1417.

Bavelas, A. (1948). A mathematical model for group structures. *Human Organization*, 7(3), 16.

Bavelas, A. (1950). Communication patterns in task-oriented groups. *J. Acoust. Soc. Am.*, 22(6), 725–730.

Benzi, M. and Klymko, C. (2015). On the limiting behavior of parameter-dependent network centrality measures. *SIAM J. Matrix Anal. Appl.*, 36(2), 686–706.

Berman, A. and Plemmons, R.J. (1994). *Nonnegative Matrices in the Mathematical Sciences*. Society for Industrial and Applied Mathematics (SIAM), Philadelphia, Pennsylvania.

Biganda, P.S., Abola, B., Engström, C., Mango, J.M., Kakuba, G., Silvestrov, S. (2018). Traditional and lazy pageranks for a line of nodes connected with complete graphs. In *Stochastic Processes and Applications*, Silvestrov, S., Malyarenko, A., Rančić, M. (eds). Springer Preceeding in Mathematics & Statistics, vol. 271, pp. 391–412.

Bloch, F., Jackson, M.O., Tebaldi, P. (2019). Centrality measures in networks [Online]. Available at: SSRN 2749124.

Boldi, P. and Vigna, S. (2014). Axioms for centrality. *Internet Math.*, 10(3–4), 222–262.

Bonacich, P. (1972). Factoring and weighting approaches to status scores and clique identification. *J. Math. Sociol.*, 2(1), 113–120.

Bonacich, P. and Lloyd, P. (2004). Calculating status with negative relations. *Soc. Netw.*, 26(4), 331–338.

Das, K., Samanta, S., Pal, M. (2018). Study on centrality measures in social networks: A survey. *Soc. Net. Anal. and Min.*, 8(1), 13.

Engström, C. (2016). PageRank in evolving networks and applications of graphs in natural language processing and biology. Doctoral Dissertation, vol. 271. Mälardalen University, Västerås.

Engström, C. and Silvestrov, S. (2019). Using graph partitioning to calculate PageRank in a changing network. In *Data Analysis and Applications 2: Utilization of Results in Europe and Other Topics*, Skiadas, C.H. and Bozeman, J.R. (ed.). ISTE Ltd, London, and John Wiley & Sons, New York.

Ghosh, R. and Lerman, K. (2011). Parameterized centrality metric for network analysis. *Phys. Rev. E*, 83(6), 066118.

Golub, B. and Jackson, M.O. (2010). Naive learning in social networks and the wisdom of crowds. *AEJ: Microeconomics*, 2(1), 112–49.

Hagan, R., Feng, Y., Bush, J. (2015). *Centrality Metrics*. University of Tennessee, Knoxville, Tennessee.

Izaac, J.A., Zhan, X., Bian, Z., Wang, K., Li, J., Wang, J.B., Xue, P. (2017). Centrality measure based on continuous-time quantum walks and experimental realization. *Phys. Rev. A*, 95(3), 032318.

Jalili, M., Salehzadeh-Yazdi, A., Asgari, Y., Arab, S.S., Yaghmaie, M., Ghavamzadeh, A. (2015). CentiServer: A comprehensive resource, web-based application and R package for centrality analysis. *PLoS One*, 10(11).

Katz, L. (1953). A new status index derived from sociometric analysis. *Psychometrika*, 18(1), 39–43.

Leavitt, H.J. (1951). Some effects of certain communication patterns on group performance. *J. Abnorm. Psychol.*, 46(1), 38.

Li, H. and Zhang, Z. (2018). Kirchhoff index as a measure of edge centrality in weighted networks: Nearly linear time algorithms. *Proceedings of the Twenty-Ninth Annual ACM SIAM Symposium on Discrete Algorithms*. SIAM, pp. 2377–2396

Liu, N.N. and Yang, Q. (2008). Eigenrank: A ranking-oriented approach to collaborative filtering. *Proceedings of the 31st Annual International ACM SIGIR Conference on Research and Development in Information Retrieval*. ACM, pp. 83–90.

Lü, L., Chen, D., Ren, X.L., Zhang, Q.M., Zhang, Y.C., Zhou, T. (2016). Vital nodes identification in complex networks. *Phys. Rep.*, 650, 1–63.

Mao, C. and Xiao, W. (2018). A comprehensive algorithm for evaluating node influences in social networks based on preference analysis and random walk. *Complexity*, volume 2018.

Noel, S. and Jajodia, S. (2005). Understanding complex network attack graphs through clustered adjacency matrices. *21st Annual Computer Security Applications Conference (ACSAC'05)*.

Pak, I. and Żuk, A. (2002). On Kazhdan constants and mixing of random walk. *Int. Math. Res. Notices*, 2002(36), 1891–1905.

Poulin, R., Boily, M.C., Mâsse, B.R. (2000). Dynamical systems to define centrality in social networks. *Soc. Net.*, 22(3), 187–220.

Roy, M., Tredan, G., Telekom, D. (2010). Sharpening the definition of centrality. *Social Networks and Distributed Systems (SNDS), the 2010 Workshop*.

Zhang, F. and Hancock, E.R. (2005). Image scale-space from the heat kernel. *Iberoamerican Congress on Pattern Recognition*. Springer, Berlin, Heidelberg.

Zhang, J. and Luo, Y. (2017). Degree centrality, betweenness centrality, and closeness centrality in social network. *2017 2nd International Conference on Modelling, Simulation and Applied Mathematics (MSAM2017)*. Atlantis Press, Paris.

Zhou, D. and Schölkopf, B. (2004). Learning from labeled and unlabeled data using random walks. *Joint Pattern Recognition Symposium*. Springer, Berlin, Heidelberg.

6

Error Detection in Sequential Laser Sensor Input

In this chapter, we build two low complexity and robust error-detection algorithms that adjust for sensor faults and inaccuracies in real time. The algorithms can detect consecutive and abrupt sensor slips and provide a paradigm for real-time corrections. In addition to reliability, the algorithms do not require prior data knowledge, nor do they rely on any assumptions about the distribution of the data. The run time is independent of the input size, which is ideal for large volumes of data, and allows for implementation at the sensor level.

6.1. Introduction

A major obstacle to progress in robotics is a machine's limited ability to interact with the environment. Fortunately, the cost of laser sensors has decreased, making them an attractive choice as detection tools. Laser sensors are low-cost devices that are readily available and simple to operate. However, laser sensors can suffer from jitter and are not very reliable, especially when they encounter reflective surfaces. The trade-off for their convenience is the rudimentary quality of the data received, which explains the motivation for research on the topic of increasing sensor accuracy.

Prior research on sensor accuracy has been in either engineering or data processing. Papers in the field of engineering focus on the physical aspects of sensor construction and data acquisition, with a goal of obtaining better data through hardware improvements (see, for example, Shirafuji and Hosoda (2011) and Morita *et al.* (2018)). This chapter focuses on data processing. Considering existing hardware, we construct two algorithms that correct sensor slips in real time, and therefore

Chapter written by Gwenael GATTO and Olympia HADJILIADIS.

refine the quality of the data gathered online. While other papers also focus on data processing (see, for example, Veiga *et al.* (2015) and James *et al.* (2018)), our algorithms do not require prior offline training of a classifier before they can be used in real time. That is, our algorithms are non-anticipative and do not rely on any distributional assumptions or prior training.

In particular, in this chapter, we construct low-complexity error-detection algorithms to make data obtained from a two-dimensional laser sensor more accurate. Our improvement relies on the ability to detect in real time both persistent and instantaneous sensor slips, without the need of prior data knowledge or the assumption of a specific distribution. The first algorithm that we construct is a consecutive change in mean algorithm, and the second is a burst-detection algorithm. Both algorithms are based on the literature on the topic of quickest detection, which is concerned with detecting changes in the distribution of data received in real time (see Basseville and Nikiforov (1993) and Poor and Hadjiliadis (2008)). This algorithm is based on Page's 1954 CUSUM algorithm applied in repetition as in Carlisle *et al.* (2016). The second burst-detection algorithm is based on modeling the data using hidden Markov models (HMM) (see Rabiner and Juang (1986) and Ghahramani (2001)) and uses a CUSUM-like algorithm to detect a change from one HMM to another (see Chen and Willett (2000) and Hadjiliadis and Stamos (2010)).

In section 6.2, we discuss the data gathered for this research, which contains jitters and spikes, two particularly common errors encountered in laser sensor data. In section 6.3, we describe the two algorithms, and in section 6.4, we showcase our results and examine real-world performance.

6.2. Data description

A SICK TiM571 scanning range finder, manufactured by Fetch, was mounted on a robot, and provided the laser sensor data. It is a two-dimensional sensor that measures distance from its origin using the time it takes for a laser beam to reflect back. The location of the sensor defines the origin of the axis of reference, and an x–y plane is defined by the sensor. The laser has a range of 25 meters and a 220 degree field of view. Its angular resolution is 1/3 of a degree with a 15-Hz update rate. This means that one scan contains 660 points that span 220 degrees, and 15 scans are executed each second. In our data, one time step corresponds to one full scan. Each of the 660 points, defined by their (x, y) coordinates, is obtained by firing a laser beam in a straight line from the sensor until the beam collides with an object that reflects it back to the sensor. The delay between the time the beam is sent and the time it is received is converted to distance and expressed in meters.

In the raw data, each time step represents one full scan. The 660 (x, y) pairs of coordinates represent distance in meters and are the result of one full scan to which the Bellotto and Huosheng (2009) clustering algorithm is applied. This algorithm is divided into three sub-steps: data pre-processing, edge detection and leg detection. In the data pre-processing sub-step, a local minimization operator is applied to reduce some of the noise from the data, such as noise from sloped surfaces, and a local maximization operator is applied to remove thin objects, such as table legs. Once the data has been pre-processed in this way, a vertical edge-detection filter is applied to obtain a sequence of vertical edges. From this sequence, the Bellotto–Hu clustering algorithm then categorizes groups of vertical edges in one of three patterns: LA (Legs Apart), FS (Forward Straddle) or SL (Single Leg/Legs Together). Each labeled grouping of vertical edges forms a cluster. Each cluster is then characterized by its centroid. In our case, the application of the Belotto–Hu algorithm results in the reduction of the 660 (x, y) points of the scan into at most 7 (x, y) cluster centroids. Each such centroid is given an index from $\{0, 1, \ldots, 6\}$.

One issue we experienced with the Bellotto–Hu clustering algorithm is its inability to reliably order the clusters it detects across scans. To remedy this defect, we implemented our own reordering algorithm (see Table 6.1 for an example of the correction of cluster 0 at time step 205) that assigns each cluster centroid at time step n the same index of its nearest neighbor at time $n - 1$, using the Pythagorean distance from the robot-mounted laser as the measure. This results in time step continuity of cluster centroids across scans. Our data is then the time series (x_n, y_n) for $n = 1, 2, \ldots, 739$, corresponding to the centroids of cluster 0.

Once the time series for cluster 0 was determined, we converted our data to polar coordinates:

$$(x_n, y_n)_{n=1,2,3,\ldots,739} \rightarrow (r_n, \theta_n)_{n=1,2,3,\ldots,739},$$

####	Before our sorting algorithm				After our sorting algorithm			
####	Cluster 0 (x,y)		Cluster 1 (x,y)		Cluster 0 (x,y)		Cluster 1 (x,y)	
Time/scan	x	y	x	y	x	y	x	y
⋮	⋮	⋮	⋮	⋮	⋮	⋮	⋮	⋮
203	0.460388	-0.422849	2.455110	-0.352946	0.460388	-0.422849	2.455110	-0.352946
204	0.460214	-0.422685	2.430700	-0.342202	0.460214	-0.422685	2.430700	-0.342202
205	2.599970	-0.433634	0.460402	-0.422850	0.460402	-0.422850	2.599970	-0.433634
206	0.460663	-0.423155	2.414260	-0.332548	0.460663	-0.423155	2.414260	-0.332548
207	0.460762	-0.423176	2.402750	-0.345458	0.460762	-0.423176	2.402750	-0.345458
⋮	⋮	⋮	⋮	⋮	⋮	⋮	⋮	⋮

Table 6.1. *Sorting example at time step 205*

where

$$r_n = \sqrt{x_n^2 + y_n^2}$$

$$\theta_n = \arctan(y_n/x_n)$$

at each time step.

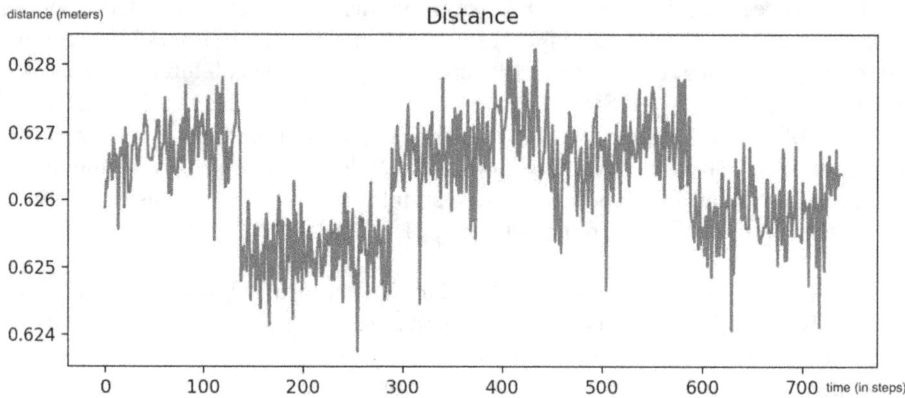

Figure 6.1. *Distance from the robot, in meters, for cluster no. 0 at each time step*

We observe two types of sensor slips. The distance plot of Figure 6.1 shows an up-and-down jitter pattern, and the angle plot of Figure 6.2 contains spikes that occur over intervals as short as one point. These inputs are known to be faulty since both the robot and the scene are static. The error shown in the distance plot pertains to a change in the mean, while the error illustrated in the angle plot can be described as a localized spike in the variance. Thus, we need to capture both types of errors to correct faults in the input. In the next section, we will build non-parametric algorithms to detect sensor slips in real time without making any assumptions about the underlying distribution of the data.

6.3. Algorithms

In section 6.2, we observed that the sensor slips related to the distance display themselves as consecutive changes in the mean of the distribution of the data, while the faulty inputs related to the angles display themselves as localized spikes in the variance or as bursts (see Figures 6.1 and 6.2). Therefore, we think of our data as the following sequences of random variables $\{R_n\}_{n=0,1,2,...}$ and $\{\Theta_n\}_{n=0,1,2,...}$, whose realizations are $\{r_n\}_{n=0,1,2,...}$ and $\{\theta_n\}_{n=0,1,2,...}$, described in section 6.2.

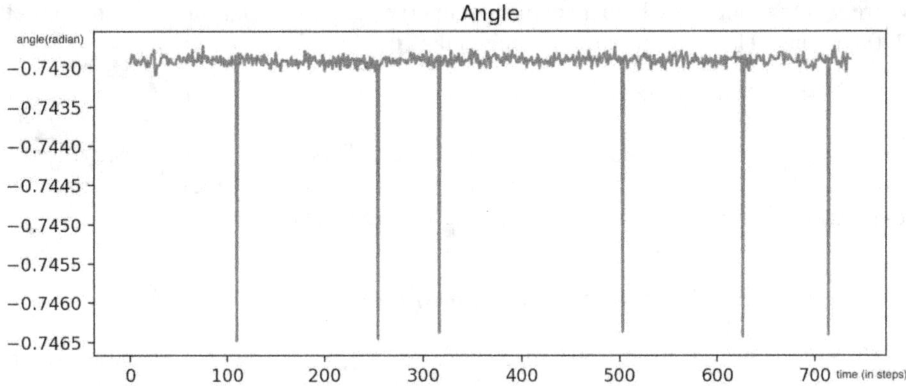

Figure 6.2. *Angle orientation from the robot, for cluster no. 0 at each time step*

First, we devise an algorithm that can detect changes in the mean of the distribution of $\{R_n\}_{n=0,1,2,...}$ when it occurs. Second, we devise an algorithm to detect the localized variance spikes in $\{\Theta_n\}_{n=0,1,2,...}$. To achieve both goals, we use *quickest change detection*.

The field of *quickest change detection* (QCD) aims to identify the times when the probability distribution of a stochastic process changes. To formalize the quickest change detection problem, consider the sequentially observed variable $\{X_n\}_{n=0,1,2,...}$. We introduce the following hypothesis testing problem:

$$H_0 : X_n \sim f_0, \qquad n = 1, 2, ..., \nu - 1$$

$$H_1 : X_n \sim f_1, \qquad n = \nu, \nu + 1, ...$$

where f_0 is the starting distribution of the data and f_1 is the distribution the data moves to at time ν. The objective is to detect ν by minimizing the delay between ν, the change point and the time the change point is declared. Detection of the change is declared by using a stopping time chosen to carefully balance the trade-off between time of detection and the frequency of false alarms (Poor and Hadjiliadis (2008)). The stopping time is usually based on the threshold crossing of an appropriately chosen statistic process.

A popular stopping time to detect a change in distribution is known as the CUSUM stopping time (see, for example, Chen and Willett (2000), Page (1954) for the original concept, or Chapter 2.6 in Siegmund (1985)). It is based on the CUSUM statistic process $\{L_k^n\}_{n=1,2,...}$, which, for each n, is the likelihood ratio from k to n, $k \leq n$, where k is the index at which a change occurs (see equation [6.4] in Chen and Willett (2000)). The CUSUM stopping time then takes the form:

$$N = \min\{n : \max_{0\leq k\leq n} L_k^n \geq h\},$$

where $h > 0$ represents the threshold parameter. This leads to the construction of two algorithms based on the CUSUM stopping time: the first focuses on detecting consecutive changes in the mean, and the second focuses on detecting bursts.

6.3.1. *Algorithm for consecutive changes in mean*

Consider the random variables $\{R_n\}_{n=0,1,2,...}$, whose realizations are $\{r_n\}_{n=0,1,2,...}$, described in section 6.2. Without making any assumptions about their distribution, we focus on detecting a change in their mean. Our hypothesis test then becomes:

$$H_0 : E_0[R_n] = \mu_0, \qquad n = 0, 1, 2, ..., \nu - 1$$

$$H_1 : E_1[R_n] = \mu_1, \qquad n = \nu, \nu + 1, ...$$

where μ_0 is the expected value of $\{R_n\}_{n=0,1,...,\nu-1}$ under the null hypothesis and μ_1 is the expected value of $\{R_n\}_{n=\nu,\nu+1,...}$ after the change occurs.

In order to detect either case in which $\mu_1 > \mu_0$ or $\mu_1 < \mu_0$, we introduce the constant k to capture the sensitivity to a change, whether upward or downward, and construct two CUSUM stopping times to detect each of the changes (see Chapter 16 in Devore (2012)).

1) $\mu_1 > \mu_0$.

$$N^+ = \min\{n : d_n \geq h\}, \qquad [6.1]$$

where

$$d_n = \max[0, d_{n-1} + (r_n - (\mu_0 + k))], \text{ with } d_0 = 0. \qquad [6.2]$$

The $\{d_n\}_{n=0,1,...}$ constitute the CUSUM statistic process, r_n is the realization of the data at time n, μ_0 is the starting mean of the data and k is the upward change that is allowed before the process is considered out of control.

2) $\mu_1 < \mu_0$.

$$N^- = \min\{n : e_n \geq h\}, \qquad [6.3]$$

where

$$e_n = \max[0, e_{n-1} - (r_n - (\mu_0 - k))], \text{ with } e_0 = 0. \qquad [6.4]$$

The $\{e_n\}_{n=0,1,...}$ is the CUSUM statistic process, r_n is the realization of the data at time n, μ_0 is the starting mean of the data and k is the downward change that is allowed before the process is considered out of control.

We note that $d_n \geq 0$ and $e_n \geq 0$ for all $n = 0, 1, 2, \ldots$ – i.e. both of the CUSUM statistics processes appearing in equations [6.2] and [6.4] are capped at below 0. We can thus select a positive threshold h in the above equations. We then compare both processes to the threshold h. A different threshold could be set for [6.2] and [6.4]. However, since we have no reason to believe that a change in the mean of the distribution is more likely to be positive or negative, we choose to set the same threshold for both CUSUM statistic processes. The first time $d_n \geq h$ or $e_n \geq h$, an alarm is raised. The threshold h can be set either with domain knowledge or as an appropriate multiple of the deviation k. In our application, we set h to be $5 * k$ and the mean μ_0 is initialized from the sample mean of the first 20 data points. In practice, the parameter k represents the sensitivity of the algorithm to changes in the mean, and can thus be set to a percentage of the original mean μ_0. In particular, in this chapter, it is set to 1% of μ_0. After an alarm is raised, both CUSUM statistic processes in equations [6.2] and [6.4] are reset to zero, and the new mean μ_1 is set to

1) $\mu_1 = \mu_0 + k$, if $N^+ \wedge N^- = N^+$ (see [6.1], [6.3]) and

2) $\mu_1 = \mu_0 - k$, if $N^+ \wedge N^- = N^-$ (see [6.1], [6.3])

Note that the algorithm is robust with respect to the choice of threshold. This is further discussed in section 6.4.

The algorithm is detailed below using pseudocode notation[1]:

1) Set the values of the following parameters:

– μ_0: the starting mean of the data;

– k: the deviation allowed from the mean (usually taken to be $\frac{1}{2}$ the allowed deviation from the process mean);

– h: the threshold which, when crossed, triggers an alarm (usually set as a multiple of k).

2) Initialize an upper and lower CUSUM statistic:

– set d_0 to 0;

– set e_0 to 0;

– set μ to μ_0.

3) Read in data of r_n's and compute, at every time step $n = 1, 2, \ldots$,:

– $d_n = \max[0, d_{n-1} + (r_n - (\mu + k))]$;

– $e_n = \max[0, e_{n-1} - (r_n - (\mu - k))]$.

4) If at time n, $d_n \geq h$:

– raise alarm;

– set d_n to 0;

1 For the implementation of the algorithm, see Gatto (2019).

– set e_n to 0;

– set μ to $\mu + k$.

5) If at time n, $e_n \geq h$:

1) raise alarm;

2) set d_n to 0;

3) set e_n to 0;

4) set μ to $\mu - k$.

6) Repeat step 2 through 5 until data stream ends.

6.3.2. *Algorithm for burst detection*

In this subsection, we build an algorithm to detect sensor slips as manifested in the angles data. The sensor slips in the angles data are manifested as bursts, or spikes, which occur over a minute time interval (see Figure 6.2). Such slips can be described as localized increases in the variance of the data. To detect such changes, we therefore must capture changes in that variance that occur over minute intervals. This makes the typical CUSUM algorithm we used previously ineffective in this case. As we can see in Figure 6.3, the typical CUSUM algorithm fails to detect the first and fourth angle spikes and further falsely detects a spike around time step 475. Even a CUSUM algorithm used to capture changes in the variance parameter itself would either be blind to such changes or be too sensitive to be practical.

Therefore, to facilitate the construction of a CUSUM algorithm, which detects the spikes, we must introduce a model that allows for spikes. This leads us to the consideration of hidden Markov models (HMM).

A discrete-time finite-state HMM is a probabilistic model of a sequence of random variables $\{X_t\}_{\{t\geq 0\}}$ whose underlying structure is a Markov chain. A key advantage of HMMs is the assumption that an observation at time t originates from a process whose inner working is hidden from the observer Ghahramani (2001). Therefore, the distribution of the X_t's for each $t = 0.1.2., \ldots$ depends on the value of a sequence of state variables $\{s_t\}_{t=0,1,\ldots}$, with $s_t \in \{1, ..., N\}$ which are governed by the N x N transition matrix A . A discrete HMM can be described by the triple

$$\lambda = (A, B, \pi), \qquad [6.5]$$

where A is the state transition matrix of the underlying Markov chain,

$$A = [a_{ij}] = [P(s_{t+1} = j | s_t = i)], \qquad i, j = 0, ..., N,$$

B is the observation matrix

$$B = [b_{ix_t}] = [P(X_t \in dx_t | s_t = i)], \qquad i = 1, ..., N$$

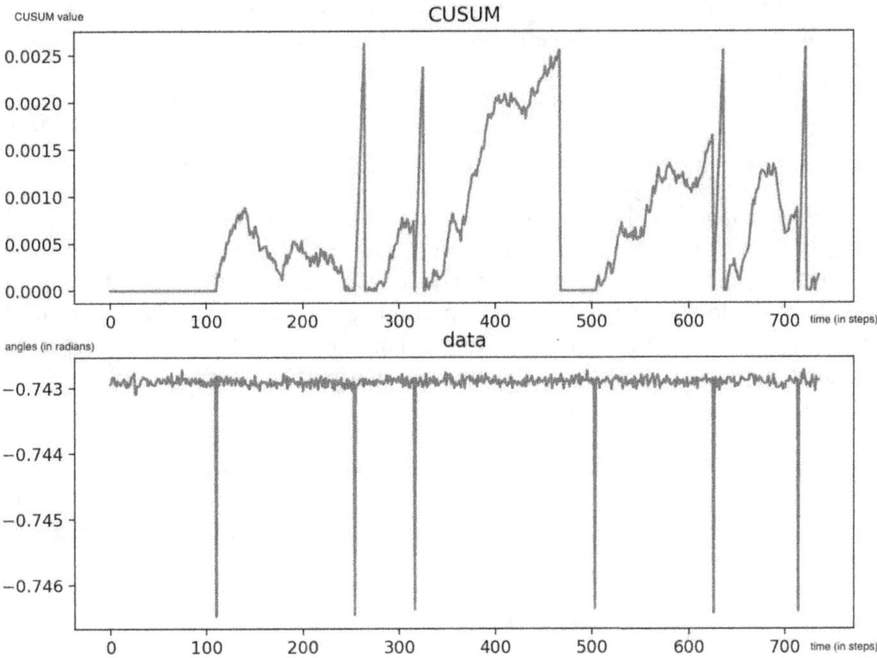

Figure 6.3. *The CUSUM algorithm of section 6.3.1 fails to detect the first and fourth spikes and falsely detects a spike around time step 475*

and π is the initial probability distribution of the underlying Markov states,

$$\pi = [\pi_i = P(s_t = i)], \qquad\qquad i = 0, ..., N.$$

In our case, we set $N = 1$. That is, there are two states, 0 and 1. We then use two HMM models to describe the probabilistic behavior of our sequence of angle random variables $\{\Theta_t\}_{\{t \geq 0\}}$: one that does not allow spikes in variance and another that does. In other words, one in which state 0 is the only possible state and one in which both states 0 and 1 are possible. To simplify the model for Θ_t's, we first normalize by subtracting the sample mean of the first few observations (the first 20). We can now represent the "low"- and "high"-variance regimes by using two Gaussian models in the observation matrix: one with a "low"-variance parameter and another with a "high"-variance parameter. Note that the Gaussian assumption in the observation matrix is redundant. All we need is a distribution with a finite second moment so that the "low"- and "high"-variance regimes can be distinguished (in our case, with a low or high variance). Arbitrarily, we chose state 0 to be the state with low variance and state 1 to be the state with high variance. Our observation matrix is:

$$B = \begin{bmatrix} \frac{1}{\sqrt{2\pi}\sigma_{\text{low}}} e^{\frac{-x}{2\sigma^2_{\text{low}}}} \\ \frac{1}{\sqrt{2\pi}\sigma_{\text{high}}} e^{\frac{-x}{2\sigma^2_{\text{high}}}} \end{bmatrix}. \qquad [6.6]$$

The low variance is the baseline variance, obtained by computing the sample variance of the first few points of data (in our case, the first 20 points). An alternative method is to set the variance by using domain knowledge. The high variance is set as a multiple of the low variance. We used a multiple of 5 in our case. The algorithm is extremely robust with respect to the multiple that is used. Although the observation matrix uses the Gaussian kernel, the data from a normal distribution is not necessary for adequate real-world performance.

Now, let A_0, π_0 be the state transition matrix and the initial probability distribution vector for the HMM that does not allow spikes in the variance, and A_1, π_1 be the state transition matrix and the initial probability distribution vector for the HMM that allows spikes in the variance.

$$A_0 = \begin{bmatrix} 1 & 0 \\ 0 & 1 \end{bmatrix}, \pi_0 = \begin{bmatrix} 1 \\ 0 \end{bmatrix}, \qquad [6.7]$$

$$A_1 = \begin{bmatrix} 0.9 & 0.1 \\ 0.9 & 0.1 \end{bmatrix}, \pi_1 = \begin{bmatrix} 0.5 \\ 0.5 \end{bmatrix}. \qquad [6.8]$$

Under model $\lambda_0 = (A_0, B, \pi_0)$, state 1 is impossible as the initial probability distribution vector π_0 begins in state 0, which is an absorbing state according to the state transition matrix A_0. On the other hand, under the model described by $\lambda_1 = (A_1, B, \pi_1)$, both state 0 and state 1 are possible and, although transitions from state 0 to itself are more likely to occur, it remains possible to transition from state 0 to state 1, which is a "high"-variance state. The choice of values for the transition probabilities in matrix A_1 reflects the possibility of transitioning from any state to another with a higher probability to remain in state 0 when in it, and a higher probability to transition to state 0 when in state 1. Such a choice of transition matrix A_1 describes the nature of the possibility of instantaneous bursts, since state 1 is a high-variance state. Our algorithm is robust with respect to the exact choice of the transition probabilities in the A_1 matrix above. A demonstration of the robustness of our algorithm to the exact choice of the A_1 matrix is seen in section 6.4.

This gives us the following hypothesis testing:

$$H_0 : \Theta_t \sim \lambda_0, \qquad t = 1, 2, \ldots$$

$$H_1 : \Theta_t \sim \lambda_1, \qquad t = 1, 2, \ldots$$

Θ_t's are the sequence of random variables representing the angles as described in section 6.2, $\lambda_0 = (A_0, B, \pi_0)$ and $\lambda_1 = (A_1, B, \pi_1)$.

Our goal is to determine which model at each point in time is most likely to describe the data. To that end, we introduce a conditional form of the log-likelihood ratio $g(n; k)$ (see equation (24) in Chen and Willett (2000))

$$g(n; k) = \sum_{t=k}^{n} \log \left(\frac{f_1(\theta_t|\theta_{t-1}, \ldots, \theta_k)}{f_0(\theta_t|\theta_{t-1}, \ldots, \theta_k)} \right), \qquad [6.9]$$

where f_0 and f_1 are the conditional distributions of the Θ_t's under H_0 and H_1, respectively, which enables the construction of the CUSUM-like statistic algorithm of equation (23) in Chen and Willett (2000):

$$S_n = \max\{0, S_{n-1} + g(n; k)\}. \qquad [6.10]$$

In the following, we will use this paradigm to build an algorithm that detects bursts in real time.

The joint probability density of the history of the stochastic process and the state at a certain time is known as the forward variable. It is defined as:

$$\alpha_t(i) = p(\theta_1, \theta_2, \ldots, \theta_t, s_t = i|\lambda), i = 0, 1, \ldots, N, t = 0, 1, \ldots, \qquad [6.11]$$

where s_t is the state occupied at time t, i is the state index and N is the total number of states. Due to the Markovian property, the forward variable can easily be computed since it depends solely on the last state, and not on the joint probability of all the states visited so far.

In our case, the number of states is 2, i.e. $N = 1$. According to equation (25) in Chen and Willett (2000), the likelihood function of an HMM with parameter triple λ can be written as

$$f(\theta_1, \theta_2, \ldots, \theta_t|\lambda) = \sum_{i=0}^{1} \alpha_t(i), \qquad [6.12]$$

where $\alpha_t(i)$'s are the sequence of forward variables that we use to compute the likelihood of the HMM described by λ in [6.5].

Furthermore, note that [6.9] is equivalent to

$$f_l(\theta_t|\theta_{t-1}, \ldots, \theta_1) = f(\theta_t|\theta_{t-1}, \ldots, \theta_1, \lambda_l) = \frac{\sum_{i=0}^{1} \alpha_t(i)}{\sum_{i=0}^{1} \alpha_{t-1}(i)}, l \in \{0, 1\}, \qquad [6.13]$$

where f_l is the conditional distribution of Θ's described by λ_l for $l = 0, 1$ corresponding to H_0 and H_1. In other words, the conditional form of the log-likelihood ratio $g(n; 1)$ of [6.9] can be expressed as the ratio of the sum of forward variables, which can easily be computed as a running sum.

Due to the limitations of machine precision, computations with increasingly small values cause numerical underflow. In order to circumvent this issue, we use the scaled forward variable α'_t as defined in equation (27) in Chen and Willett (2000). That is, $\alpha'_1(i) = \alpha_1(i)$ and for $t > 1$:

$$\alpha'_{t+1}(j) = \frac{\left[\sum_{i=0}^{1} \alpha'_t(i) a_{ij}\right] b_{j\theta_{t+1}}}{\sum_{i=0}^{1} \alpha'_t(i)}, j = 0, 1 \qquad [6.14]$$

where a_{ij} is the state transition probability to state i from state j, and $b_{j\theta_{t+1}}$ is the first entry in the matrix B of [6.6] for $j = 0$ and the second entry of the matrix B of [6.6] for $j = 1$, evaluated at the observation θ_{t+1}. Note that there are two sets of α'_t's corresponding to each state for each of the hypotheses H_0 and H_1, resulting in four arrays of α'_t's indexed by the time variable t.

The algorithm is given below in pseudocode notation[2] :

1) Set the values of the following parameters:

– t to 1;

– l_0 to 0, where l_t denotes the LLR at time t.

2) Initialize the scaled forward variable α'_t:

$$\alpha'_t(j) = \pi_j b_{j\theta_t} \qquad [6.15]$$

for each state $j = 0, 1$ and for both hypotheses H_0 and H_1.

3) Update the log-likelihood ratio process $\{l_t\}_{t \geq 0}$ as follows:

$$l_t = l_{t-1} + \log\left(\frac{\sum_{i=0}^{1} \alpha'_t(i|H_1)}{\sum_{i=0}^{1} \alpha'_t(i|H_0)}\right) \qquad [6.16]$$

4) If $l_t > h$, declare detection of high variance.

If $l_t < 0$:

– l_t to 0;

– t to $t + 1$;

– go to 2.

2 For the implementation of the algorithm, see Gatto (2019).

5) Else (if $0 < l_t < h$):
   – set t to $t + 1$;
   – update α'_t using [6.14];
   – go to 3.

Note that $g(t; 1)$ of [6.9] is

$$g(t;1) = \log\left(\frac{\sum_{i=1}^{N} \alpha'_t(i|H_1)}{\sum_{i=1}^{N} \alpha'_t(i|H_0)}\right)$$

of step 3 and the CUSUM-like algorithm is constructed by the sequence of l_t's above by setting the left-hand threshold to 0 (see Siegmund (1985)), i.e. $\{l_t\}_{t\geq 0}$ represents the log-likelihood ratio process LLR.

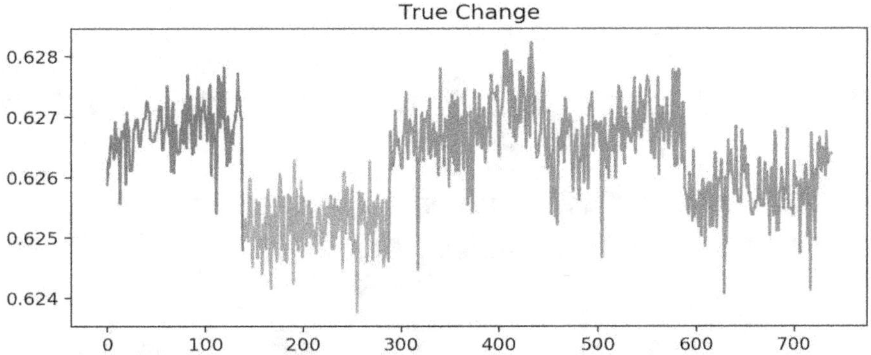

Figure 6.4. *Boundaries of the highlighted sections mark when a change happens. For a color version of this figure, see www.iste.co.uk/dimotikalis/analysis1.zip*

6.4. Results

We applied the consecutive changes in mean CUSUM algorithm of section 6.3.1 to detect consecutive changes in the mean of the distances of the data. Figure 6.4 is the graph of Figure 6.1 presented in section 6.2 with the changes we aim to capture as highlighted sections. We ran our algorithm on the data of distances and compiled the results in Figure 6.5 for different thresholds ($h = 2.5k$, $h = 5k$ and $h = 10k$). Our algorithm managed to capture the three changes we intended to capture, and it did so for various values of h, demonstrating robustness of the consecutive changes in mean CUSUM algorithm with respect to the threshold (i.e. the threshold does not need to be perfectly tuned). There is a slight lag in detection when the threshold is increased, which is to be expected.

We ran the burst-detection algorithm to detect instantaneous sensor slips on the angle component of the data, shown in Figure 6.6. As demonstrated in the graph, the

LLR statistic spikes up whenever there is a sensor slip in the angles data. Therefore, almost any threshold would be sufficient in detecting a sensor slip. To test the robustness of the algorithm with respect to the choice of transition matrix A_1 (see [6.8]), we ran the algorithm with the matrix in [6.8] and the following matrices:

$$A_1' = \begin{bmatrix} 0.1 & 0.9 \\ 0.1 & 0.9 \end{bmatrix}, A_1'' = \begin{bmatrix} 0.5 & 0.5 \\ 0.5 & 0.5 \end{bmatrix}. \quad [6.17]$$

Regardless of which transition matrix was used, our algorithm detected the spikes at the same locations. Therefore, the burst-detection algorithm is very robust with respect to the choice of parameters in the models of section 6.3.2.

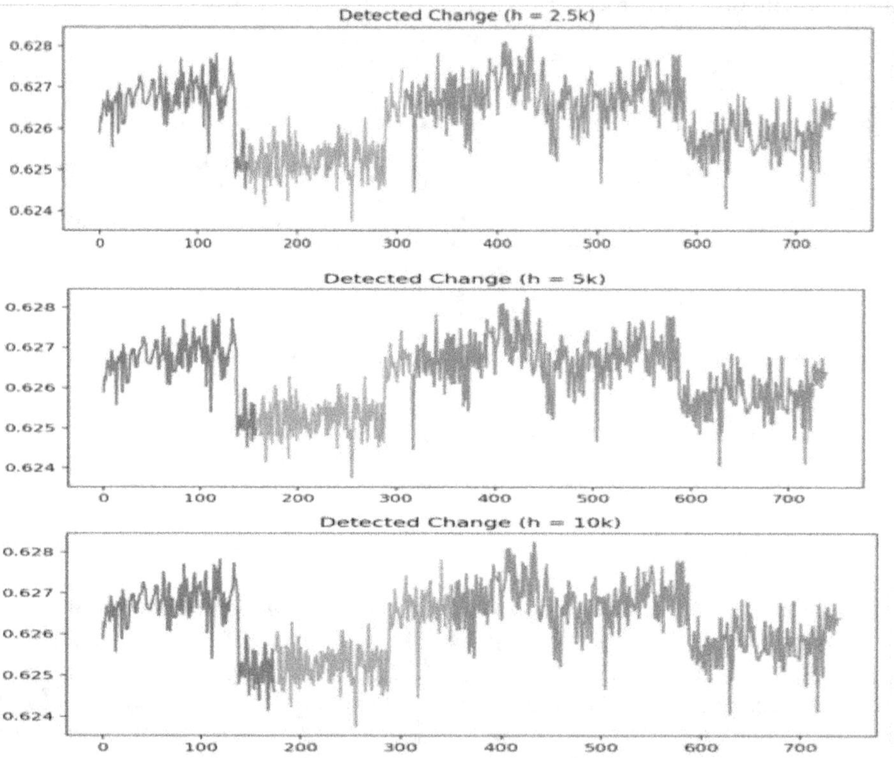

Figure 6.5. *Boundaries of the highlighted sections represent the different levels in mean detected by our algorithm with thresholds set to $h = 2.5k$, $h = 5k$ and $h = 10k$, respectively. For a color version of this figure, see www.iste.co.uk/dimotikalis/analysis1.zip*

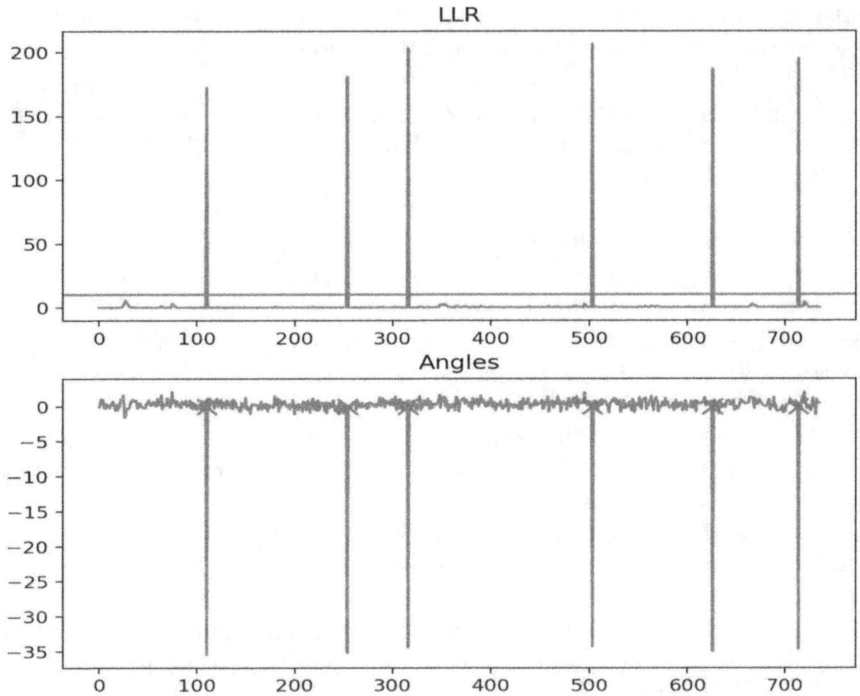

Figure 6.6. *Comparison of LLR values to the plot of angles. The transition matrix is the same as in [6.8]. For a color version of this figure, see www.iste.co.uk/dimotikalis/analysis1.zip*

6.5. Acknowledgments

We would like to acknowledge Professor Ioannis Stamos for sharing his data and giving access to his lab, along with his graduate students Jaspal Singh and Jamie Canizales. We would also like to acknowledge Andrew J. Pearson, ALM, for his thorough proofreading of this chapter.

6.6. References

Basseville, M. and Nikiforov, I. (1993). *Detection of Abrupt Changes: Theory and Practice*. Prentice Hall, Upper Saddle River.

Bellotto, N. and Huosheng, H. (2009). Multisensor-based human detection and tracking for mobile service robots. *IEEE Transactions on Systems, Man, and Cybernetics, Part B (Cybernetics)*, 1(39), 167–181. DOI: 10.1109/TSMCB.2008.2004050.

Carlisle, M., Hadjiliadis, O., Stamos, I. (2016). Trends and trades. In *Handbook of High-Frequency Trading and Modeling in Finance*, Viens, F., Mariani, M.C., Florescu, I. (eds). John Wiley and Sons, Hoboken, New Jersey.

Chen, B. and Willett, P. (2000). Detection of hidden Markov model transient signals. *IEEE Transactions on Aerospace and Electronic Systems*, 4(36), 1253–1268. DOI: 10.1109/7.892673.

Devore, J. (2012). *Probability and Statistics for Engineering and the Science*, 8th edition, Brooks/Cole, Boston, Massachusetts.

Gatto, G. (2019). Error detection algorithm [Online]. Available at: https://github.com/Revidyloh/error-correction, v.1.0, 2019. DOI: https://zenodo.org/badge/latestdoi/194529788.

Ghahramani, Z. (2001). An introduction to hidden Markov models and Bayesian networks. *International Journal of Pattern Recognition and Artificial Intelligence*, 1(15), 9–42. DOI: 10.1142/S0218001401000836.

Hadjiliadis, O. and Stamos, I. (2010). Sequential classification in point clouds of urban scenes. *Proceedings of the 5th International Symposium on 3D Data Processing, Visualization and Transmission*, May 17–20, Paris, France.

James, J.W., Pestell, N., Lepora, N.F. (2018). Slip detection with a biomimetic tactile sensor. *IEEE Robotics and Automation Letters*, 4(3), 3340–3346. DOI: 10.1109/LRA.2018.2852797.

Morita, N., Nogami, H., Higurashi, E., Sawada, R. (2018). Grasping force control for a robotic hand by slip detection using developed micro laser Doppler velocimeter. *Sensors*, 2(18), 326–343. DOI: 10.3390/s18020326.

Page, E.S. (1954). Continuous inspection schemes. *Biometrika*, 1–2(41), 100–115. DOI: 10.2307/2333009.

Poor, H.V. and Hadjiliadis, O. (2008). *Quickest Detection.* Cambridge University Press, Cambridge.

Rabiner, L. and Juang, B. (1986). An introduction to hidden Markov models. *IEEE ASSP Magazine*, 1(3), 4–16. DOI: 10.1109/MASSP.1986.1165342.

Shirafuji, S. and Hosoda, K. (2011). Detection and prevention of slip using sensors with different properties embedded in elastic artificial skin on the basis of previous experience. *Proceedings of the 15th International Conference on Advanced Robotics (ICAR)*, Jun 20–23, Estonia. DOI: 10.1109/ICAR.2011.6088598.

Siegmund, D. (1985). *Sequential Analysis.* Springer-Verlag, New York.

Veiga, F., Van Hoof, H., Peters, J., Hermans, T. (2015). Stabilizing novel objects by learning to predict tactile slip. *Proceedings of IEEE/RSJ International Conference on Intelligent Robots and Systems*, 5065–5072, September 28–October 2, Hamburg, Germany.

7

Diagnostics and Visualization of Point Process Models for Event Times on a Social Network

Point process models have been used to analyze interaction event times on a social network, in the hope to provide valuable insights for social science research. However, the diagnostics and visualization of the modeling results from such an analysis have received limited discussion in the literature. In this chapter, we develop a systematic set of diagnostic tools and visualizations for point process models fitted to data from a network setting. We analyze the residual process and Pearson residual on the network by inspecting their structure and clustering structure. Equipped with these tools, we can validate whether a model adequately captures the temporal and/or network structures in the observed data. The utility of our approach is demonstrated using simulation studies and point process models applied to a study of social interactions between animals.

7.1. Introduction

Interaction event times observed on a social network provide valuable information for social scientists to gain insight into the dynamics and dependence structure among actors on this network. Williamson *et al.* (2016) studied the social interaction patterns of group-housed male mice over long time periods. In Figure 7.1(a), we plot the univariate event times from one pair of mice, which show irregular event-intense intervals and heterogeneity in event densities. Figure 7.1(b) displays the observed social interactions among one cohort of mice over time. The interaction patterns appear to be heterogeneous and structured across the social network. We will apply

---

Chapter written by Jing WU, Anna L. SMITH and Tian ZHENG.

the proposed diagnostic tools to analyze this dataset as real data examples throughout this chapter.

A number of continuous-time social network event times models have been recently developed. Saito *et al.* (2009) studied the cross-reference network of blogs and proposed a model for topic propagation diffusion dynamics. Fan and Shelton (2009) have considered a continuous-time Bayesian network with time-varying nodal attributes. Point process models with network dependence structure have become common practice for such studies, with a focus on intensity function estimation (e.g. Perry and Wolfe 2013; Linderman and Adams 2014; Zipkin *et al.* 2016; Yang *et al.* 2017). In section 7.2, we provide a short review of point process models and existing strategies for incorporating network structure.

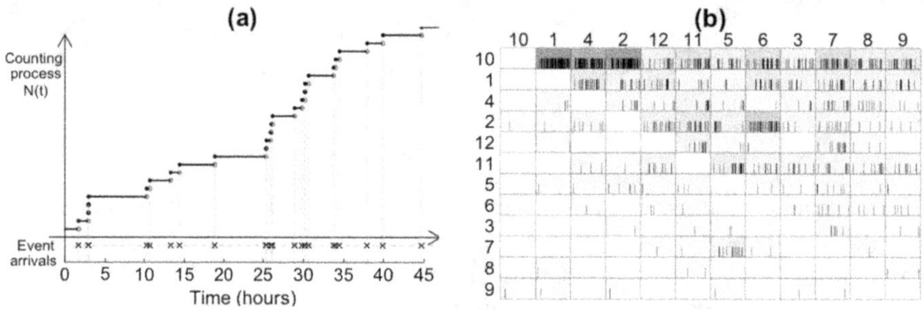

Figure 7.1. *Interaction event times on a social network of mice from Williamson et al. (2016). (a) The univariate point process between one pair of mice. The crosses at the bottom indicate the observed event times. The counting process $N(t)$, defined as the number of events observed up to time t, is plotted. (b) Network point processes. Each row represents the initiator (sender), and each column indicates the recipient (receiver) for a social interaction. The rows (columns) are reordered by the I&SI rank (Schmid and de Vries 2013), where the top row (left column) is the most dominant mouse. The I&SI ranking is a widely used dominance ranking method from the animal behavior literature. For the square at i-th row and j-th column, the observed event times from i to j are plotted as line segments at the bottom. The color shade reflects the total number of events for each pair, with darker color representing more events. For a color version of this figure, see www.iste.co.uk/dimotikalis/analysis1.zip*

Despite the amount of effort that has been devoted to developing such models, there is limited work on the assessment and diagnostics of those models. To summarize and compare a model's performance, the likelihood of the model with respect to the observed data is usually evaluated. As an overall assessment of model fit, however, it does not provide details for detecting *when*, *where* and *how* the proposed model inadequately accounts for patterns and variations in the observed data. Especially, for interaction dynamics on a social network, the timing often exhibits a bursty and heavy-tailed pattern (Barabasi 2005), while the structure of interdependence among network actors can be heterogeneous and complicated. Further model developments for addressing these challenges call for the development of better tools that allow researchers to systematically examine and identify the lack of fit of existing models against observed data.

The goal of this chapter is to propose diagnostic statistics and visualization tools for network event times models, which we develop as extensions of evaluation techniques for univariate point process models. We investigate the proposed techniques using simulated studies and real data. In section 7.2, we introduce the notation for point processes, network point processes and related models. Section 7.3 focuses on time-domain diagnoses by applying the time rescaling theorem and inspecting residual processes. The diagnostic tools for detecting network heterogeneity and network structure in residual processes are developed and demonstrated in section 7.4.

7.2. Background

7.2.1. *Univariate point processes*

Let $(0, T]$ be the time interval of observation. We denote the history of arrival times of observed events up to T as $\mathcal{H}(T) = \{t_m\}_{m=0}^{M}$, where $t_0 = 0$, $t_M = T$ and M is the total number of events. Figure 7.1(a) is a point process of interaction event times between one pair of mice from the animal behavior study in Williamson *et al.* (2016). The associated univariate point process is defined via a *counting process*, $N(t), t \in (0, T]$, where $N(t)$ is a right-continuous function that records the number of events observed during the interval $(0, t]$. The stochastic properties of a point process is usually specified by its conditional intensity function $\lambda(t|\mathcal{H}(t))$ at any time t,

$$\lambda(t|\mathcal{H}(t)) = \lim_{\Delta t \to 0} \frac{Pr([N(t+\Delta t) - N(t)] = 1|\mathcal{H}(t))}{\Delta t}.$$

Inference on the intensity function is conducted by evaluating the likelihood function,

$$\prod_{m=1}^{M} \lambda(t_m|\mathcal{H}(t_m)) \exp\Big\{ - \int_0^T \lambda(s|\mathcal{H}(s))ds \Big\}.$$

A *homogeneous Poisson process* is the simplest model that assumes a constant intensity $\lambda(t) \equiv \lambda, \lambda > 0$. It cannot accommodate situations where event densities vary as shown in Figure 7.1(a). A *Hawkes process* (Hawkes 1971) is a self-exciting process whose intensity has the form

$$\lambda(t) = \lambda_1 + \sum_{t_m < t} \phi_\theta(t - t_m),$$

where $\phi_\theta(t)$ is a self-exciting kernel with parameter θ. The exponential kernel $\phi(t) = \alpha \exp(-\beta t)$ is most widely used, where $\alpha > 0$ calibrates the instantaneous boost to the event intensity at each arrival of an event and $\beta > 0$ controls the decay of past events' influence over time.

When event dynamics display a bimodal pattern, for example, an alternation between long waiting times and intervals of more intensive events, some models assume that the intensity is modulated by a latent continuous-time Markov chain, $Z(t)$. Here, a $Z(t)$ with S-states is parameterized by its infinitesimal generator matrix $Q \in \mathbb{R}^{S\times S}$ (Rabiner 1989). To accommodate these bimodal patterns, Fischer and Meier-Hellstern (1993) proposed a *Markov-modulated Poisson process* (MMPP) model whose intensity function is $\lambda_{Z(t)}$. When the Markov process $Z(t)$ is in state s ($s = 1, ..., S$), arrivals occur according to a homogeneous Poisson process of rate λ_s. Instead of simply using a constant rate λ_s in the MMPP, Wang *et al.* (2012) assume Hawkes processes with piecewise constant kernel functions as *Markov-modulated Hawkes processes with stepwise decay* (MMHPSD). The inference procedure heavily relies on the piecewise constant assumption. As a result, MMHPSD suffers from the problem that the inferred latent state is highly sensitive to single events and hard to interpret. Wu *et al.* (2019) introduce the more widely used exponential kernel for $\lambda_s(t)$ in the *Markov-modulated Hawkes process* (MMHP) model under the assumption that $S = 2$ and utilizes a variational approximation to overcome computational challenges. When the underlying Markov process is in the *active* state ($Z(t) = 1$), the events occur in bursty patterns as in a Hawkes process with intensity $\lambda_1(t) = \lambda_1 + \sum_{t_m<t} \alpha e^{-\beta(t-t_m)}$, while in the *inactive* state ($Z(t) = 0$), the dynamics switch to a quieter period following a homogeneous Poisson process with constant rate λ_0. The parameter set for an MMHP is then $\Theta := \{\lambda_0,\ \lambda_1,\ \alpha,\ \beta,\ Q\}$, where Q is defined as $\begin{pmatrix} -q_1 & q_1 \\ q_0 & -q_0 \end{pmatrix}$. We choose the MMHP as the paradigm model, as suggested in Wu *et al.* (2019) that this model can best address the modeling of sporadic inter-event waiting times shown in the real data example in Williamson *et al.* (2016).

7.2.2. *Network point processes*

Consider a network consisting of a fixed set of N nodes, $V = \{1, 2, ..., N\}$. The observation of event arrival times on a network is defined on $\mathbb{R} \times V \times V$, where $N(t, i, j) := N^{i,j}(t)$ is the number of interactions between node i and node j during the time interval $(0, t]$. This is essentially a marked point process with finite mark space and is also called a multivariate point process (Cox and Lewis 1972). Throughout this chapter, we will consider a directed network; hence, the sequence i, j conveys information that i is the sender and j is the receiver. We also assume that there are no self-loops, i.e. $i \neq j$. Define the history of interactions up to time T as $\mathcal{H}_V(T) = \{(t_1, i_1, j_1), ..., (t_M, i_M, j_M)\}$, where M is the total number of events on the network. $\mathcal{H}_V(T)$ can also be represented by $\bigcup_{i,j\in V, i\neq j} \mathcal{H}^{i,j}(T)$, where $\mathcal{H}^{i,j}(T)$ is the history for pair (i,j) containing $M^{i,j}$ event times, i.e.$\{t_m^{i,j}, m = 1, 2, ..., M^{i,j}\}$.

The conditional intensity for the marginal counting process $N^{i,j}(t)$ between a pair (i,j) is defined as the instantaneous expected rate of events occurring around a time t given the history:

$$\lambda^{i,j}(t|\mathcal{H}_V(t)) = \lim_{\Delta t \to 0} \frac{Pr([N^{i,j}(t+\Delta t) - N^{i,j}(t)] = 1|\mathcal{H}_V(t))}{\Delta t}.$$

Following proposition 7.3.III in Daley and Vere-Jones (2003), the likelihood function is given by

$$\prod_{i=1}^{N}\prod_{j\neq i}^{N}\Big[\prod_{m=1}^{M^{i,j}} \lambda^{i,j}(t_m^{i,j}|\mathcal{H}_V(t_m^{i,j}))\Big] \exp\Big\{-\int_0^T \lambda^{i,j}(s|\mathcal{H}_V(s))ds\Big\}.$$

The simplest model for point processes on a network is a *homogeneous* network point process model, where $\lambda^{i,j}(t_m^{i,j}|\mathcal{H}_V(t_m^{i,j}))$ does not depend on i, j. As dynamic interactions on a social network are known to be heterogeneous, clustered and structured by the underlying social distance among actors, a few models have been developed recently to take into account network heterogeneity and structure. However, the discussion on model assessment and diagnosis remains limited in the literature. To adjust for higher-degree actors in a network, and pair-specific covariates, Perry and Wolfe (2013) propose a multivariate point process with the model assumption that the intensities between pairs are decided by a function of a sender-specific baseline rates. By treating the sender baseline intensity rates as nuisance parameters, the model estimation is carried out by maximizing the log partial likelihood. For model checking, they calculate and visualize normalized residuals. These residuals do not account for the sender-specific baseline rates and are not used to examine how well the proposed model explains network dependence in event arrivals among pairs of actors. Zipkin *et al.* (2016) model the arrival times of interaction events between pairs of actors of known identity using independent Hawkes processes and use the fitted pairwise processes to *resolve* the actors' identities of an interaction event given only the event's arrival time and no actor identities. They validate the fitted model by comparing the resolved actor identities for a hold-out set of event times against the ground truth. Both Linderman and Adams (2014) and Yang *et al.* (2017) assume a latent network structure and consider a reciprocity effect, using a Hawkes process model. The models are validated by prediction of hold-out links and interpretation of the latent space structure. The above validation procedures are inadequate to inspect the proposed models' lack of fit in terms of temporal trends and/or network structure. We will show, in sections 7.3 and 7.4, a systematic approach for model checking and diagnostics, with theoretically well-grounded diagnostic statistics and visualization tools.

7.3. Model checking for time heterogeneity

We first consider univariate point processes. The goal is to check that the fitted model can capture the variability in temporal trends. Wu *et al.* (2019) show that the *Markov-modulated Hawkes process* (MMHP) can model the patterns of event dynamics that are sporadic with bursts and long wait times. In our simulated case studies, we will generate event times according to a MMHP and compare the model fit of a homogeneous point process model, a Hawkes process model, an MMPP model and an MMHPSD model. In our real data example, we fit the above five models to the interaction events between one pair of mice as in Figure 7.1(a). The inference result is shown in Table 7.1. For the MMHP model, the inference is carried out following the Bayesian estimation in Wu *et al.* (2019); hence, the posterior means are reported. Maximum likelihood estimates are reported for the other models.

Model	λ_0	λ_1	α	β	q_1	q_0
Homogeneous Poisson	0.56					
Hawkes process	0.39		0.89	2.78		
MMPP	0.25	0.53			0.14	0.16
MMHPSD	6.7×10^{-7}	2.63	0.39	0.06	0.49	2.38
MMHP	0.24	0.28	0.49	0.60	0.18	0.19

Table 7.1. *Estimation of the parameters in the five models after fitting to our real data example*

7.3.1. *Time rescaling theorem*

One approach to test the goodness of fit of point process models is to apply the time rescaling theorem (Brown *et al.* 2002). It states that if $\{t_m\}_{m=1:M}$ is a realization of events from a point process with conditional intensity $\lambda(t|\mathcal{H}(t))$, then the rescaling transformation $\int_0^{t_m} \lambda(s|\mathcal{H}(s))ds$ over $m = 1, ..., M$ yields a Poisson process with rate 1. Hence, the *rescaled inter-event times*, defined as $\{\Lambda_m := \int_{t_{m-1}}^{t_m} \lambda(s|\mathcal{H}(s))ds\}_{m=1:M}$, are independently and identically distributed as exponential random variables with rate 1.

Case study I: simulated examples. We simulate an MMHP using thinning algorithm (Lewis and Shedler 1979) with parameter value $\Theta = \{\lambda_0 = 1,\ \lambda_1 = 1.1,\ \alpha = 1.6,\ \beta = 1.9,\ Q = \left(\begin{smallmatrix}-0.2 & 0.2\\ 0.4 & -0.4\end{smallmatrix}\right)\}$ and termination time $T = 100$. We fit the five models to the simulated data separately. Figure 7.2 shows the estimated intensity versus true intensity, and Q–Q plots to test for goodness of fit. The MMHP model can recover the true intensity precisely and reveals a nearly exponential distributed *rescaled inter-event times*, i.e. $\{\Lambda_m\}_{m=1:M}$.

The Q–Q plot for the homogeneous Poisson process model jumps nearly vertically at larger quantiles, since it fails to capture the intensity at state 0 and thus has a heavier tail than the exponential distribution. For the Hawkes process and MMHPSD models, the Q–Q plots show that the empirical distribution has a spike concentrated on a larger value than the theoretical median. Both models fail to capture the high intensity during state 1 because they compromise their parameter estimation between state 1 and state 0. The MMPP model yields an empirical distribution with a lighter right tail and is not flexible enough to capture the intensity during state 1.

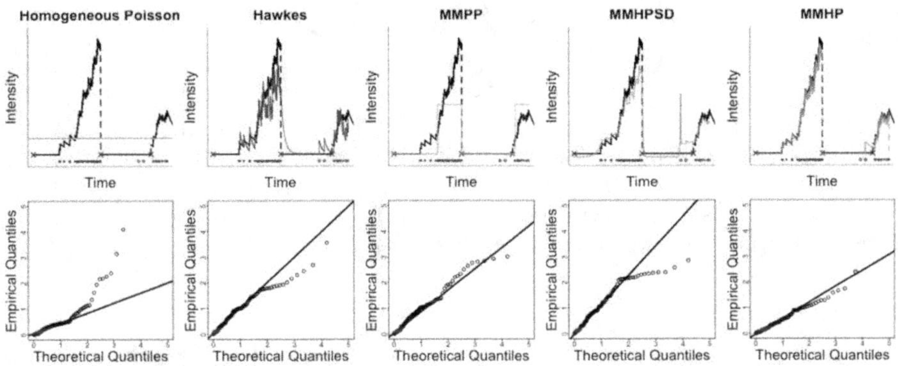

Figure 7.2. *Intensity estimation and Q–Q plots for testing the distribution of rescaled inter-event times after fitting the five models to synthetic data. The top panels plot intensity functions versus true intensity, where the blue dots are events, black lines indicate the true intensity function and colored lines are the estimations. Knowing the true intensity, the red crosses are state transitions. The bottom panels give Q–Q plots for testing whether the rescaled inter-event times are distributed as exponential random variables with rate 1. For a color version of this figure, see www.iste.co.uk/dimotikalis/analysis1.zip*

Case study II: an application to Williamson *et al.*'s (2016) data. Figure 7.3 shows the intensity estimation and Q–Q plots after fitting the five models to observed event times for the pair of mice shown in Figure 7.1(a). MMHP fits the data most reasonably according to the Q–Q plot, whereas the other four models show some indications of lack of fit. The fitted homogeneous Poisson process, Hawkes process and MMPP models all show a lighter right tail than the exponential distribution, while the fitted MMHPSD model has a heavier right tail. By using Q–Q plots to examine the distribution of $\{\Lambda_m\}_{m=1:M}$, we can compare model fit across competing point process models and can conclude that the MMHP model captures the sporadic and bursty event dynamics of mice social interactions.

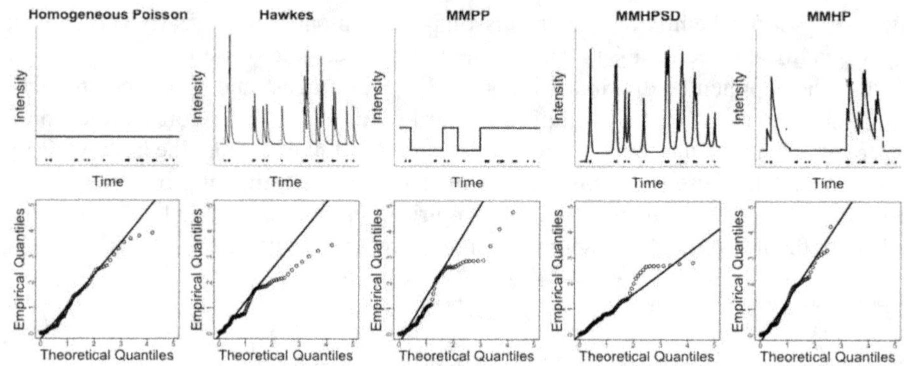

Figure 7.3. *Model fit diagnostics for Williamson et al.'s (2016) data: intensity estimation and Q–Q plots after fitting the five models to real event time data between one pair of mice as shown in Figure 7.1(a). The top panels plot estimated intensity functions after fitting the five models, which are indicated by the black lines. The black dots are the events occurring over time. The bottom panels are Q–Q plots to test goodness of fit*

7.3.2. *Residual process*

According to the Doob–Meyer decomposition theorem (Andersen *et al.* 2012), given a counting process $N(t)$ with its conditional intensity function $\lambda(t|\mathcal{H}(t))$, the *residual process* $M(t)$, defined as $M(t) = N(t) - \int_0^t \lambda(s|\mathcal{H}(s))ds$, is a martingale. Hence, when a point process model is fitted to data and gives an estimated intensity function $\hat{\lambda}(t)$, the *raw residual process*

$$R(t) = N(t) - \int_0^t \hat{\lambda}(s)ds$$

can be used to inspect the fit of the model by measuring its discrepancy from 0. Here, we evaluate the raw residual process at the final time point, T.

The variance of the above residual process $M(T)$ depends on the intensity function, var$(M(t)) = \int_0^t \lambda(s)ds$. To compare model fit between scenarios with different intensity functions, it is desirable to compute a *standardized* residual (Baddeley *et al.* 2005), similar to the Pearson residuals for linear regression. Clements *et al.* (2011) propose the *Pearson residual process* for a counting process as an intensity-weighted version of the raw residual process,

$$\text{PR}(t) = \sum_{t_m<t} \frac{1}{\sqrt{\lambda(t_m)}} - \int_0^t \sqrt{\lambda(s)}ds.$$

Under the true model specification, the variance of $PR(t)$ does not depend on the intensity, i.e. var$(\text{PR}(t)) = t$ with mean 0 (Clements *et al.* 2011).

Case study I: simulated examples. We simulate 50 independent MMHPs with the same parameter settings as in section 7.3.1. After fitting the Hawkes process, MMPP, MMHPSD and MMHP models to the data, the raw residual $R(T)$ and Pearson residual $\mathrm{PR}(T)$ are calculated. Due to the formulation of the Pearson residual process, the integral part cannot be calculated explicitly, and hence, is approximated by numerical integration. Boxplots for 50 raw residuals and Pearson residuals are shown in Figure 7.4(a) and (b) separately. We can see that the Hawkes process and MMPP models yield positive residuals most of the time, which means that the models tend to underestimate the intensity. The Hawkes process biased its parameter estimation in order to compromise to the low intensity in state 0, and underestimated the intensity in state 1. The MMPP model cannot capture the volatility of the intensity function in state 1 and yield an underestimated intensity. On the contrary, the MMHPSD model overestimates the intensity, mostly due to its over-sensitive intensity function with the stepwise kernel.

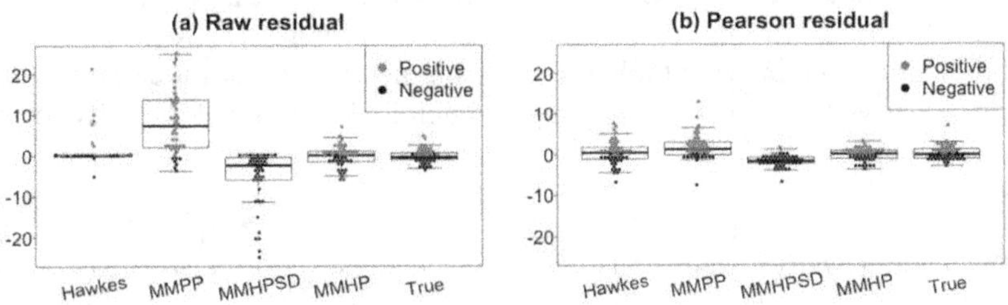

Figure 7.4. *Beeswarm boxplots for raw residual and Pearson residual processes after fitting the four models to synthetic data from 50 independent MMHPs. The residuals calculated from the true intensities are also plotted for comparison. For a color version of this figure, see www.iste.co.uk/dimotikalis/analysis1.zip*

Case study II: an application to Williamson *et al.*'s (2016) data. Using the data from Williamson *et al.* (2016), we fit the above four models to all pairs of mice in a cohort separately and then evaluate the raw residuals and Pearson residuals at final observation time T for all pairs. Figure 7.5 plots the residuals versus number of events that occurred between a pair, with smoothing regression lines using the LOWESS method (Cleveland 1979). The MMHP model has residuals closest to 0 and is not sensitive to the number of events (in a sense, the number of events is a rough indicator of intensity), while the other models fail. Comparing the raw residual of the MMPP and MMHPSD models with their Pearson residual, we also observe that the raw residuals have more variability over the number of events. This suggests that the Pearson residual is more appropriate in the network setting, since it is important to have a measurement that is comparable across pairs with varying numbers of pairwise interactions.

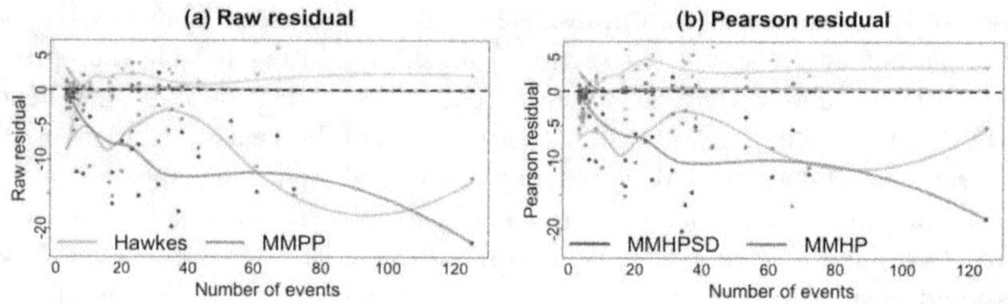

Figure 7.5. *Raw residuals and Pearson residuals after fitting the four models to all pairs in the cohort of mice as in Figure 7.1(b) from Williamson et al. (2016). Each model is represented by a different color, with residual values (dots) and a smoothed LOWESS line. For a color version of this figure, see www.iste.co.uk/dimotikalis/analysis1.zip*

7.4. Model checking for network heterogeneity and structure

In this section, we provide tools for assessing network point process models. We will focus on MMHP in a network setting and define the marginal intensity for a pair (i, j) as

$$\lambda^{i,j}(t|\mathcal{H}_V(t)) = \begin{cases} \lambda_1^{i,j} + \sum_{t_m^{i,j}<t} \alpha^{i,j} e^{-\beta^{i,j}(t-t_m^{i,j})}, & Z^{i,j}(t) = 1 \\ \lambda_0^{i,j}, & Z^{i,j}(t) = 0, \end{cases} \quad [7.1]$$

where $Z^{i,j}(t)$ is an independent CTMC across all pairs (i, j). We denote the parameters for (i, j) as $\Theta^{i,j} = \{\lambda_0^{i,j},\ \lambda_1^{i,j},\ \alpha^{i,j},\ \beta^{i,j},\ Q^{i,j}\}$.

7.4.1. *Kolmogorov–Smirnov test*

For each pair (i, j), we wish to use the time rescaling theorem presented in section 7.3.1 and test whether the *rescaled inter-event times*, $\{\Lambda_m^{i,j}\}_{m=1:M^{i,j}}$, are distributed as exponential random variables with rate 1, where $\Lambda_m^{i,j} = \int_{t_{m-1}^{i,j}}^{t_m^{i,j}} \lambda^{i,j}(s|\mathcal{H}_V(s))ds$. To summarize the model fit across pairs in the network, it is important to quantify the test results with a unified measurement. We use the Kolmogorov–Smirnov statistics, $\text{KS}^{i,j} = \sup_x |\hat{F}^{i,j}(x) - F(x)|$ for each pair (i, j), where $\hat{F}^{i,j}(x)$ is the empirical distribution of the *rescaled inter-event times* and $F(x)$ is the cumulative distribution function of the exponential distribution. Larger values of the K–S statistics indicate a larger discrepancy between the model and the network point pattern data.

Case study I: simulated examples. We simulate network point process data with 10 nodes until time $T = 500$ as follows: for each pair (i, j), the marginal intensity

$\lambda^{i,j}(s|\mathcal{H}_V(s))$ follows [7.1]. All the pairs (i,j) share the same parameters that $\lambda_0^{i,j} = 0.05$, $\lambda_1^{i,j} = 0.08$, $\beta^{i,j} = 22$, $Q^{i,j} = \begin{pmatrix} -0.01 & 0.01 \\ 0.04 & -0.04 \end{pmatrix}$. The matrix $\boldsymbol{\alpha} = [\alpha^{i,j}]_{N\times N}$ is block-structured,

$$\alpha_{ij} = \begin{cases} 20, & i,j \in B_k \\ 0.5, & i \in B_k, j \in B_l, k \neq l \end{cases}.$$

where class $B_1 = \{1,2,3,4\}$ and $B_2 = \{5,6,...,10\}$. The total number of events between pair (i,j) until time T, $N^{i,j}(T)$, is shown in Figure 7.6(a), which has a clear block structure.

We compare the fit of the following two models: (1) across all pairs, the parameter set $\Theta^{i,j}$ shares a same set of values (referred to as the *homogeneous* network model), and (2) $\alpha^{i,j}$ varies across pairs such that $\boldsymbol{\alpha}$ has a block structure with the same classes as in the simulation (all other elements of $\Theta^{i,j}$ are the same across all pairs, as in the first model). This is referred to as the *block* network model.

Figure 7.6(b) shows the K–S statistics over the network after fitting the *homogeneous* network model to the simulated data, whereas Figure 7.6(c) shows that for the *block* network model. Figure 7.6(d) displays results from the true intensity function, which has the lowest K–S statistics. *The homogeneous* network model (Figure 7.6(b)) fits the data the worst, since it has higher K–S statistics values across the network compared to the other models. Figure 7.6(b) also shows block-structured K–S statistics and demonstrates that the *homogeneous* network model performs worse for the between-block pairs, since within-block pairs generate more events and contribute more to the likelihood function.

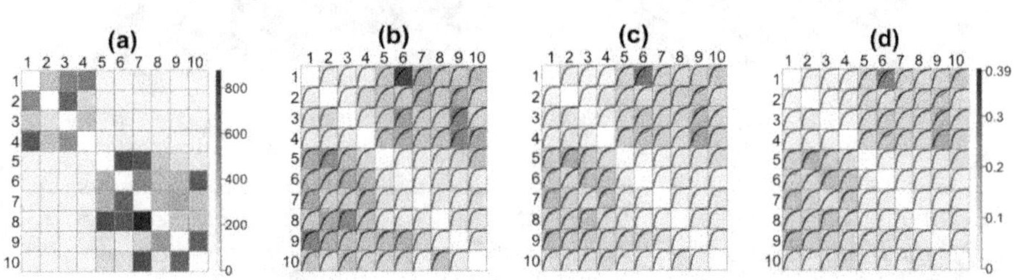

Figure 7.6. *(a) Heatmap of simulated network point process data: the total number of events occurring between each pair. (b), (c) Matrices of K–S statistics after fitting the homogeneous and block network models to the simulated data in (a). The rows and columns of each matrix correspond to senders and receivers. The color shades reflect the values of the K–S test statistics. (d) Matrix of K–S statistics calculated from true intensity function. For a color version of this figure, see www.iste.co.uk/dimotikalis/analysis1.zip*

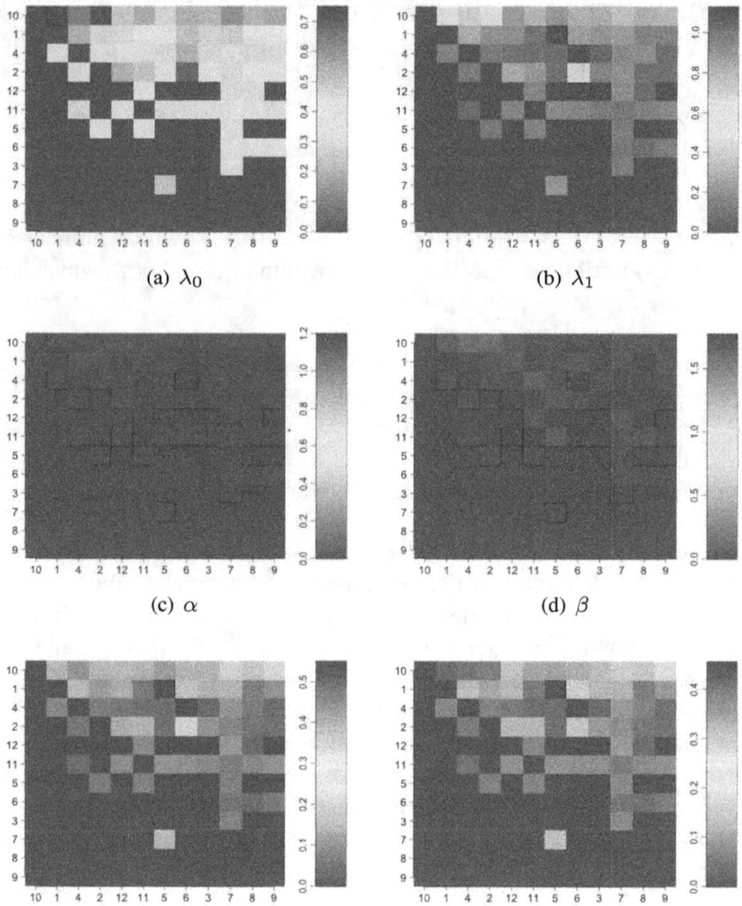

(a) λ_0 (b) λ_1

(c) α (d) β

(e) q_0 (f) q_1

Figure 7.7. *Estimation of the parameters in the heterogeneous network model after fitting to network event data, where each panel represents the matrix of one type of parameter. For a color version of this figure, see www.iste.co.uk/dimotikalis/analysis1.zip*

Case study II: an application to Williamson *et al.*'s (2016) data. For the interaction event times among a group of mice as shown in Figure 7.1(b), we fit the following two models: (1) the parameter set $\Theta^{i,j}$ for MMHP is the same across pairs (referred to as the *homogeneous* network model) and (2) $\Theta^{i,j}$ is allowed to vary freely across pairs (referred to as the *heterogeneous* network model).The inference result of the *homogeneous* network model) is shown in Table 7.2. The estimates of the parameters of the *heterogeneous* network model are visualized in Figure 7.7.

λ_0	λ_1	α	β	q_1	q_0
0.18	0.21	0.83	0.99	0.09	0.14

Table 7.2. *Estimation of the parameters in the homogeneous network model after fitting to network event data*

Figure 7.8 plots the matrices of K–S statistics after fitting the above two models. Since the *heterogeneous* network model has more flexibility across pairs and can adapt well on various pairs' dynamics, its K–S statistics are smaller and the model fits the data better. Especially for the pair $(7, 5)$, they started engaging in interactions during the last period of observations, which sets them apart from the other pairs, as shown in Figure 7.8 (b). The *homogeneous* network model fits this pair poorly, whereas the *heterogeneous* network model better captures this pattern.

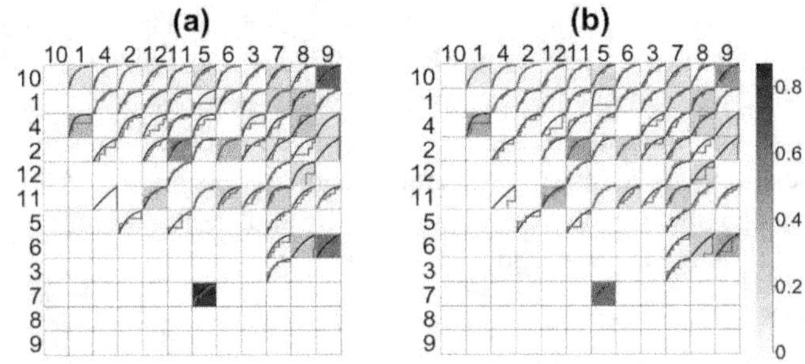

Figure 7.8. *Matrices of K–S statistics after fitting the homogeneous network model (in (a)) and the heterogeneous network model (in (b)) to the mice interaction data. The color shades reflect the values of K–S test statistics. For a color version of this figure, see www.iste.co.uk/dimotikalis/analysis1.zip*

7.4.2. *Structure score based on the Pearson residual matrix*

The Pearson residual is more valuable for diagnosing model fit in the network setting, since it is comparable across pairs and not influenced by intensity function. For a pair (i, j), the Pearson residual is

$$\mathrm{PR}_{i,j}(T) = \sum_{t_m^{i,j}} \frac{1}{\sqrt{\hat{\lambda}^{i,j}(t_m^{i,j})}} - \int_0^T \sqrt{\hat{\lambda}^{i,j}(s)}ds.$$

To assess model fit, we propose to quantify the network structure in the Pearson residual matrix, $\mathrm{PR} := [\mathrm{PR}_{i,j}]_{N \times N}$. In order to spot systematic overestimation or underestimation issues, we separate the Pearson residual matrix into two matrices for the positive and negative residual values, respectively. We preserve the positive residuals in the *underestimation* matrix PR^+, such that $\mathrm{PR}^+_{i,j} = \mathrm{PR}_{i,j}$ if $\mathrm{PR}_{i,j} > 0$ and

$\text{PR}^+_{i,j} = 0$ otherwise. Similarly, we create the *overestimation* matrix PR^- using the absolute values of negative residuals. We conduct non-negative matrix factorization (NMF) (Lee and Seung 2001) on each of the matrices, $A \approx WH$, where $A \in \mathbb{R}^{N\times N}$ is PR^+ or PR^- and $W \in \mathbb{R}^{N\times K}, H \in \mathbb{R}^{K\times N}$. K is usually set to be much smaller than N, so that WH is a lower-rank approximation of A. We introduce a *matrix structure score* as $\frac{||A-WH||_F}{||A||_F}$, where $||\cdot||_F$ is the matrix Frobenius norm. If NMF can recover the residual matrix well, the score is larger, suggesting that the model fits the data worse and yields a more structured residual matrix.

Case study I: simulated examples. Using simulated data as in section 7.4.1, we calculate the Pearson residual matrix after fitting the *homogeneous* and *block* network models and plot them in Figure 7.9(a) and (b), respectively. For a better comparison, we also use the true model intensity to calculate the residuals and plot in Figure 7.9(c). Table 7.3 compares the matrix structure score of the underestimation PR^+ and overestimation matrix PR^- across models by using $K = 2$. The *homogeneous* model exhibits larger positive residuals within-block. This demonstrates that the estimation of the intensities for within-block pairs are biased towards lower values because the model is network homogeneous and not adequate to capture intense interactions of within-block pairs. The matrix structure score of this model is higher in both overestimation and underestimation matrices, as shown in Table 7.3, which is a quantified validation for its lack of fit.

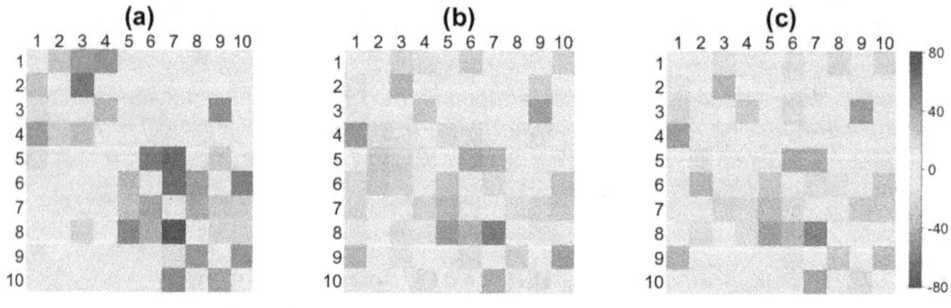

Figure 7.9. Matrices of the Pearson residuals after fitting models to simulated data. (a), (b) Results after fitting the homogeneous and block network models to the simulated data. (c) From the true model that is used to generate simulation. Red colors represent positive residuals, indicating that the model underestimates the intensity function, whereas blue colors represent negative residuals and hence overestimation. For a color version of this figure, see www.iste.co.uk/dimotikalis/analysis1.zip

Case study II: an application to Williamson *et al.*'s (2016) data. Figure 7.10 shows the Pearson residual matrix after fitting the *homogeneous* and *heterogeneous* network models to the mice interaction data. The models' assumptions are defined in section 7.4.1. The *homogeneous* network model tends to overestimate the intensity function when individuals 4 and 11 are the senders, because the inference is biased by a large number of interactions initiated by active individuals. Thus, this pattern in the residual matrix yields a larger matrix structure score, especially for the negative residuals, as shown in Table 7.4, where K is set to be 2. It suggests that the *homogeneous* network model is not adequate to capture the network dynamics, while the *heterogeneous* network model has more flexibility and is superior in terms of Pearson residual and structure score.

	Homogeneous	Block	True
Positive	0.46	0.41	0.41
Negative	0.93	0.43	0.36

Table 7.3. *Structure scores for simulated examples*

	Homogeneous	Heterogeneous
Positive	0.79	0.69
Negative	0.77	0.55

Table 7.4. *Structure scores for the Williamson et al. (2016) example*

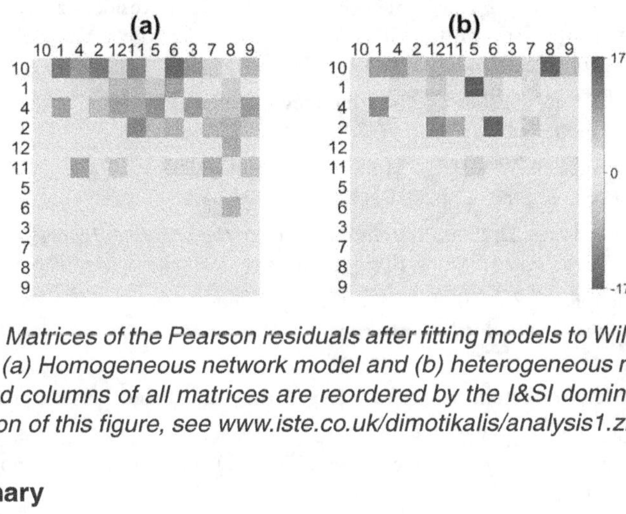

Figure 7.10. *Matrices of the Pearson residuals after fitting models to Williamson et al.'s (2016) data. (a) Homogeneous network model and (b) heterogeneous network model. The rows and columns of all matrices are reordered by the I&SI dominance rank. For a color version of this figure, see www.iste.co.uk/dimotikalis/analysis1.zip*

7.5. Summary

In this chapter, we propose diagnostic statistics and visualization tools for network point process models. The evaluation techniques are theoretically well grounded and can be applied to intensity-based models for network point pattern data generally. We use simulations and real data examples to demonstrate the utility of our approaches. By inspecting lack of fit in terms of both temporal dependence and network structures, the proposed suite of diagnostic statistics and visualizations can reveal deficiencies in network point process models and provide important insights that can lead to model improvements.

7.6. Acknowledgments

This work was based on research sponsored by DARPA agreement number D17AC00001. The content of the information does not necessarily reflect the position or the policy of the Government, and no official endorsement should be inferred.

7.7. References

Andersen, P.K., Borgan, O., Gill, R.D., and Keiding, N. (2012). *Statistical Models Based on Counting Processes*. Springer Science & Business Media, New York.

Baddeley, A., Turner, R., Møller, J., and Hazelton, M. (2005). Residual analysis for spatial point processes (with discussion). *Journal of the Royal Statistical Society: Series B (Statistical Methodology)*, 67(5), 617–666.

Barabasi, A.-L. (2005). The origin of bursts and heavy tails in human dynamics. *Nature*, 435(7039), 207–211.

Brown, E.N., Barbieri, R., Ventura, V., Kass, R.E., and Frank, L.M. (2002). The time-rescaling theorem and its application to neural spike train data analysis. *Neural Computation*, 14(2), 325–346.

Clements, R.A., Schoenberg, F.P., and Schorlemmer, D. (2011). Residual analysis methods for space-time point processes with applications to earthquake forecast models in California. *The Annals of Applied Statistics*, pp. 2549–2571.

Cleveland, W.S. (1979). Robust locally weighted regression and smoothing scatterplots. *Journal of the American Statistical Association*, 74(368), 829–836.

Cox, D.R. and Lewis, P.A.W. (1972). Multivariate point processes. *Proc. 6th Berkeley Symp. Math. Statist. Prob*, volume 3, pp. 401–448.

Daley, D.J. and Vere-Jones, D. (2003). *An Introduction to the Theory of Point Processes: Volume I: Elementary Theory and Methods*. Springer Science & Business Media, New York.

Fan, Y. and Shelton, C.R. (2009). Learning continuous-time social network dynamics. *Proceedings of the Twenty-Fifth Conference on Uncertainty in Artificial Intelligence*. AUAI Press, Arlington.

Fischer, W. and Meier-Hellstern, K. (1993). The Markov-modulated Poisson process (MMPP) cookbook. *Performance Evaluation*, 18(2), 149–171.

Hawkes, A.G. (1971). Spectra of some self-exciting and mutually exciting point processes. *Biometrika*, 58(1), 83–90.

Lee, D.D. and Seung, H.S. (2001). Algorithms for non-negative matrix factorization. *Advances in Neural Information Processing Systems*, 556–562.

Lewis, P.W. and Shedler, G.S. (1979). Simulation of nonhomogeneous Poisson processes by thinning. *Naval Research Logistics Quarterly*, 26(3), 403–413.

Linderman, S. and Adams, R. (2014). Discovering latent network structure in point process data. *International Conference on Machine Learning*, PMLR, 32(2), 1413–1421.

Perry, P.O. and Wolfe, P.J. (2013). Point process modelling for directed interaction networks. *Journal of the Royal Statistical Society: Series B (Statistical Methodology)*, 75(5), 821–849.

Rabiner, L.R. (1989). A tutorial on hidden markov models and selected applications in speech recognition. *Proceedings of the IEEE*, 77(2), 257–286.

Saito, K., Kimura, M., Ohara, K., and Motoda, H. (2009). Learning continuous-time information diffusion model for social behavioral data analysis. *Asian Conference on Machine Learning*. Springer, Nanjing.

Schmid, V.S. and de Vries, H. (2013). Finding a dominance order most consistent with a linear hierarchy: An improved algorithm for the i&si method. *Animal Behaviour*, 86(5), 1097–1105.

Wang, T., Bebbington, M., and Harte, D. (2012). Markov-modulated Hawkes process with stepwise decay. *Annals of the Institute of Statistical Mathematics*, 64(3), 521–544.

Williamson, C.M., Lee, W., and Curley, J.P. (2016). Temporal dynamics of social hierarchy formation and maintenance in male mice. *Animal Behaviour*, 115, 259–272.

Wu, J., Zheng, T., and Curley, J. (2019). Markov-modulated Hawkes processes for sporadic and bursty event occurrences. arXiv:1903.03223.

Yang, J., Rao, V., and Neville, J. (2017). Decoupling homophily and reciprocity with latent space network models. Document, UAI.

Zipkin, J.R., Schoenberg, F.P., Coronges, K., and Bertozzi, A.L. (2016). Point-process models of social network interactions: Parameter estimation and missing data recovery. *European Journal of Applied Mathematics*, 27(3), 502–529.

PART 2

Data Analysis Methods and Tools

8

Exploring the Distribution of Conditional Quantile Estimates: An Application to Specific Costs of Pig Production in the European Union

This communication uses symbolic data analysis tools to visualize conditional quantile estimation intervals, applying it to the problem of cost allocation in agriculture. After recalling the conceptual framework of the estimation of agricultural production costs, the first part presents the empirical model, the quantile regression approach and the interval data processing techniques used as symbolic data analysis tools. The second part presents the comparative analysis of the econometric results of pig production between 12 European Member States, using principal component analysis and the divisive hierarchical clustering of the estimation intervals, by discussing the relevance of the exploratory graphs obtained for the international comparisons.

"*Applied economists increasingly want to know what is happening to an entire distribution, to the relative winners and losers, as well as to averages.*"

Angrist and Pischke (2009)

Chapter written by Dominique DESBOIS.
This communication develops some of author's works done during the preparation of his PhD dissertation (Desbois 2015), co-directed by Y. Surry and J.C. Bureau, supported by the *Farm Accountancy Cost Estimation and Policy Analysis* project (FACEPA) of the 7th Framework Program of the European Community (FP7 / 2007 2013, Approval No. 212292). This mention does not imply any approval by the persons and organizations mentioned hereby, the author assuming full responsibility for the text.

8.1. Introduction

The integration of agriculture in the 28 Member States resulting from the enlargement of the European Union (EU) has raised both in the context of competitive markets as markets subject to regulation, recurring needs for estimating production costs for major agricultural products, following the successive reforms of the Common Agricultural Policy (CAP). The analysis of agricultural production costs is a tool for analyzing economic results of farmers: it allows us to assess the price competitiveness of farmers, one of the major elements for development and sustainability of food chains in the European regions. To meet the needs of simulations and impact assessment in the various common market organizations, we must be able to provide information on the entire distribution of production costs for the assessment of public agricultural policy options. Based on the observation of asymmetry and heterogeneity within the empirical distribution of agricultural inputs, we have proposed a methodology adapted to the problem of estimating the empirical distributions of specific production costs for the main agricultural products in a European context where agricultural holdings remain mainly oriented towards multiple productions (Desbois *et al.* 2017).

We first present the empirical model for estimating the specific production costs, derived from an econometric cost allocation approach inspired by Divay and Meunier (1980) using microeconomic data to build an input–output matrix. Then, we introduce the estimation methodology according to the conditional quantiles proposed by Koenker and Bassett (1978). Next, we present the symbolic data analysis procedures used to explore the empirical estimates of conditional quantile distribution intervals based on the concepts and methods provided by the symbolic approach of Billard and Diday (2006). Then, we present graphs of results from the analysis tools of the symbolic data applied to the estimation intervals of the conditional quantiles. Eventually, we conclude on the relevance of this approach applied to pig production proposing an extension of this type of analysis at the regional level.

8.2. Conceptual framework and methodological aspects

Firstly, we present the methodology for estimating specific costs. Secondly, we introduce factorial analysis and the clustering procedure of the estimation intervals in the formalism of the symbolic data.

8.2.1. *The empirical model for estimating the specific production costs*

Inspired by Divay and Meunier (1980), the allocation of the sum x_i of the input costs for farm holding i is made by linear decomposition along the gross products Y_i^j of farm holding i for each production j, where u_i is a random vector with a zero mathematical expectation:

$$x_i = \sum_{j=1}^{p} \beta_j Y_i^j + u_i \qquad [8.1]$$

Like Cameron and Trivedi (2005), we assume that the data generator process is a linear model with multiplicative heteroscedasticity characterized in matrix form by:

$$x = Y'\beta + u \text{ with } u = Y'\alpha \times \varepsilon \text{ and } Y'\alpha > 0 \qquad [8.2]$$

where $\varepsilon \sim iid[0, \sigma]$ is a random-vector identically and independently distributed with zero mean and constant variance σ^2. Under this assumption, $\mu_q(x|Y, \beta, \alpha)$, the $q^{t□}$ conditional quantile of the production cost x, conditioned by Y and the α and β parameters, is derived analytically as follows:

$$\mu_q(x|Y, \beta, \alpha) = Y'[\beta + \alpha \times F_\varepsilon^{-1}(q)] = Y'\gamma. \qquad [8.3]$$

where F_ε is the cumulative distribution function (CDF) of the errors.

The technical coefficient for the jth product of the qth quantile of the specific cost is defined by the jth component of the multivariate slope vector:

$$\gamma^j(q) = [\beta + \alpha \times F_\varepsilon^{-1}(q)]^j. \qquad [8.4]$$

Following D'Hautfoeuille and Givord (2014), three models can be derived:

i) $x = Y'\beta + u$ with $u = K\epsilon$, homoscedastic errors $V(\epsilon|Y) = \sigma^2$, denoted as the *location shift* model, i.e. the linear model of conditional quantile with homogeneous slopes; while $Y'\alpha = K$ is constant, the conditional quantiles $\mu_q(x|Y, \beta, \alpha) = Y'\beta + KF_e^{-1}(q)$ get all the same β slope, but differ only by a constant gap, growing as q, the quantile order, increases;

ii) $x = Y'\beta + (Y'\alpha)\varepsilon$ and $Y'\alpha > 0$ with heteroscedastic residuals, referred to as the *location-scale shift* model, i.e. the linear model of heterogeneous conditional quantile slopes;

iii) $X = Y'\gamma_\zeta$ with ζ a random variable independent of Y following a uniform distribution over the interval [0,1] such as $\zeta \longrightarrow Y'\gamma_\zeta$ be strictly increasing whatever

Y, designated as the random coefficient model. ζ corresponds to a random component determining the rank of the individual within the distribution of X. Under the strong distributional hypothesis of rank invariance, the random coefficient γ_q represents the effect of a marginal change in Y for agricultural holdings located at the qth quantile of the ζ distribution. This distributional assumption of rank invariance means that median farms in terms of input productivity would maintain the $q = 0.5$ rank, regardless of the different levels of production Y_i registered for the ith farm holding.

8.2.2. *The procedures for estimating and testing conditional quantiles*

The quantile regression is defined for each quantile of order q as the solution of a problem of minimization of the sum of the deviations in absolute value (L_1 norm):

$$\hat{\beta}(q) = \underset{\beta\in\Re^p}{\arg\min}\left\{\sum_{i\in\{i/x_i\geq y_i'\beta\}} q\left|x_i - y_i'\beta\right| + \sum_{i\in\{i/x_i\leq y_i'\beta\}} (1-q)\left|x_i - y_i'\beta\right|\right\} \qquad [8.5]$$

can be written in matrix form:

$$\hat{\beta}(q) = \underset{\beta\in\Re^p}{\arg\min}\left\{\begin{matrix} qe'(X - Y'\beta \geq 0)\delta^1[X - Y'\beta] \\ +(1-q)e'(Y'\beta - X \geq 0)\delta^1[Y'\beta - X]\end{matrix}\right\} \qquad [8.6]$$

with $e'(X - Y'\beta \geq 0)$, index of farms i such as $x_i - y'_i\beta \geq 0$, and δ^1, vector of absolute deviations.

Thus, the linear optimization problem solving methods developed for the L1 (median) regression easily extend to quantile regression (Koenker and d'Orey 1994). Although the simplex method (Dantzig 1949) has an algorithmic complexity in $O(n^6)$, the Karmarkar (1984)'s method of the "interior-point" is in practice preferably as soon as the sample size becomes large, because of its reduced algorithmic complexity to $O(n^{3,5})$. For large samples, Portnoy and Koenker (1977) have shown that a combination of the "interior-point" algorithm and a smoothing algorithm for the objective function by Madsen and Nielsen (1993) makes quantile regression calculations competitive with those of least squares regression.

The weighted conditional quantiles have been proposed by Koenker and Zhao (1994) as L-estimates[1] in linear heteroscedastic models. The $W = \{w_i, i = 1, \ldots, n\}$

1 An L-estimate is an estimate defined by a linear combination of ordinal statistics.

weighting of the observations leads to a quantile regression scheme solving the following minimization problem:

$$\hat{\beta}_{\omega}(q) = \arg\min_{\beta \in \Re^p} \left\{ \sum_{i \in \{i/x_i \geq y_i'\beta\}} w_i q \left| x_i - y_i'\beta \right| + \sum_{i \in \{i/x_i \leq y_i'\beta\}} w_i (1-q) \left| x_i - y_i'\beta \right| \right\} \qquad [8.7]$$

The weighted estimation procedure uses the "predictor–corrector" implementation of the primal-dual algorithm proposed by Lustig *et al.* (1992).

The regularity conditions are specified as follows:

i) the distributions $F_i(x)$ input expenditures for a given product mix are absolutely continuous with densities $f_i(x)$ continuous and uniformly bounded over $]0, +\infty[$ in $\xi_i = \mu_q(x|y_i)$;

ii) $\Sigma_0 = \lim_{n\to\infty} \frac{1}{n}\sum_{i=1}^{n} y_i y_i'$ exists and is positive definite;

iii) $\Sigma_1 = \lim_{n\to\infty} \frac{1}{n}\sum_{i=1}^{n} f_i(\xi_i) y_i y_i'$ exists and is positive definite;

iv) the product mix is bounded in absolute value by $Sup_{i=1,\dots,n}\|y_i\| \sim O(\sqrt{n})$.

Pollard (1993) shows that under conditions i) and ii), the estimator, $\hat{\beta}_q \xrightarrow{p} \beta_q$, converges in probability; in addition, under the conditions i), ii), iii) and iv), we obtain the asymptotic normality, i.e.

$$\sqrt{n}\left(\hat{\beta}_q - \beta_q\right) \underset{loi}{\rightarrow} N\left(0, q(1-q){\Sigma_1}^{-1}\Sigma_0\Sigma_1\right) \qquad [8.8]$$

Finally, under the assumption of equality and independence of distributions $f_i(\xi_i) = f_\varepsilon(0)$, this result is simplified by:

$$\sqrt{n}\left(\hat{\beta}_q - \beta_q\right) \underset{loi}{\rightarrow} N\left(0, \sigma^2(q){\Sigma_0}^{-1}\right) \text{ with } \sigma(q) = \frac{\sqrt{q(1-q)}}{f_\varepsilon(0)} \qquad [8.9]$$

In addition, under conditions iii) and iv), if the hazards attached to the ith observation $\varepsilon_i = x_i - y_i{}'\beta$ of distribution F_i identical and independent admitting a density $f = \dot{F}$ such as $f\left(F^{-1}(q)\right) > 0$ in the neighborhood of q, as established by Koenker and Bassett (1982):

$$\sqrt{n}\left(\hat{\beta}_q - \beta_q\right) \underset{loi}{\rightarrow} N(0, \omega^2(q,F)\Omega^{-1}) \qquad [8.10]$$

with $\omega(q,F) = \frac{\sqrt{q(1-q)}}{f(F^{-1}(q))}$ and $\Omega = \lim_{n\to\infty} \frac{1}{n}\sum_{i=1}^{n} y_i y_i'$

These results can be used to construct confidence intervals for estimates using three procedures: inverse density function, rank method or resampling algorithm. The inverse density function estimation is the most direct and the fastest method, but it is sensitive to the hypothesis of identically and independently distributed data (iid). For data that are not iid, the rank method, which calculates confidence intervals by reversing the rank score test, is preferred. However, based on the simplex method, the rank method generates significant computation times for large datasets. The resampling method, based on the bootstrap technique, makes it possible to overcome all hypotheses but is unstable for small samples. Given the size of the Farm Accounting Data Network (FADN) sample, its non-random selection and the existence *a priori* of distinct sub-populations (e.g. specialized types of farming), we opted for the resampling method, based on the Markov Chain Marginal Bootstrap (MCMB) technique. Without distributional assumption, this method yields robust empirical confidence intervals in a reasonable computation time (He and Hu 2002).

For a given product j_0 such as a pig and the lth European country, the estimation interval of technical coefficients for $q^{t\square}$ conditional quantile of the specific costs

$$z_l^q = \left[Inf\_\hat{\gamma}_l^{j_0}(q); Sup\_\hat{\gamma}_l^{j_0}(q)\right] = \left[\underline{z_l^q}; \overline{z_l^q}\right] \qquad [8.11]$$

is obtained by the MCMB.

8.2.3. *Symbolic PCA of the specific cost distributions*

As symbolic objects (SO), the L national distributions $\Omega = \{\omega_1, \cdots, \omega_l, \cdots \omega_L\}$ are described by a set of $Q = 5$ descriptors[2], which are the estimation intervals of technical coefficients for conditional quantiles $Z = \{z^1, \dots, z^q, \dots, z^Q\} = \{z^{0.10}, z^{0.25}, z^{0.50}, z^{0.75}, z^{0.90}\}$, coding for the D1 and D9 deciles combined with the three quartiles Q1, Q2 and Q3.

Proposed by Lauro and Palumbo (2000), the Symbolic Object Principal Component Analysis (SO-PCA) is a mixed strategy for the principal component analysis (PCA) of symbolic objects designed in a three-step approach that we think is best suited to account for differences in scale and shape between empirical distributions of specific costs.

---

2 This choice of a small number of descriptors was made for comparative convenience with some more classical graphic approaches (Desbois *et al.* 2017); however, like this earlier work, it could be extended without disadvantage to sets of descriptors of cardinality *Q=9* (deciles) or even *Q=99* (percentiles) if the analysis objectives required it.

8.2.3.1. *The PCA of the interval extrema for quantile estimates*

In the extension of the PCA to the interval data proposed by Cazes *et al.* (1997), called Vertex Principal Component Analysis (V-PCA), a normalized PCA of the $\overline{\underline{Z}}$ array is carried out according to the vertices of the intervals.

As an example for quantile estimates with $Q = 2$, i.e. the bivariate case with Q1 ($Z^{0.25}$) and Q3 ($Z^{0.75}$) quartiles, the submatrix associated with the lth country is defined by:

$$\overline{\underline{Z}}_l = \begin{array}{c} \begin{array}{cc} Z^{0.25} & Z^{0.75} \end{array} \\ \begin{Bmatrix} \overline{z_l^{0.25}} & \overline{z_l^{0.75}} \\ \underline{z_l^{0.25}} & \underline{z_l^{0.75}} \\ \underline{z_l^{0.25}} & \overline{z_l^{0.75}} \\ \overline{z_l^{0.25}} & \underline{z_l^{0.75}} \end{Bmatrix} \end{array}. \qquad [8.12]$$

In this way, the vertices of hyper-rectangles $\mathcal{H}_l$ are vectors of $\mathbb{R}^Q$, while the estimates of the conditional quantiles are elements of $\mathbb{R}^N$, with $N = 2^Q L$. Thus, the V-PCA provides a dual representation of the specific empirical cost distributions represented by their estimation intervals, which are the symbolic objects, and conditional quantiles which are the descriptors.

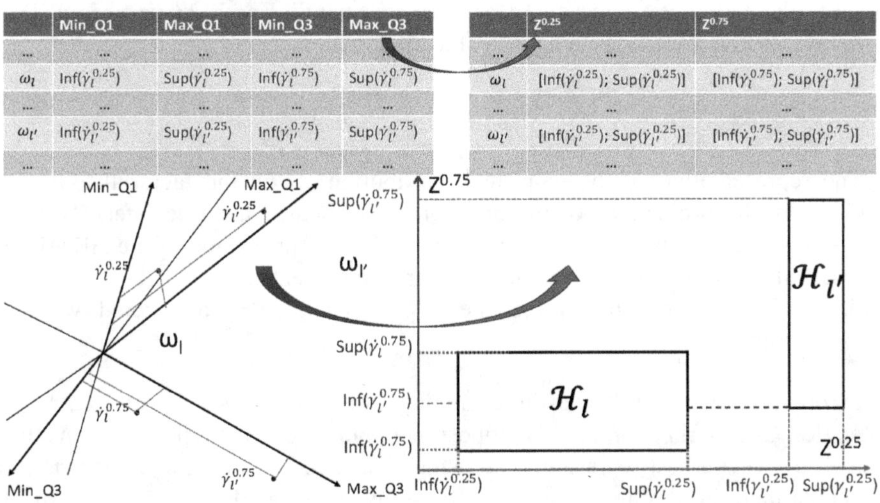

Figure 8.1. *The symbolic coding of the estimation intervals for the technical coefficients of the lower (Q1) and higher (Q3) quartiles of specific costs (graphic adapted from Lauro and Palumbo (2000) and tables from Diday (2006)). For a color version of this figure, see www.iste.co.uk/dimotikalis/analysis1.zip*

As in the classical PCA, the proper subspace (optimal for dual representation) is structured by orthonormal axes v_m $(1 \leq m \leq Q)$, maximizing the sum of squares of vertex coordinates $\psi_m = \underline{\overline{Z}} v_m$ and satisfying in $\mathbb{R}^N$ the characteristic eigenvector equation v_m and eigenvalues λ_m of the matrix $\frac{1}{N}\underline{\overline{Z}}'\underline{\overline{Z}}$:

$$\frac{1}{N}\underline{\overline{Z}}'\underline{\overline{Z}} v_m = \lambda_m v_m \qquad [8.13]$$

The dual analysis in $\mathbb{R}^Q$ leads to a similar equation

$$\frac{1}{N}\underline{\overline{Z}}\,\underline{\overline{Z}}' w_m = \lambda_m w_m \qquad [8.14]$$

having the same non-zero eigenvalues but eigenvectors w_m such as:

$$v_m = \lambda_m^{-1/2} \underline{\overline{Z}}' w_m$$

The interpretation of the axes of the V-PCA is based on the conditional quantiles (variables of the V-PCA) presenting the strongest contributions. In normalized PCA, the contributions to the inertia of the Q quantiles to the axis m are calculated as the square of the correlation between the factorial axis and the quantile factorial coordinates. For the lth land empirical distribution ω_l of the technical coefficients of specific costs as a symbolic object, the coordinates of the projections of the extremes linked to the estimation intervals of conditional quantiles with 2^Q vertices $s\,(l)$ on the main factorial axes are provided by the relationship:

$$\psi_{s(l)}^{m} = \underline{\overline{Z}}_{s(l)} v_m \qquad [8.15]$$

The representation of the empirical distribution ω_l on the factorial axis m is provided by the projections of the ends of the estimated quantile intervals as the hyper-rectangle of maximum inaccuracy (HRMI). The projection of the HRMI on a factorial plane provides a maximum imprecision rectangle (MIR) for the lth empirical distribution representing the hyper-rectangle $\mathcal{H}_l$ associated with the symbolic object ω_l.

In order to avoid the over-dimensioning projection bias of projected hyper-rectangles, Chouakria *et al.* (1998) propose to retain among the representations those whose vertices are best represented by using as a criterion, the relative contribution (CTR) defined in cosine terms, i.e. for a hyper-rectangle vertex $s\,(l)$ of the estimation intervals of the lth national quantile distribution:

$$CRT_{s(l)}^{m} = \Sigma_{q=1}^{Q}\left(\underline{\overline{Z}}_{s(l)}^{q} v_m\right)^2 / \Sigma_{q=1}^{Q} \underline{\overline{Z}}_{s(l)}^{q\;2} \qquad [8.16]$$

The problems of representation of symbolic objects are studied by Verde and De Angelis (1997) in terms of better adjustment of convex envelopes. Hyper-rectangle representations $\mathcal{H}_l$ of empirical distributions of conditional quantiles ω_l as symbolic objects are constructed by the Maximum Covering Area Rectangular (MCAR) method. However, in order to coherently process the estimation extrema (vertices) belonging to the same empirical distribution of estimated quantiles (symbolic object) and to optimally represent the size and shape of these distributions, we will use the PCA constrained by interval extrema, proposed by Lauro and Palumbo (2000).

8.2.3.2. *The constrained PCA of estimation interval extrema*

In V-PCA, the vertices are considered as independent statistical units. In order not to lose information on the size and shape of the hyper-rectangles, Lauro and Palumbo (2000) introduce a constraint of cohesion between the vertices. The method is based on maximizing the variance between symbolic inter-objects, constituting a particular application of PCA constrained to a reference space defined by D'Ambra and Lauro (1982).

Let A be the Boolean matrix that indicates the membership of the N estimation interval vertices to the L empirical distributions. The expression of the variance between symbolic objects is given by:

$$\frac{1}{N}\underline{\overline{Z}}'A(A'A)^{-1}A'\underline{\overline{Z}} \qquad [8.17]$$

In our case study, all the empirical distributions have the same number of conditional quantiles, then the same number of estimation intervals, hence the same number of vertices $\frac{N}{L} = 2^Q$. It follows that $A'A = 2^Q I_L$.

Let P_A be the orthogonal projector associated with the A matrix on the sub-space of reference:

$$P_A = A(A'A)^{-1}A' \qquad [8.18]$$

In the space $\mathbb{R}^N$, the factorial axes of inertia are obtained as a solution of the following eigenvalue equation:

$$\frac{1}{N}\underline{\overline{Z}}'P_A\underline{\overline{Z}}\tilde{v}_m = \tilde{\lambda}_m\tilde{v}_m \qquad [8.19]$$

where $\tilde{\lambda}_m$ and $\tilde{v}_m$ are the eigenvalues and the eigenvectors associated with the inertia operator $\frac{1}{N}\underline{\overline{Z}}'P_A\underline{\overline{Z}}$, respectively.

The coordinates of the hyper-rectangle $\mathcal{H}_l$ associated with the lth empirical country distribution are then computed as follows:

$$\tilde{\psi}_l^m = \underline{Z}_l \tilde{v}_m \qquad [8.20]$$

The analysis in $\mathbb{R}^Q$ is equivalent to solving the following equation for eigenvalues:

$$(A'A)^{-1/2}\left(A'\overline{\underline{Z}\underline{Z}}'A\right)(A'A)^{-1/2}\widetilde{w}_m = \tilde{\lambda}_m \widetilde{w}_m \qquad [8.21]$$

where $\widetilde{w}_m = (A'A)^{-1/2}A'\overline{\underline{Z}}\tilde{v}_m$, modulo the $\frac{1}{N}$ constant.

The absolute contribution (CTA) of the qth conditional quantile of the empirical distribution to the mth factor is defined in the same way as for the V-PCA:

$$CTA_q^m = \frac{\left(\tilde{\lambda}_m^{1/2}\tilde{v}_q^m\right)^2}{\tilde{\lambda}_m} = \left(\tilde{v}_q^m\right)^2 \qquad [8.22]$$

The CRTs are also used to select the lth empirical distribution to be represented on the factorial graphs:

$$CRT_l^m = \sum_{s(l)\in\omega_l}\sum_{q=1}^{Q}\left(z_s^q \tilde{v}_j^m\right)^2 / \sum_{s(l)\in\omega_l}\sum_{q=1}^{Q}\left(z_s^q\right)^2 \qquad [8.23]$$

The processing of interval data is similar to Escofier and Pages (2014)'s multiple PCA in the following point: the extrema of each technical coefficients as instances of the symbolic object are centered on the general mean of each conditional quantile as a symbolic variable.

As in V-PCA, representations of empirical distributions as a symbolic object are constructed by the MCAR method. However, the imposition of extrema membership constraints on distributions makes their factor representation more compact.

8.2.3.3. *PCA ranges of estimation intervals*

Partial PCA can also be used to better emphasize the differences between symbolic objects. The following section defines a partial PCA where the vertices of the hyper-rectangles are centered relative to the *Inf* value.

In order to take into account only the sizes and shapes of the hyper-rectangles associated with the descriptions, Lauro and Palumbo (2000) have proposed a PCA based on scaling interval data that summarizes useful information to describe the size and shape of symbolic objects, with the following range transformation

$$r_l^q = \overline{z_l^q} - \underline{z_l^q}. \quad [8.24]$$

The *description potential – DP,* introduced by Carvalho (1997), is the hyper-volume associated with the description of the lth empirical distribution, domain defined by the Cartesian product of the ranges $R_l = r_l^1 \times \cdots \times r_l^q \cdots r_l^Q$ of the Q conditional quantiles associated with the description of the symbolic object ω_l, representing the lth country distribution.

The transformation of the data into a range with an affine translation where the minima $\left\{\underline{z_l^1}, \cdots, \underline{z_l^q}, \cdots, \underline{z_l^Q}\right\}$ are all located at the origin, as in Figure 8.2.

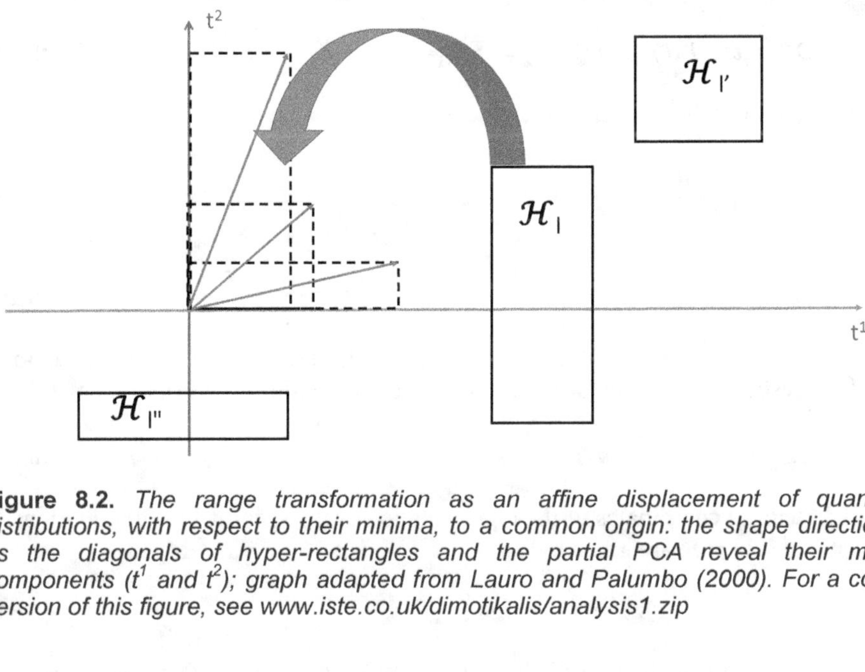

Figure 8.2. *The range transformation as an affine displacement of quantile distributions, with respect to their minima, to a common origin: the shape directions as the diagonals of hyper-rectangles and the partial PCA reveal their main components (t^1 and t^2); graph adapted from Lauro and Palumbo (2000). For a color version of this figure, see www.iste.co.uk/dimotikalis/analysis1.zip*

The range measurement is defined by:

$$\pi(\omega_l) = \prod_{q=1}^{Q} \mu\left(\widetilde{r_l^q}\right) \text{ where} \quad [8.25]$$

$\widetilde{r_l^q} = r_l^q / s_q$ is the normalized range relative to the domain $D^q = \left\{r_l^q; l \in L\right\}$ of the conditional quantile z^q as a descriptor.

However, if the measure of one of the descriptors tends to zero, then the description potential tends to zero also. To overcome this drawback, Lauro and Palumbo (2000) use the linear measure of the potential of description defined by Carvalho (1997), i.e. for the instance a_l of the symbolic object ω_l:

$$\sigma(a_l) = \sum_{q=1}^{Q} \mu\left(\widetilde{r_l^q}\right). \qquad [8.26]$$

Let all the instances of the empirical distributions of specific costs $\{a_1, \cdots, a_l, \cdots, a_L\}$ and $\boldsymbol{U}$, the matrix $L \times Q$ of general term $u_l^q = \sqrt{\widetilde{r_l^q}}$, then the PCA in the range of estimation ranges is defined by the factor decomposition of the total linear description potential $LDP = \sum_{l=1}^{L} \sigma(a_l)$, allowing a different geometrical representation of the vertices than in the V-PCA.

The Range Transformation PCA (RT-PCA) breaks down the criterion

$$LDP = tr(U'U) = tr(UU') = \sum_{l=1}^{L} \sigma(a_l) \qquad [8.27]$$

according to the following characteristic eigenvector equations:

$$U'Ut_m = \mu_m t_m \qquad [8.28]$$

and

$$UU'u_m = \mu_m u_m \qquad [8.29]$$

Thus, the sum of eigenvalues μ_m associated with eigenvectors t_m in $\mathbb{R}^L$ and u_m in $\mathbb{R}^Q$ corresponds to the factorial decomposition of the linear description potential:

$$\sum_{m=1}^{Q} \lambda_m = \sum_{l=1}^{L} \sigma(a_l). \qquad [8.30]$$

The factorial coordinates of the representation of the specific cost distributions in the optimal subspace are given by:

$$\phi_m = Ut_m \qquad [8.31]$$

The absolute contribution (CTA), as the ratio between the factorial coordinate and the eigenvalue, measures the contribution of the empirical distribution of specific costs to the potential of description of the mth factorial axis; it is defined by:

$$CTA_l^m = \frac{(\phi_l^m)^2}{\mu_m} \qquad [8.32]$$

The relative contribution (CTR) measures the representation quality of the empirical distribution in the chosen representation factorial subspace:

$$CTR_l^m = \frac{\sum_{m=1}^{M*}(\phi_l^m)^2}{\sum_{q=1}^{Q}(u_l^m)^2} \quad [8.33]$$

The interpretation of the factorial axes is performed according to the absolute contributions (factorial coordinates) of the estimated quantiles estimated for the empirical distributions of specific costs, as descriptors of the symbolic objects:

$$CTA_q^m = (t_q^m)^2. \quad [8.34]$$

The RT-PCA can be represented by the projection of the factorial coordinates of the maxima. The distributions described by conditional quantile estimates share representations in hyper-rectangles similar in size and shape if they are projected in the same neighborhood.

If all the terms of matrix U are positive, then the first eigenvector u_1 and the associated factor

$$\phi_1 = Ut_1 \quad [8.35]$$

are positive. According to Lauro and Palumbo (2000), the first major component can be interpreted as a size factor, while the higher rank factors order the empirical distributions according to their shape characteristics.

8.2.3.4. *Mixed strategy PCA of estimation intervals*

The mixed strategy in principal component analysis of symbolic objects (SO-PCA*)* combines the vertex PCA (V-PCA) and the range PCA (RT-PCA) in a three-step approach to account for differences in scale and shape between empirical distributions of specific costs:

i) PCA ranges to extract the main axes that best represent the scales and forms of empirical distributions of conditional quantiles;

ii) projection from $\underline{\overline{Z}}$ to $\hat{Z} = P_A\underline{\overline{Z}}$ in order to take into account the relations between the different extrema, given the order relationships between the different conditional quantiles of the distribution of specific costs;

iii) PCA line projections $\hat{Z}_i$ on the sub-space of optimal representation $\Phi = \{\phi_1, \cdots, \phi_m, \cdots, \phi_{M^*}\}$ on the sub-space of optimal representation $P_\Phi = \Phi(\Phi'\Phi)^{-1}\Phi'$.

The mixed analysis strategy is therefore based on the solution of the following characteristic equation:

$$\hat{Z}'P_{\Phi}\hat{Z} = \underline{Z}'A(A'A)^{-1/2}P_{\Phi}(A'A)^{-1/2}A'\underline{Z}s_m = \rho_m S_m \quad [8.36]$$

where the diagonal matrix $(A'A)^{-1}$ is broken down in the product $(A'A)^{-1/2}(A'A)^{-1/2}$ for reasons of symmetry, with, respectively, ρ_m and S_m, eigenvalues and eigenvectors associated with the usual ortho-normality conditions.

The interpretation of the results of the analysis depends on the choice of the projection operator P_{Φ}, whose l^{th} diagonal term, interpretable as a normalized weight associated with the lth country distribution, is equal to:

$$\phi_l(\phi'_l\phi_l)^{-1}\phi'_l = \sum_{m=1}^{M^*}(\phi_l^m)^2/\mu_m \quad [8.37]$$

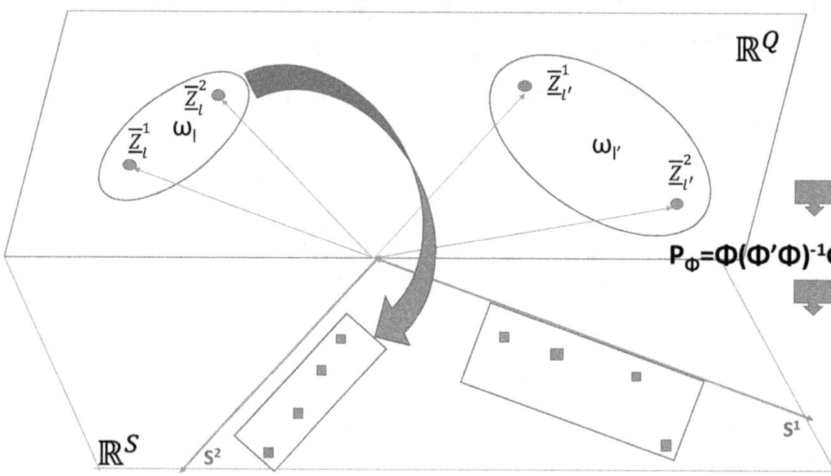

Figure 8.3. *Extrema estimation intervals of the technical coefficients of lower and upper quartile estimates for country l and country l' are projected on the optimal subspace (S^1, S^2) for representation of the size and shape of their MCARs (adapted from Lauro and Palumbo (2000)). For a color version of this figure, see www.iste.co.uk/dimotikalis/analysis1.zip*

8.2.4. *Symbolic clustering analysis of the specific cost distributions*

For a given product j_0 such as a pig and the lth European country, the estimation interval of technical coefficients for the specific costs

$$z_l^q = \left[Inf\_\hat{\gamma}_l^{j_0}(q); Sup\_\hat{\gamma}_l^{j_0}(q)\right] \quad [8.38]$$

is obtained by Markov chain marginal bootstrap (He and Hu 2002). As symbolic objects, the L national distributions $\Omega = \{\omega_1, \cdots, \omega_l, \cdots \omega_L\}$ are described by a set of $Q = 5$ descriptors[3], which are the estimation intervals of technical coefficients for conditional quantiles:

$$Z = \{z^1, \ldots, z^q, \ldots, z^Q\} = \{z^{0.10}, z^{0.25}, z^{0.50}, z^{0.75}, z^{0.90}\}. \qquad [8.39]$$

The local dissimilarities between country l and country l', associated with these estimation intervals of technical coefficients for the qth conditional quantile, are computed according to the Euclidean distance metric:

$$\delta_M(z_l^q, z_{l'}^q) = \sqrt{\left(Inf\_\hat{y}_l^{j_o}(q) - Inf\_\hat{y}_{l'}^{j_o}(q)\right)^2 + \left(Sup\_\hat{y}_l^{j_o}(q) - Sup\_\hat{y}_{l'}^{j_o}(q)\right)^2}. \qquad [8.40]$$

For this metric M, a global dissimilarity between country l and country l' based on the differences over the national distributions of estimation intervals for the technical coefficients is computed according to the following quadratic criterion:

$$d(\omega_l, \omega_{l'}) = \left(\sum_{q=1}^{Q} \delta_M^2(z_l^q, z_{l'}^q)\right)^{1/2}. \qquad [8.41]$$

Given a matrix of dissimilarities between national empirical distributions of specific costs issued from the previous computations, we can use the methods of unsupervised clustering. In a way similar to the Ward method, Chavent *et al.* (2007) propose a divisive hierarchical clustering algorithm on symbolic data (DIVCLUS-T), valid for both interval data and categorical data. Subsequently, we detail for interval data the principles on which the operations of this unsupervised clustering procedure are based.

The divisive hierarchical clustering algorithm recursively splits each cluster into two sub-clusters, starting from the whole set of countries as symbolic objects

$$\Omega = \{\omega_1, \cdots, \omega_l, \cdots \omega_L\}.$$

At each partition in k symbolic clusters $P_K = \{C_1, \cdots, C_k \cdots, C_K\}$, a cluster has to be divided in order to get a partition P_{K+1}, with $K + 1$ clusters, optimizing the selected adequacy criterion based on the inertia.

3 This choice of a small number of descriptors was made for comparative convenience with some more classical graphic approaches (Desbois 2015); however, it could be extended to sets of descriptors with cardinality Q=9 (deciles), or even Q=99 (percentiles).

The inertia of the kth cluster is defined by $I(C_k) = \sum_{l \in C_k} \mu_l d_M^2\left(z_l, g(C_k)\right)$, where μ_l is the weight of the lth country and $g(C_l)$ is the cluster centroid defined as:

$$g(C_k) = \frac{1}{\sum_{l \in C_k} \mu_l} \sum_{l \in C_k} \mu_l z_l. \qquad [8.42]$$

The intra-inertia is defined by the sum of the inertias of the clusters to their centroids:

$$W(P_K) = \sum_{k=1,\ldots,K} I\left(C_k\right). \qquad [8.43]$$

The inter-inertia is defined by the inertia of the centroids with regard to the g overall centroid of Ω, as follows:

$$B(P_K) = \sum_{k=1,\ldots,K} \mu_k d_M^2\left(g(C_k), g\right) \text{ where } \mu_k = \sum_{l=1,\ldots,k} \mu_l. \qquad [8.44]$$

For a partition P_K, the total inertia sums the intra-inertia with the inter-inertia:

$$I(\Omega) = W(P_K) + B(P_K). \qquad [8.45]$$

Hence, minimizing the inertia unexplained by the partition (measured by *W*) is equivalent to maximizing the inertia explained by the partition (measured by *B*).

Generated by the logical binary choice (*yes/no*) to a numerical binary question $\Psi = [Is\ z^q\ \leq c\ ?]$, let us denote $\{A_k, \overline{A_k}\}$ the induced bipartition of a cluster C_k formed of n_k objects. In order to choose among the $n_k - 1$ possible bipartitions of the C_k cluster, a discriminating criterion is defined by the following ratio:

$$D(\Psi) = \frac{B^q(A_k, \overline{A}_k)}{I^j(C_k)} = 1 - \frac{W^j(A_k, \overline{A}_k)}{I^q(C_k)}, \qquad [8.46]$$

where the inter-inertia $B^q\left(A_k, \overline{A}_k\right)$ and the inertia $I^q(C_k)$ are computed with regard to the qth conditional quantile. Hence, minimizing the intra-inertia $\mathrm{W}\{A_k, \overline{A_k}\}$ is equivalent to maximizing the inter-inertia $\mathrm{B}\{A_k, \overline{A_k}\}$ and, as a result, the $D(\Psi)$ discriminating criterion.

As in the Ward method, the "*upper hierarchy*" (quoting Mirkin 2005) of partition P_K is indexed by the height h of a cluster C_K, defined by its inter-inertia as follows:

$$h(C_k) = B\left(A_k, \overline{A_k}\right) = \frac{\mu(A_k)\mu(\overline{A_k})}{\mu(A_k)+\mu(\overline{A_k})} d^2\left(g(A_k), g\left(\overline{A_k}\right)\right) \qquad [8.47]$$

The DIVCLUS-T algorithm splits the cluster C_K^* that maximizes $h(C_K)$, ensuring that the next partition $P_{K+1} = P_K \cup \{A_K, \overline{A_K}\} - C_K^*$ has the minimum intra-inertia value, with respect to the rule

$$W(P_{K+1}) = W(P_K) - h(C_K^*). \qquad [8.48]$$

8.3. Results

Based on the gross product, the estimation according to the quantiles provides a conditional allocation of the specific costs by main products, within the framework of a multi-product exploitation. In the framework of the Farm Accountancy Cost Estimation and Policy Analysis project (FACEPA) research project, the managers in charge of the Knowledge Based Bio-Economy project of the 7th EU Framework Program of Research chose to focus on the three main agricultural commodities that are wheat, milk and pig, produced at a level sufficiently broad at the European level to allow meaningful cross-country comparisons for the 12 European Member States which are the main producers (EU12), choosing 2006 as a baseline for comparison convenience. Thus, we will analyze the results obtained in particular for pigs, one of the three main commodities selected in the framework of the FACEPA project. The pig-specific inputs mainly encompass feed, veterinary, water and energy inputs.

Country	D1 [Min ; Max]	Q1 [Min ; Max]	Q2 [Min ; Max]	Q3 [Min ; Max]	D9 [Min ; Max]
Austria	[347.2 ; 369.2]	[397.3 ; 409.1]	[425.6 ; 447.4]	[463.5 ; 485.3]	[523.1 ; 562.5]
Belgium	[539.9 ; 566.1]	[561.2 ; 579.8]	[591.5 ; 608.7]	[642.7 ; 674.7]	[684.6 ; 707.4]
Denmark	[445.2 ; 458.6]	[503.0 ; 515.4]	[558.9 ; 570.9]	[617.7 ;632.1]	[654.7 ; 671.3]
France	[470.9 ; 493.5]	[509.9 ; 527.7]	[547.9 ; 561.7]	[577.6 ; 594.2]	[610.8 ; 644.6]
Germany	[444.3 ; 454.7]	[475.6 ; 485.4]	[514.8 ; 526.4]	[567.7 ; 582.7]	[593.4 ; 618.8]
Hungary	[369.0 ; 451.8]	[459.3 ; 568.1]	[589.2 ; 662.2]	[633.3 ; 681.5]	[647.8 ; 737.0]
Italy	[116.5 ; 170.1]	[162.2 ; 245.2]	[325.1 ; 386.5]	[559.6 ; 633.0]	[627.9 ; 718.3]
Netherlands	[487.4 ; 506.2]	[528.2 ; 550.4]	[584.6 ;602.4]	[639.9 ; 661.5]	[676.0 ; 721.6]
Poland	[471.3 ; 483.3]	[541.8 ; 552.2]	[603.0 ; 618.0]	[655.1 ; 674.7]	[704.5 ; 727.3]
Spain	[191.5 ; 285.3]	[369.8 ; 441.6]	[552.3 ; 638.5]	[743.8 ; 802.2]	[824.6 ; 893.4]
Sweden	[396.3 ; 443.9]	[507.2 ; 533.1]	[533.1 ; 578.1]	[547.5 ; 619.5]	[641.7 ; 722.9]
United-Kingdom	[376.8 ; 559.2]	[548.4 ; 596.0]	[599.2 ; 629.6]	[641.3 ; 712.9]	[723.2 ; 805.4]

Table 8.1. *Pig, estimation intervals for technical coefficients of quantile-specific costs for €1,000 of gross product, EU12. Source: author's processing, from EU-FADN 2006*

Table 8.1 presents for pigs the estimation intervals of conditional quantiles (lower decile D1, lower quartile Q1, median Q2, upper quartile Q3, upper decile D9) of the specific inputs of agricultural production. The figures are estimated from a quantile regression of the specific inputs[4] on a decomposition of the gross product into 15 aggregates, for the set of 12 European countries (UE12) selected regarding 2006.

The pre-visualization of the specific cost estimates is done according to the graph in Figure 8.4, showing the point estimates of conditional quantiles in ascending order for each country. This graph of point estimates displaying the conditional quantiles of specific costs for pigs by country highlights some distributional facts: first, Italy (ITA) and Spain (ESP) having a similar shape (high inter-quantile growth with an inter-decile difference D1-D9 greater than €400) despite distinct locations (the minimum of the differentials between respective quantiles is greater than €200); second, Austria (OST) opposes the previous model with an inter-decile gap D1-D9 of €100; third, the United Kingdom presenting a distributional scale with significant inter-quantile growth (inter-decile gap D1–D9 greater than €300); and fourth, a subset of countries with moderate inter-quantile growth (inter-decile range of between €200 and €300). The conditional median (Q2) estimation levels are also helpful in distinguishing between these different distributional scales: on the one hand, Italy (ITA) and Austria (OST) with median estimates less than €450; on the other hand, all other countries whose median estimates are between €500 and €600.

Among the differences that can be identified regarding 2006, let us first note the level differences between two similar distributional scales with heterogeneous slopes: Spain (ESP)[5] and Italy (ITA)[6]. Table 8.1 confirms this separation of distributional scales for all conditional quantiles; these two countries illustrate the *location-scale shift* linear model of conditional quantile with heterogeneous slopes. Less easy to observe, we secondly note the absence of overlapping between the distributional scales of Belgium (BEL), Denmark (DAN) and Austria (OST); apart from some differences in precision for the estimation of the upper conditional decile (D9) between Belgium and Austria, on the one hand, and Denmark, on the other hand, these three countries illustrate the *location shift* linear model of conditional quantile with homogeneous slopes.

4 The specific inputs are given by the SE281 accounting aggregate of the European FADN.

5 For Spain, the differences between extreme conditional quantiles overlap those between *Comunitat Valenciana* (PDO *Jamon de Teruel*) at the highest costs and *Extremadura* at the lowest costs.

6 For Italy, the highest quantile estimates correspond to those recorded in *Emilia-Romagna* (PDO *Prosciutto di Parma*) or *Venetto* (PDO *Veneto Berico-Eugeano*), which oppose the lowest quartile and decile estimates in *Lombardia*.

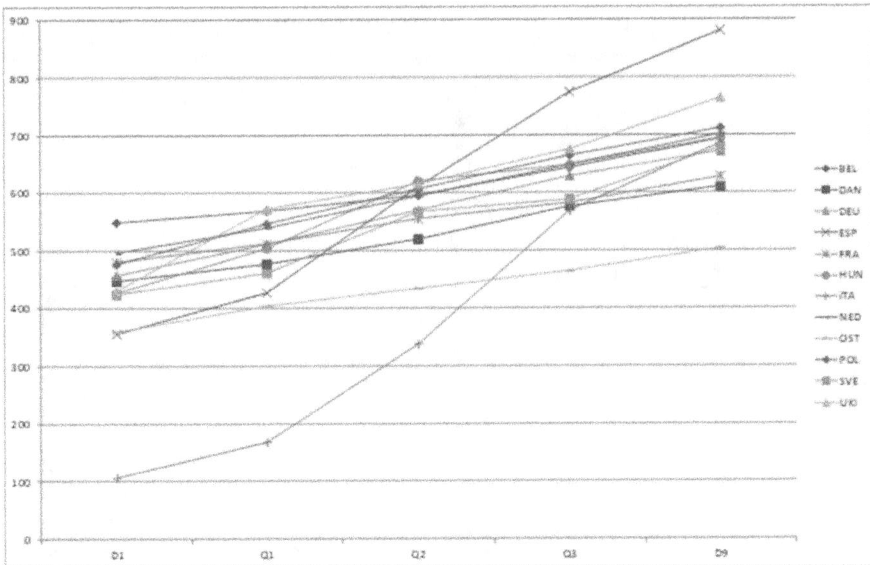

Figure 8.4. *Pigs, point estimation of technical coefficients of conditional quantiles for 12 EU member states. Source: author's processing, from EU-FADN 2006. For a color version of this figure, see www.iste.co.uk/dimotikalis/analysis1.zip*

8.3.1. *The SO-PCA of specific cost estimates*

The SO-PCA interval estimates of conditional quantiles allow us to specify this distributional structure in Figure 8.5. The first principal component $F1$ (Axis 1) representing 64% of the inertia is negatively correlated with the first conditional quantiles (decile D1 and quartile Q1 highly correlated). The second principal component $F2$ (Axis 2), representing 35% of the inertia, is positively correlated with the upper decile (D9) and the third quartile (Q3). The median Q2 is also correlated with the first two major components. The SO-PCA $F1 \times F2$ first factorial plane, representing 99% of the country variability, makes it possible to identify two distinct groups of countries differentiating according to the level of the conditional estimate of the first quantiles (D1 and Q1) strongly correlated with the first axis $F1$: on one hand, in $F1 > 0$, Italy with the first quantiles lower than €205 and, on the other hand, in $F1 < 0$, all other countries for which the first quantiles are greater than €235. The second main component allows us to distinguish three groups: on the one hand, Austria in $F2 > 0$ to the most homogeneous quantile estimates situated between €350 and €430; on the other hand, Spain with the highest estimates (from €770 for Q3 to € 860 for D9); and all other countries in the quadrant ($F1 < 0$, $F2 < 0$) or close to it.

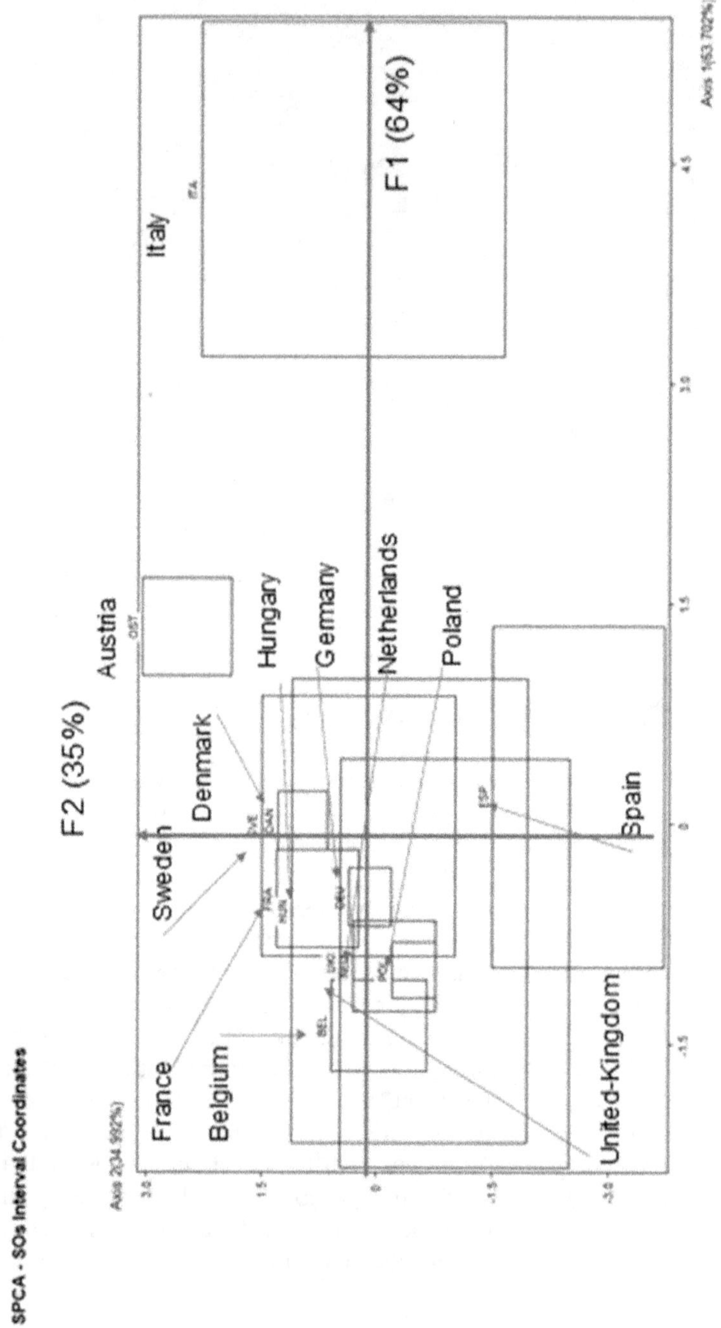

Figure 8.5. *Pigs, SO-PCA of quantile estimation intervals, factorial plane F1xF2 of EU12 countries. Source: author's processing, from EU-FADN 2006. For a color version of this figure, see www.iste.co.uk/dimotikalis/analysis1.zip*

Coherently with Figure 8.6, the SO-PCA identifies Austria's cost homogeneity model, clearly apart from the other national distributions of Northern and Central Europe. Moreover, it distinguishes two models of cost heterogeneity belonging to Southern Europe, one for lower cost quantiles (D1 and Q1) illustrated by Italy, and the second by the higher costs (D9 and Q3) represented by Spain.

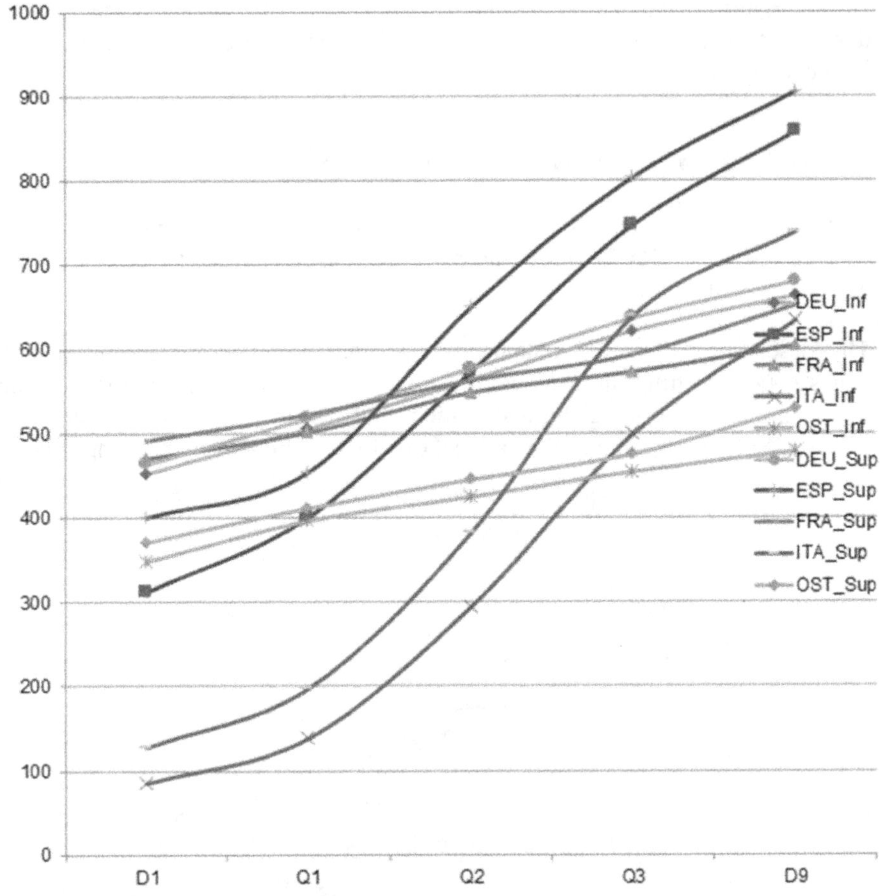

Figure 8.6. *Pigs, estimation intervals of technical coefficients (€) by conditional quantiles for €1,000 of gross product: the location shift model (FRA-DEU/OST) versus the location-scale shift model (ITA/ESP). Source: author's processing, from EU-FADN 2006. For a color version of this figure, see www.iste.co.uk/dimotikalis/analysis1.zip*

Thus, Austria illustrates the *location shift* model, while Italy represents the *location-scale shift* model. The first axis of the SO-PCA is a gradient of the central level of the cost distributions and the second one is a gradient of quantile heterogeneity within the cost distributions, both of them with an inverted sign of variation.

However, in order to better communicate the observed statistical realities, we need to define the groupings observed in terms that can be directly mobilized for the economic analysis.

8.3.2. *The divisive hierarchy of specific cost estimates*

In this section, we analyze the results obtained from the symbolic divisive clustering procedure DIVCLUS-T for pigs the meat commodity selected in the framework of the FACEPA project.

The descending hierarchy obtained shows that most of the set of quantile estimates is used to fully discriminate the countries studied, which implies keeping all the parameters describing the distribution, and possibly extending it by a finer quantile scale allowing some of the national distributions to be better distinguished.

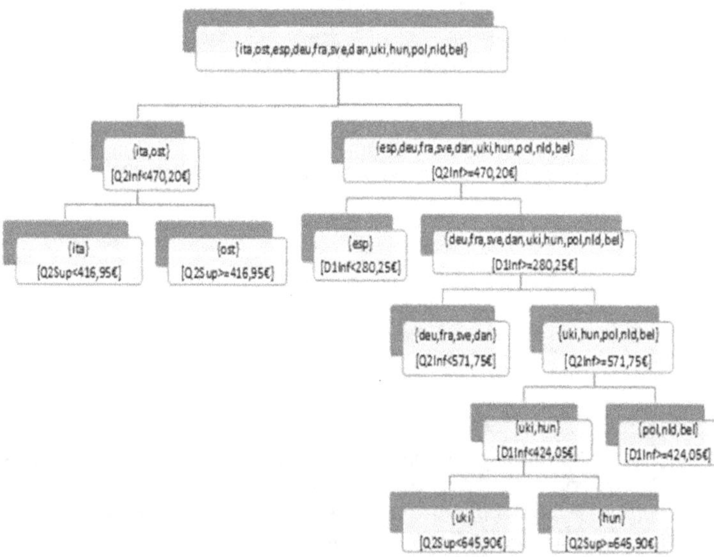

Figure 8.7. *The EU12 country clustering based on specific costs for €1,000 of pig gross product. Source: author's processing, according to EU-FADN 2006. For a color version of this figure, see www.iste.co.uk/dimotikalis/analysis1.zip*

At the top of the divisive hierarchy, the clustering procedure allows us to identify two countries with contrasted models for empirical distributions of the pig technical coefficients for specific production costs: grouped by their infimums of median (*Q2Inf*) level which are lower than the €470.20 threshold, Italy (*ita*) and Austria (*ost*) are split in the following step by the threshold of €416.95, to which the median supremum (*Q2Sup*) of Italy is lower while Austria one is higher.

Among the group of other countries for which the median infimum is greater than the €470.20 threshold, Spain presents an infimum of the first decile (*D1Inf*) lower than the €280.25 threshold. Then, the other countries of this group split into two main subgroups: the first one (Denmark, France, Germany, Sweden) with a median infimum lower than the €571.75 threshold; the second one (Belgium, Hungary, Netherlands, Poland, United Kingdom) with a median infimum higher than €571.75. This latter group is split into two subgroups by a €424.05 threshold for the infimum of the first decile (*D1Inf*) which is lower for United Kingdom and Hungary while it is greater for Belgium, Netherlands and Poland. Lastly, with a level of median supremum (*Q2Sup*) lower than the €645.90 threshold, the United Kingdom is distinguished from the Hungary which has a higher median supremum level.

These specific cost thresholds, highlighted by the interval-based divisive clustering algorithm, provide the parameters of subsequent analyses on the cost-effectiveness of European pig producers, in order to design proper regulations on this very competitive market.

8.4. Conclusion

Combining quantile regression with symbolic data analysis, this chapter has presented a global methodology that aims to keep as much as possible relevant information for public policy design, during the econometric process of estimating and analyzing agricultural specific costs for pig production. The principal component analysis of symbolic objects identifies different models of distributional scale, notably that of the *location shift* model opposite that of the *location-scale shift* one. In addition, differences and similarities between estimate intervals are used by the divisive hierarchical clustering algorithm to produce country classes delineated through *cost thresholds*. The differences between these groups of countries are delimited by thresholds expressed according to the conditional quantiles in unitary terms of the gross product. These thresholds can be used for segmenting farm populations to later analyze the differential impacts of agricultural policy measures envisioned in order to regulate the European pig market. We plan to apply this methodology at the second level of the European Nomenclature of Territorial Units for Statistics (i.e. NUTS 2, based on 281 regions).

8.5. References

Angrist, J. and Pischke, J. (2009). *Mostly Harmless Econometrics: An Empiricist's Companion*. Princeton University Press, Princeton.

Billard, L. and Diday, E. (2006). *Symbolic Data Analysis: Conceptual Statistics and Data Mining*. Wiley Interscience, Chichester.

Cameron, A.C. and Trivedi, P.K. (2005). *Microeconometrics. Methods and Applications*. Cambridge University Press, Cambridge.

Carvalho, F.A.T. (1997). Clustering of constrained symbolic objects based on dissimilarity functions. Indo–French Workshop on Symbolic Data Analysis and its Applications, Paris IX University, Paris.

Cazes, P., Chouakria, A., Diday, E., Schektman, Y. (1997). Extension de l'analyse en composantes principales à des données de type intervalle. *Revue de statistique appliquée*, 45(3), 5–24.

Chavent, M., Lechevalier, Y., Briant, O. (2007). DIVCLUS-T: A monothetic divisive hierarchical clustering method. *Computational Statistics & Data Analysis*, 52(2), 687–701.

Chouakria, A., Diday, E., Cazes, P. (1998). An improved factorial representation of symbolic objects. *Studies and Research, Proceedings of the Conference on Knowledge Extraction and Symbolic Data Analysis (KESDA'98)*. Office for Official Publications of the European Communities, Luxembourg, 276–289.

D'Ambra, L. and Lauro, C.N. (1982). Analisi in componenti principali in rapporto a un sottospazio di riferimento. *Rivista di Statistica Applicata*, 15(1), 51–67.

Dantzig, G.B. (1949). Programming in a linear structure. *Econometrica*, 17, 73–74.

Desbois, D. (2015). Estimation des coûts de production agricoles : approches économétriques. PhD dissertation directed by Bureau J.C. and Surry, Y. (eds). ABIES-AgroParisTech, Paris.

Desbois, D., Butault, J.-P., Surry, Y. (2017). Distribution des coûts spécifiques de production dans l'agriculture de l'Union européenne : une approche reposant sur la méthode de régression quantile. *Économie rurale*, 361, 3–22.

Diday, E. (2006). Thinking by classes in data science: The symbolic data analysis paradigm. *WIREs Computational Statistics*, 8, 171–205.

Divay, J.F. and Meunier, F. (1980). Deux méthodes de confection du tableau entrées-sorties. *Annales de l'INSEE*, 37, 59–109.

D'Haultfoeuille, X. and Givord, P. (2014). La régression quantile en pratique. *Economie et statistique*, 71, 85–111.

Escofier, B. and Pagès, J. (1988). *Analyse factorielles multiples*. Dunod, Paris.

He, X. and Hu, F. (2002). Markov chain marginal bootstrap. *Journal of the American Statistical Association*, 97, 783–795.

Karmarkar, R. (1984). A new polynomial-time algorithm for linear programming. *Combinatorica*, 4, 373–395.

Koenker, R. and Bassett, G. (1978). Regression quantiles. *Econometrica*, 46, 33–50.

Koenker, R. and Bassett, G. (1982). Robust tests for heteroscedasticity based on regression quantiles. *Econometrica*, 50(1), 43–61.

Koenker, R. and d'Orey, V. (1994). Remark AS R92: A remark on algorithm AS 229: Computing dual regression quantiles and regression rank scores. *Applied Statistics*, 43, 410–414.

Koenker, R. and Zhao, Q. (1994). L-estimation for linear heteroscedastic models. *Journal of Nonparametric Statistics*, 3, 223–235.

Lauro, C.N. and Palumbo, F. (2000). Principal component analysis of interval data: A symbolic data analysis approach, *Computational Statistics*, 15(1), 73–87.

Lustig, I.J., Marsden, R.E., Shanno, D.F. (1992). On implementing Mehrotra's predictor-corrector interior-point method for linear programming. *SIAM Journal on Optimization*, 2, 435–449.

Madsen, K. and Nielsen, H.B. (1993). A finite smoothing algorithm for linear estimation. *SIAM Journal on Optimization*, 3, 223–235.

Mirkin, B. (2005). Clustering for data mining. *A Data Recovery Approach*. Chapman & Hall, London.

Pollard, D. (1991). Asymptotics for least absolute deviation regression estimators. *Econometrics Theory*, 7, 186–199.

Portnoy, S. and Koenker, R. (1977). The Gaussian hare and the Laplacian tortoise: Computation of squared-errors vs. absolute-errors estimators. *Statistical Science*, 1, 279–300.

Verde, R. and De Angelis, P. (1997). Symbolic objects recognition on a factorial plan. *NGUS'97*, Bilbao.

9

Maximization Problem Subject to Constraint of Availability in Semi-Markov Model of Operation

The semi-Markov (SM) decision processes theory delivers the methods that allow us to control the operation processes of systems. The infinite duration SM decision processes are presented in this chapter. We discuss the gain maximization problem subject to an availability constraint for the infinite duration of the SM model of the operation in the reliability aspect. The problem is transformed into a linear programming maximization problem.

9.1. Introduction

In many articles and books, we can find applications of semi-Markov (SM) processes in reliability theory. We consider the most interesting and important books on these issues to be Gertsbakh (1969), Mine and Osaki (1970), Howard (1971), Silvestrov (1980) and Grabski (2015). The SM decision processes theory was developed by Howard (1969, 2015, 2018), Main and Osaki (1971), Gercbakh (2001) and Jewell (2018). These processes are also discussed in Feinberg (1994) and Boussemart and Limnios (2004), as well as by Beutler and Ross (1986), Boussemart *et al.* (2001) and Boussemart and Limnios (2004). Feinberg (1994) presented the Markov decision processes with a constraint on the average asymptotic failure rate. We should mention that the extended abstract of the discussed problem has been published in the AIP Conference Proceedings (1969).

Chapter written by Franciszek GRABSKI.

The gain maximization problem subject to an availability constraint for the SM model of the operation in reliability aspect is also discussed in those papers. The problem is transformed into a linear programing maximization problem.

9.2. Semi-Markov decision process

The semi-Markov (SM) decision process is an SM process with a finite states space $S = \{1,2, \dots N\}$, such that its trajectory depends on decisions that are made at an initial instant and at the moments of the state changes. We assume that a set of decisions in each state i, denoted by D_i, is finite. To make a decision $k = \delta_i \in D_i$ means to select k-th row among the alternating rows of the SM kernels $\left\{Q_{ij}^{(k)}(t): \ t \geq 0, \ k \in D_i, \ i,j \in S \ \right\}$, where $Q_{ij}^{(k)}(t) = p_{ij}^{(k)} F_{ij}^{(k)}(t)$. A number $p_{ij}^{(k)}$ denotes a transition probability of the so-called *embedded Markov chain* if a decision $k = \delta_i \in D_i$ is made. We assume that the strategy has the Markov property. It means that for every state $i \in S$, a decision $\delta_i(n)$ does not depend on the process evolution until the moment $\tau_n^{(k)}$. If $\delta_i(n) = \delta_i = const$, then it is called a *stationary decision*. This means that the decision does not depend on n. The policy consisting of the stationary decisions is called a *stationary policy*. Hence, a stationary policy is defined by the sequence $\delta = (\delta_1, \delta_2, \dots, \delta_N)$. The strategy that is a sequence of stationary policies is called a *stationary strategy*.

To formulate an optimization problem, we have to introduce the reward structure for the process. We assume that an object, which occupies the state i when a successor state is j, earns a gain at a rate $r_{ij}^{(k)}(x), i,j \ \in S, \ k \in D_i$ and at a moment x of the entering state i for decision $k \in D_i$. A value of a function $R_{ij}^{(k)}(t) = \int_0^t r_{ij}^{(k)}(x)\, dx, \ i,j \ \in S, \ k \in D_i$ denotes the profit that the object bears by spending time t in state i before making a transition into the state j, for the decision $k \in D_i$. We suppose that $r_{ij}^{(k)}(x) = r_{ij}^{(k)}$. Here, we also suppose that $r_{ij}^{(k)} = r_i^{(k)}$ and $m_{ij}^{(k)} = m_i^{(k)}, \ i,j \ \in S, \ k \in D_i$. Under these assumptions, we obtain an expected value of the gain that is generated by the process in the state i in a certain unit of time, at one interval of its realization, for the decision k:

$$u_i^{(k)} = \sum_{j \in S} \int_0^\infty r_{ij}^{(k)} t \, d\, Q_{ij}^{(k)}(t) = m_i^{(k)} r_i^{(k)} \sum_{j \in S} p_{ij}^{(k)} = m_i^{(k)} r_i^{(k)}. \qquad [9.1]$$

Recall that

$$m_i^{(k)} = E\left(T_i^{(k)}\right) = \lim_{t \to \infty} \int_0^\infty t \, dG_i^{(k)}(t), \qquad i \in S, \ k \in D_i. \qquad [9.2]$$

9.3. Semi-Markov decision model of operation

9.3.1. *Description and assumptions*

The object (device) works by performing two types of tasks: 1 and 2. The duration of task r is a non-negative random variable $\xi_r,\ r = 1{,}2$ governed by a CDF $F_{\xi_r}(x),\ x \geq 0,\ r = 1{,}2$. The working object may be damaged. The time to failure of the object executing a task r is also a non-negative random variable $\zeta_r,\ r = 1{,}2$ with a probability density function $f_{\zeta_r}(x),\ x \geq 0,\ r = 1{,}2$. A repair time of the object performing task r is a non-negative random variable $\eta_r,\ r = 1{,}2$ governed by a probability density function $f_{\eta_r}(x),\ x \geq 0,\ r = 1{,}2$. Each repair renews the object. This is a strong assumption, but in the case of sudden damage, the expected repair time is a number that is independent of the shape of the lifetime distribution. After completing the task, the inspection and renewal takes place. A duration after task r is a non-negative random variable γ_r, determined by a PDF $f_{\gamma_r}(x),\ x \geq 0,\ r = 1{,}2$. After the inspection is completed, the object starts the execution of task 1 with the probability p_1 or task 2 with the probability $q_1 = 1 - p_1$. Furthermore, we assume that all random variables and their copies are mutually independent, and have finite and positive second moments.

9.3.2. *Model construction*

We start with the model construction by introducing the operation process states:

1) check the object technical condition and renewal after executing task 1;

2) check the object technical condition and renewal after executing task 2;

3) object operation – perform task 1;

4) object operation – perform task 2;

5) repair after failure during the execution of task 1;

6) repair after failure during the execution of task 2.

To construct a decision stochastic process, we need to determine the sets of decision (alternatives) for every state.

D_1: 1) normal inspection after performing task 1;

2) expensive inspection after performing task 1;

3) cheap inspection after performing task 1.

D_2: 1) normal inspection after performing task 2;

2) expensive inspection after performing task 2;

3) cheap inspection after performing task 2.

D_3: 1) normal profit per unit of time for executing task 1;

2) higher profit for executing task 1.

D_4: 1) normal profit per unit of time for executing task 1;

2) higher profit for executing task 2.

D_5: 1) normal repair after failure during the execution of task 1;

2) expensive repair after failure during the execution of task 1.

D_6: 1) normal repair after failure during the execution of task 2;

2) expensive repair after failure during the execution of task 2.

Possible state changes of the process are shown in Figure 9.1.

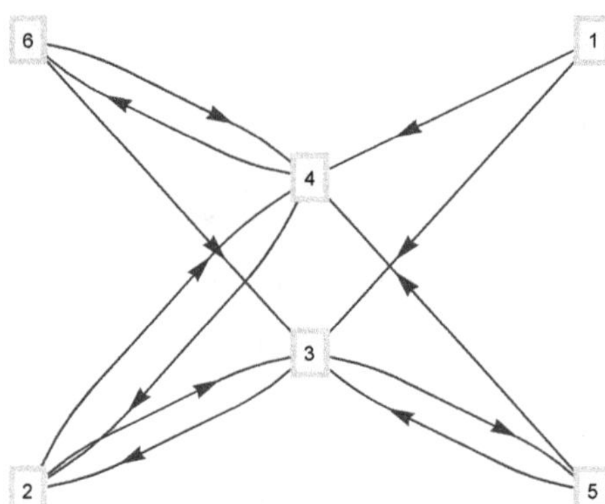

Figure 9.1. *Possible state changes of the process in the semi-Markov model of operation. For a color version of this figure, see www.iste.co.uk/dimotikalis/analysis1.zip*

9.4. Optimization problem

The gain optimization problem under the availability limitation for the infinite duration SM reliability model is considered here. The problem is transformed into a linear programing problem.

A number $u_i^{(k)}$ means an expected value of the gain that is generated by the process in the state i at one interval of its realization for the decision $k \in D_i$. Assume that the considered SM decision process with a state space $S = \{1,\dots, N\}$ satisfies the assumptions of the limiting theorem (Gertsbakh 1969; Mine and Osaki 1970; Howard 1971; Korolyuk and Turbin 1976; Silvestrov 1980; Grabski 2015).

The criterion function

$$\boldsymbol{g}(\delta) = \frac{\sum_{j\in S}\pi_j(\delta)u_j^{(k)}}{\sum_{j\in S}\pi_j(\delta)m_j^{(k)}} = \frac{\sum_{j\in S}\pi_j(\delta)m_j^{(k)}r_j^{(k)}}{\sum_{j\in S}\pi_j(\delta)m_j^{(k)}} \qquad [9.3]$$

denotes the gain per unit of time as a result of a long operating system. The numbers $\pi_i(\delta)$ and $i \in S$ represent the stationary distribution depending on the stationary strategy of the embedded Markov chain of the SM process defined by the kernel $Q^{(\delta)}(t) = \left[Q_{ij}^{(k)}(t):\ t \geq 0,\ \ i,j \in S,\ \ k \in D_i\right]$. Recall that the stationary distribution of the embedded Markov chain means that probabilities $\pi_i(\delta),\ i \in S$ satisfy the system of linear equations:

$$\sum_{i\in S} \pi_i(\delta)p_{ij}^{(k)} = \pi_j(\delta), \qquad \sum_{i\in S} \pi_i(\delta) = 1,\ \pi_j(\delta) > 0,\ \ j \in S, \qquad [9.4]$$

A number $m_i^{(k)}$ denotes an expectation value of the waiting time in state i if the decision k is chosen.

The function

$$\boldsymbol{k}(\delta) = \frac{\sum_{j\in S_+}\pi_j(\delta)m_j^{(k)}}{\sum_{j\in S}\pi_j(\delta)m_j^{(k)}} \qquad [9.5]$$

denotes the limiting availability.

We wish to find a strategy δ that maximizes the gain $\boldsymbol{g}(\delta)$ subject to the availability constraint $k(\delta) > \alpha$ for $0 < \alpha \leq 1$.

9.4.1. *Linear programing method*

Mine and Osaki (1963) used the linear programing method to solve the problem of optimization without any additional constraints. In this chapter, the problem of optimization with the availability constraint of a system is considered.

The stationary probabilities $\pi_j(\delta),\ j \in S,$ for every decision $k \in D_i$ satisfy the following linear system of equations:

$$\sum_{i\in S} \pi_i(\delta) p_{ij}^{(k)} = \pi_j(\delta), \quad \sum_{i\in S} \pi_i(\delta) = 1,\ \pi_j(\delta) > 0, \quad j \in S, \qquad [9.6]$$

where

$$p_{ij}^{(k)} = \lim_{t\to\infty} Q_{ij}^{(k)}(t), \quad i,j \in S.$$

A number $a_j^{(k)}$ denotes the probability of a decision k in state j. It is obvious that

$$\sum_{k\in D_j} a_j^{(k)} = 1, \quad 0 \le a_j^{(k)} \le 1,\ j \in S. \qquad [9.7]$$

The criterion function [9.1] and the constraint can be rewritten as

$$\boldsymbol{g}(\delta) = \frac{\sum_{j\in S}\sum_{k\in D_j} a_j^{(k)} \pi_j(\delta) m_j^{(k)} r_j^{(k)}}{\sum_{j\in S}\sum_{k\in D_j} a_j^{(k)} \pi_j(\delta) m_j^{(k)}},\ \boldsymbol{k}(\delta) = \frac{\sum_{j\in S_+}\sum_{k\in D_j} a_j^{(k)} \pi_j(\delta) m_j^{(k)}}{\sum_{j\in S}\sum_{k\in D_j} a_j^{(k)} \pi_j(\delta) m_j^{(k)}} > \alpha \qquad [9.8]$$

Substituting $a_j^{(k)} \ge 0,$ for $j \in S$ and $k \in D_j,$ and taking advantage that $\pi_j(d) = \sum_{k\in D_i}^{(k)} a_{(j),}^{k}$ we obtain [9.9]

$$\sum_{i\in S} \sum_{k\in D_i} a_i^{(k)} \pi_i(\delta) p_{ij}^{(k)} = \pi_j(\delta),\ \sum_{i\in S} \pi_i(\delta) = 1,\ \pi_j(\delta) > 0,\ j \in S \qquad [9.9]$$

Substituting

$$a_i^{(k)} \pi_i(\delta) = x_j^{(k)} \ge 0, \quad j \in S,\ k \in D_j \qquad [9.10]$$

and taking into account that

$$\sum_{k\in D_i} a_i^{(k)} \pi_i(\delta) = \pi_i(\delta) = \sum_{k\in D_i} x_j^{(k)} \ge 0,\ j \in S,\ k \in D_j. \qquad [9.11]$$

From [9.7], [9.8] and [9.9], we obtain the following optimization problem: find a stationary strategy δ by maximizing the function

$$\mathbf{g}(\delta) = \frac{\sum_{i\in S}\sum_{k\in D_i} u_i^{(k)} x_i^{(k)}}{\sum_{i\in S}\sum_{k\in D_i} m_i^{(k)} x_i^{(k)}} \qquad [9.12]$$

under constraints

$$\frac{\sum_{j\in S_+}\sum_{k\in D_j} m_j^{(k)} x_j^{(k)}}{\sum_{j\in S}\sum_{k\in D_j} m_j^{(k)} x_j^{(k)}} > \alpha \qquad [9.13]$$

$$\sum_{k\in D_j} x_j^{(k)} - \sum_{i\in S}\sum_{k\in D_i} p_{ij}^{(k)} x_i^{(k)} = 0,\ x_j^{(k)} \geq 0,\ j \in S,\ k \in D_j \qquad [9.14]$$

$$\sum_{j\in S}\sum_{k\in D_j} x_j^{(k)} = 1. \qquad [9.15]$$

Note that $\sum_{i\in S}\sum_{k\in D_i} m_i^{(k)} x_i^{(k)} > 0$.

We introduce some new variables

$$y_j^{(k)} = \frac{x_j^{(k)}}{\sum_{i\in S}\sum_{k\in D_i} m_i^{(k)} x_i^{(k)}},\quad y = \frac{1}{\sum_{i\in S}\sum_{k\in D_i} m_i^{(k)} x_i^{(k)}} \qquad [9.16]$$

We now obtain the following problem of linear programming: Find the stationary strategy δ by maximizing the function

$$\mathbf{g}(\delta) = \sum_{i\in S}\sum_{k\in D_i} u_i^{(k)} y_i^{(k)} \qquad [9.17]$$

under the constraints

$$\sum_{j\in S_+}\sum_{k\in D_j} m_j^{(k)} y_j^{(k)} \geq \alpha, \qquad [9.18]$$

$$(\text{or } \sum_{j\in S_-}\sum_{k\in D_j} m_j^{(k)} y_j^{(k)} \leq 1-\alpha,\ S_- = S - S_+) \qquad [9.19]$$

$$\sum_{k\in D_j} y_j^{(k)} - \sum_{i\in S}\sum_{k\in D_i} p_{ij}^{(k)} y_i^{(k)} = 0, \qquad [9.20]$$

$$\sum_{j\in S}\sum_{k\in D_j} m_j^{(k)} y_j^{(k)} = 1, \qquad y_j^{(k)} \geq 0,\ j \in S,\ k \in D_j \qquad [9.21]$$

$$\sum_{j\in S}\sum_{k\in D_j} y_j^{(k)} = y. \qquad [9.22]$$

Finally, we obtain the linear programing problem defined by [9.17]–[9.22].

From [9.14], [9.20] and [9.22], we have

$$x_j^{(k)} = y\, y_j^{(k)} \quad \text{for all } j \in S \text{ and } k \in D_j. \qquad [9.23]$$

The probabilities

$$a_j^{(k)} = \frac{x_j^{(k)}}{\sum_{k \in D_j} x_j^{(k)}} = \frac{y\, y_j^{(k)}}{\sum_{k \in D_j} y\, y_j^{(k)}} = \frac{y_j^{(k)}}{\sum_{k \in D_j} y_j^{(k)}}, \qquad [9.24]$$

are independent of y, which **determine the optimal stationary strategy.**

A model of an object operation is the SM decision process (SMDP) with the state space $S = \{1,2,3,4,5,6\}$, sets of actions (decisions) D_1, D_2, D_3, D_4, D_5, D_6 and the family of kernels $Q^{\delta}(t)$. The model is constructed if all the elements of the kernel are determined. The transition probability matrix of the embedded Markov chain $\{X(\tau_n): n \in \mathbb{N}_0\}$ has the following form:

$$P^{\delta} = \begin{bmatrix} 0 & 0 & p_{13}^{(k)} & p_{14}^{(k)} & 0 & 0 \\ 0 & 0 & p_{23}^{(k)} & p_{24}^{(k)} & 0 & 0 \\ p_{31}^{(k)} & & 0 & 0 & p_{35}^{(k)} & 0 \\ 0 & p_{42}^{(k)} & 0 & 0 & 0 & p_{46}^{(k)} \\ 0 & 0 & p_{53}^{(k)} & p_{54}^{(k)} & 0 & 0 \\ 0 & 0 & p_{63}^{(k)} & p_{64}^{(k)} & 0 & 0 \end{bmatrix}. \qquad [9.25]$$

where

$$p_{53}^{(k)} = p_5, \quad p_{54}^{(k)} = q_5 = 1 - p_5, \quad p_{63}^{(k)} = p_6, \quad p_{64}^{(k)} = q_6 = 1 - p_6.$$

In this model, the set of the "up" states is $S_+ = \{1, 2, 3, 4\}$, while the set of the "down" states is $S_- = S - S_+ = \{5,6\}$.

9.5. Numerical example

Let us consider the decision variables

$$y_1^{(1)},\ y_1^{(2)},\ y_1^{(3)},\ y_2^{(1)},\ y_2^{(2)},\ y_2^{(3)},\ y_3^{(1)},\ y_3^{(2)},\ y_4^{(1)},\ y_4^{(2)},\ y_5^{(1)},\ y_5^{(2)},\ y_6^{(1)},\ y_6^{(2)}$$

and some known parameters

$$u_1^{(1)},\ u_1^{(2)},\ u_1^{(3)},\ u_2^{(1)},\ u_2^{(2)},\ u_2^{(3)},\ u_{3,}^{(1)}\ u_{3,}^{(2)}\ u_4^{(1)},\ u_4^{(2)},\ u_5^{(1)},\ u_5^{(2)},\ u_6^{(1)},\ u_6^{(2)},$$

$m_1^{(1)}$, $m_1^{(2)}$, $m_1^{(3)}$, $m_2^{(1)}$, $m_2^{(2)}$, $m_2^{(3)}$, $m_3^{(1)}$, $m_3^{(2)}$, $m_4^{(1)}$, $m_4^{(2)}$, $m_5^{(1)}$, $m_5^{(2)}$, $m_6^{(1)}$, $m_6^{(2)}$.

State i	*Decision k*	$p_{i1}^{(k)}$	$p_{i2}^{(k)}$	$p_{i3}^{(k)}$	$p_{i4}^{(k)}$	$p_{i5}^{(k)}$	$p_{i6}^{(k)}$	$m_i^{(k)}$	$r_i^{(k)}$	$u_i^{(k)}$
1	1	0	0	0.64	0.36	0	0	2	-150	-300
	2	0	0	0.45	0.55	0	0	2.5	-160	-400
	3	0	0	0.72	0.28	0	0	2	-120	**-240**
2	1	0	0	0.24	0.76	0	0	2.5	-120	-360
	2	0	0	0.65	0. 35	0	0	3	-150	**-450**
	3	0	0	0.40	0.60	0	0	2.5	-104	**-260**
3	1	0.99	0	0	0	0.01	0	10.2	260	**2660**
	2	0.98	0	0	0	0.02	0	9.4	380	3572
4	1	0	0.98	0	0	0	0.02	9.2	350	**3220**
	2	0	0.96	0	0	0	0.04	8.8	460	4048
5	1	0	0	0.80	0.20	0	0	72	-120	**-8640**
	2	0	0	0.38	0.62	0	0	66	-135	-8910
6	1	0	0	0.58	0.42	0	0	80	-122	-9760
	2	0	0	0.66	0.34	0	0	70	-140	-9800

Table 9.1. *Transition probabilities and gain parameters of the SMDP*

The question that arises is as follows: is it possible to specify the alpha parameter arbitrarily? Computer simulations show that this is impossible. There is the greatest availability parameter for which the problem has no solution. In this case

$$\boldsymbol{\alpha} = 0.929.$$

Using a Mathematica computer program, we obtain the solution of the problem:

$$y_1^{(1)} = 0, \quad y_1^{(2)} = 0, \quad y_1^{(3)} = 0.0524145,$$

$$y_2^{(1)} = 0, \quad y_2^{(2)} = 0.021061, \quad y_2^{(3)} = 0.0019544,$$

$$y_3^{(1)} = 0.0529439, \quad y_3^{(2)} = 0, \quad y_4^{(1)} = 0.023486, \quad y_4^{(2)} = 0,$$

$$y_5^{(1)} = 0.000529, \quad y_5^{(2)} = 0, \quad y_6^{(1)} = 0, \quad y_6^{(2)} = 0.000469.$$

From [9.24], we obtain the probabilities

$$a_1^{(1)} = 0, \quad a_1^{(2)} = 0, \quad a_1^{(3)} = 1,$$

$$a_2^{(1)} = 0, \quad a_2^{(2)} = 0{,}91509 \quad a_2^{(3)} = 0{,}08490$$

$$a_3^{(1)} = 1, \quad a_3^{(2)} = 0, \quad a_4^{(1)} = 1,$$

$$a_4^{(2)} = 0, \quad a_5^{(1)} = 1 \; a_5^{(2)} = 0,$$

$$a_6^{(1)} = 0, \quad a_6^{(2)} = 1$$

The vector of the optimal action in each step has the form $(3,\ \gamma,\ 1, 1,1, 2)$, where

$$\gamma = \begin{cases} 2 & \text{with prob. } 0.915099 \\ 3 & \text{with prob. } 0.084901 \end{cases}$$

In this case, the maximum expected gain from the unit of time at one step of the operation is:

$$g = -240 * 0.0524145 - 450 * 0.021061 - 260 * 0.0019544 +$$

$$2660 * 0.0529439 +$$

$$+ 3220 * 0.023486 - 8640 * 0.000529 - 9800 * 0.000469 \cong 184.732$$

For the availability parameter $\alpha \geq 0.930$, there is no solution subject to the above constraints.

9.6. Conclusion

– The SM decision processes theory offers us the possibility of formulating and solving the optimization problems that can be modeled by SM processes. In this type of problems, we choose the process that brings us the largest profit.

– If the SM process describing the evolution of the real system over a long period of time satisfies the assumptions of the limit theorem for the SM process, we can use the results of the infinite duration SM decision processes theory.

– The algorithm that allows the best strategy to be determined is equivalent to a linear programing problem.

– The gain optimization problem subject to the availability constraint for the SM model of operation is considered and solved.

– Using Theorem 5.5 (Korolyuk and Turbin 1976) for the problem without additional constraints, we came to the conclusion that for every $j \in S$, only one $k \in D_j$ exists such that $y_j^{(k)} > 0$.

– This theorem is not true for the gain optimization problem subject to the constraint of availability. The optimal stationary strategy may contain the vectors with mixed decisions. This fact extends the previously known results.

– "The new state will contain a mixture of systems from different ages, depending on how each system returned to this state." However, this would require the use of one of the methods of approximation of the density of the waiting time distribution, for example the approximation using splines or the estimator with the Gaussian kernel (Grabski 2015). Finally, the results will be almost identical.

9.7. References

Andrzejczak, K. and Selech, J. (2020). An aggregate criterion for selecting a distribution for times to failure of components of rail vehicles. *Maintenance and Reliability*. 22(1), 102–111.

Bernaciak, K. (2005). Multicriterial optimization of semi-Markov processes with discount. *Advances in Safety and Reliability*, 171–178.

Beutler, and Ross, K. (1986). Time-average optimal constrained semi-Markov decision processes. *Advances in Applied Probability*, 18, 341–359.

Boussemart, M. and Limnios, N. (2004). Markov decision processes with a constraint on the average asymptotic failure rate. *Communication in Statistics – Theory and Methods*, 33(7), 1689–1714.

Boussemart, M., Bicard, T., Limnios, N. (2001). Markov decision processes with a constraint on the average asymptotic failure rate. *Methodology and Computing in Applied Probability*. 3(2), 199–214.

Feinberg, E. (1994). Constrained semi-Markov decision processes with average rewards, *ZOR – Mathematical Methods of Operations Research*, 257–288.

Gertsbakh, I.B. (1969). *Models of Preventive Service* (in Russian). Sovetskoe Radio, Moscow.

Grabski, F. (2015). *Semi-Markov Processes: Applications in Systems Reliability and Maintenance*. Elsevier, Amsterdam.

Grabski, F. (2018a). Optimization problem subject to constraint of availability in semi-Markov models of operation. *AIP Conference Proceedings 2116*, 450091-1–450091-7.

Grabski, F. (2018b). Optimization problem of reliability subject to constraint of availability in semi-markov models of operation. In *AMSDA Conference Proceedings*. Data analysis and applications 4: Financial data analysis and methods; Volume 6; Big Data, Artificial Intelligence and Mata Analysis SET, Makrides, A., Karagrigoriou, A., Skidas, C.H. (eds). ISTE Ltd, London, and John Wiley & Sons, New York.

Howard, R.A. (1960). *Dynamic Programing and Markov Processes*. MIT Press, Cambridge, Massachusetts.

Howard, R.A. (1964). Research of semi-Markovian decision structures. *Journal of Operations Research Society*, Japan, 6, 163–199.

Howard, R.A. (1971). Dynamic probabilistic system. *Semi-Markov and Decision Processes*. Vol II, Wiley, New York.

Jewell, W.S. (1963). Markov-renewal programming. *Operation Research*, 11, 938–971.

Korolyuk, V.S. and Turbin, A.F. (1976). *Semi-Markov Processes and their Applications* (in Russian). Naukova Dumka, Kiev.

Mine, H. and Osaki, S. (1970). *Markovian Decision Processes*. AEPCI, New York.

Silvestrov, D.S. (1980). *Semi-Markov Processes with a Discrete State Space* (in Russian). Sovetskoe Radio, Moscow.

10

The Impact of Multicollinearity on Big Data Multivariate Analysis Modeling

The purpose of this work is to address, discuss and attempt to resolve some of the issues that appear in the presence of multicollinearity, such as the overfitting in regression analysis, the accuracy of the impact of the parameter estimation in the dependent variable and the inconsistent results during the analysis of variance, in order to properly model the public pension expenditures (PPE). For this purpose, we proceed to locate, collect and analyze the factors which may have an impact on the shaping of the PPE in the short term or long term. The analysis focuses on 20 European countries for which a large amount of data is available, including a set of 20 possible explanatory variables for the period 2001–2015.

10.1. Introduction

The tremendous increase in the development of technology, as well as the creation of new databases on a variety of topics, makes Big Data Analytics more efficient to work with. However, more is not always better. Large amounts of data may sometimes fail to perform properly in data analytic applications. Indeed, when it comes to modeling, a multitude of explanatory variables for extensive time periods can cause inconsistencies in the interpretation of statistical results. The most important obstacle that we have to overcome is the existence of multicollinearity between the covariates. To deal with this, among other issues, special techniques called dimension reduction techniques, such as principal component analysis (PCA), as well as its predecessor, Beale *et al.* (1967), will be used separately and in combination with data standardization.

Chapter written by Kimon NTOTSIS and Alex KARAGRIGORIOU.

In order to address, discuss and attempt to resolve possible issues that the existence of multicollinearity can cause, this work tries to properly model PPE. The analysis focuses on 20 European countries for which a large amount of data is available in Knoema and in the Organisation for Economic Co-operation and Development (OECD), including a set of 20 possible explanatory variables (covariates, factors) for the period 2001–2015.

OECD states that public pension expenditures (PPE) are "all cash expenditures (including lump-sum payments) on old-age and survivors pensions. Old-age cash benefits provide an income for persons retired from the labour market, or guarantee income when a person has reached a standard pensionable age or fulfilled the necessary contributory requirements. This category also includes early retirement pensions: pensions paid before the beneficiary has reached the standard pensionable age relevant to the programme. It excludes programmes concerning early retirement for labour market reasons. Old-age pensions includes supplements for dependants paid to old-age pensioners with dependants under old-age cash benefits. Old age also includes social expenditures on services for the elderly people, services such as day care and rehabilitation services, home-help services and other benefits in kind. It also includes expenditures on the provision of residential care in an institution. This indicator is measured by the percentage of gross domestic product (GDP) broken down by public and private sector".

Most of the studies for the modeling of PPE focus on a single country or a few factors of importance. de la Fuente (2011) analyzes the pension system of Spain as a function of workers' Social Security contribution histories, while Pereira *et al.* (2010) study and analyze the macroeconomic effects of public pension reforms. Marcinkiewicz and Chybalski (2014) discuss pension expenditures as one of the main indicators of pension system sustainability, propose a model based on GDP and old-age dependency ratio and apply the resulting model to countries with very different population structures. The same authors, later Marcinkiewicz and Chybalski (2016), suggest a new typology of pension regimes between OECD countries. The interested reader can see Bonoli (2003), Bonoli and Shinkawa (2005), Franco *et al.* (2006) and Lachowska and Myck (2018) for additional information and results concerning expenditures.

This work relies on multivariate dimension reduction techniques including PCA, Beale *et al.* (1967) and linear models (McCullagh and Nelder 1989) for the modeling of PPE, by identifying the appropriate set of covariates that affect the PPE. For relevant approaches, see Barr and Diamond (2006) and Farrell and Shoag (2017).

10.2. Multicollinearity

Nowadays, researchers involved in statistical modeling are able to collect data on multiple factors for a long time horizon, which may contain millions of observations. Trying to model a variable with such a large dataset can cause additional problems (such as inaccurate and disorderly databases, computational complexity and insufficient analytical skills) as opposed to a modeling with smaller datasets. Due to such problems, there was a need for the creation of a particular category of data analysis, named Big Data Analysis, which is a rapidly growing branch of data science. "Big data analysis is the use of advanced analytic techniques against very large, diverse datasets that include structured, semi-structured and unstructured data in different sizes. Big data is a term applied to datasets whose size or type is beyond the ability of traditional relational databases to capture, manage, and process the data with low-latency. And it has one or more of the following characteristics – high volume, high velocity, or high variety" (IBM N/A).

The most common consequence of such a big dataset modeling in multivariate analysis is the existence of multicollinearity, which is usually caused by the correlation among the predictor variables. It is a phenomenon which occurs when high intercorrelation among the covariates exists, which means that one or more covariates can be linearly predicted from other covariates with a notable degree of accuracy. Not so often, it can be caused by the false use of dummy variables in the modeling, the repetition of the same kind of variable, or when the data is insufficient, which can be resolved by collecting more data.

In the presence of multicollinearity, the proper interpretation of the data analysis may not be reliable. Due to this fact, multicollinearity has reached the point of being an additional "fifth" assumption in the case of multivariate regression, among the already existing assumptions – normality, homoscedasticity and independence of the residuals and linearity between the dependent and each independent variable. There are two types of multicollinearity, namely structural and data-based multicollinearity.

Structural multicollinearity, also known as perfect multicollinearity, occurs when a byproduct variable exists in the modeling along with the original variable. In other words, it is a mathematical artifact caused by generating predictors with the use of already existing ones. Due to the fact that the perfect collinear relationship between the variables included in the model exists, the researcher is unable to use the ordinary least squares (OLS) regression to estimate the value of the parameters. Therefore, perfect multicollinearity violates one of the linear regression model assumptions.

Data-based multicollinearity, also known as high multicollinearity, occurs between the variables in the original unprocessed dataset and is the most common type when it comes to observational experiments.

Perfect multicollinearity is highly uncommon and is the most easy to handle and avoid by a thorough examination of the covariates of the model. However, high multicollinearity is the most common and can cause severe estimation and interpretation problems.

The most common consequence which appears in the presence of multicollinearity is the overfitting in regression analysis modeling, due to the redundancy of variables, which reduces the power of the model in order to identify the statistically significant factors. This means that the model is too complex and the model's measures like the coefficient of determination (R-squared) are misleading, because instead of describing the proportion of the variance in the dependent variable that is predictable from the independent variables, they describe the random error in the data. It is also possible that parameter estimates may not accurately describe the impact of the associated covariates to the dependent variable. It can also result in the alteration of the indicator and the immensity of the partial regression coefficients from one sample to another.

Furthermore, although that phenomenon seems quite paradoxical, reports have been made for non-identical results between an F-test and T-test when multicollinearity exists. This phenomenon has been thoroughly examined and explained by Geary and Leser (1968), as they provide two reasons for the occurrence. The first one is the existence of multicollinearity in the model, in which the existence of a relationship can be established, but not the individual influence of each factor. The second reason stems from the value of the degrees of freedom of the residuals (DF). If the residuals DF are $\geqslant 3$, then the significance point of $F_{(k,\ n-k-1)}$ is lower than the significance point of $F_{(1,\ n-k-1)}$, which corresponds to the significance point of the t-statistics. Hence, when all t-statistics are equal or approximately so, they may all be non-significant while F is significant. The explanation is that a significant F-ratio does not indicate the significance of any given regression coefficient, but merely the existence of at least one linear combination which is significantly different from zero (Largey and Spencer 1996).

There are several ways to detect the existing multicollinearity. The most commonly used is the variance inflation factor (VIF). It expresses the rate at which the variance of the estimator increases when collinearity exists. A concrete way of estimating the proportion of multicollinearity with the use of VIF does not exist; however, several empirical rules exist as cut-offs. VIF_j of the $j-th$ factor is given by:

$$\mathrm{VIF_j} = \frac{1}{1 - \mathrm{R_j^2}}$$

where R_j^2 is the coefficient of determination of a regression of factor j on all other explanatory variables. One of the most common empirical rules states that if VIF_j exceeds 5 and the cut-off often used, then multicollinearity is considered high (Sheather 2009). (*Note that 10 is also a frequently used threshold for evaluation.*)

Another way to locate multicollinearity is with the use of the condition index (CI) (Belsley 1991), which is the multicollinearity of combinations of variables in the dataset (Hair *et al.* 2010), as well as the regression coefficient variance-decomposition matrix. The following empirical rule is commonly applied in the CI interpretation:

– if CI <10, multicollinearity is weak;

– if 10 $\leq$ CI $\leq$ 100, multicollinearity is moderate to strong;

– if CI >100, multicollinearity is severe.

Other ways to detect multicollinearity include (see WMD) the following:

– a simple correlation matrix can be used as the first index for multicollinearity detection. Relative high correlations indicate the possible existence of multicollinearity;

– in the presence of multicollinearity, it is possible for a variable to be statistically significant when it comes to simple regression between this particular variable and the dependent one, while in a multiple regression it is possible for the same variable to be statistically insignificant;

– the subjoin or subtraction of covariates that cause major shifts in the estimated regression coefficients are indicators for the existence of multicollinearity.

There are several ways to deal with the existing multicollinearity. The most common solution when perfect multicollinearity exists is to remove the byproduct explanatory variables, and in the case of high multicollinearity, remove the highly correlated factors from the model. Stepwise regression can be used for the identification of less significant variables among the highly correlated ones. The use of partial least squares (PLS) regression, PCA, or ridge regression, as well as methods that reduce the number of predictors to a smaller set of uncorrelated components, can also be considered.

10.3. Dimension reduction techniques

As mentioned above, large amounts of data may sometimes fail to properly perform in data analytics applications and can cause inconsistencies in the interpretation of the results. In order to overcome this problem, which most likely comes as a result of the existence of multicollinearity, various dimension reduction techniques were developed. Dimension reduction is the process of reducing the number of random variables under consideration by creating a new smaller set of variables based on the original one. Through this process, it is easier to interpret different statistical tests without losing the accuracy of the original variables in the sense that these techniques are intended to retain the unchanged variation as much as possible. In this work, we make use of PCA, as well as its predecessor, the Beale *et al.* (1967) dimension reduction technique, which are well established, easily implemented and highly applicable in various scientific fields.

10.3.1. *Beale et al.*

This technique is a very simple three-step procedure proposed by Beale *et al.* (1967) for discarding variables in multivariate analysis. Their technique can be summed up as follows:

– locate the minimum eigenvalue and the corresponding eigenvector of the variance–covariance or correlation matrix;

– locate the element of the eigenvector with the highest absolute value. This value corresponds to a variable which will be removed from the model;

– repeat the above steps until *p-k* variables have been removed.

Note that p is the number of all variables and k is the number of eigenvalues which are larger than one.

10.3.2. *Principal component analysis*

The PCA technique was proposed independently by Pearson (1901) and Hotelling (1933, 1936). The idea behind the PCA (Artemiou and Li 1557, 2013; Jolliffe 1972; Smallman *et al.* 2018) is the conversion of a dataset with interdependent variables into a new one with uncorrelated variables (principal components), which will be arranged in such a way so that the first variables maintain the greater part of the variance that exists in all of the original variables. With this procedure, the reduction of the dimension of the original dataset is achieved while leaving as much of the variation as possible unchanged (Jolliffe 2002).

PCA relies on the covariance or correlation matrix of the original dataset in order to obtain the eigenvalues and the eigenvectors which are essential in this procedure. The covariance matrix is used when all variables have the same measure, while the correlation matrix is used when a different unit of measurement occurs. The analysis consists of the following:

10.3.2.1. *Covariance matrix*

The *nxn* matrix for n variables, $X_1, X_2, ..., X_n$, is typically presented as follows

$$\Sigma = \begin{bmatrix} \mathrm{E}[(X_1 - \mathrm{E}(X_1))(X_1 - \mathrm{E}(X_1))] & \cdots & \mathrm{E}[(X_1 - \mathrm{E}(X_1))(X_n - \mathrm{E}(X_n))] \\ \mathrm{E}[(X_2 - \mathrm{E}(X_2))(X_1 - \mathrm{E}(X_1))] & \cdots & \mathrm{E}[(X_2 - \mathrm{E}(X_2))(X_n - \mathrm{E}(X_n))] \\ \vdots & \ddots & \vdots \\ \mathrm{E}[(X_n - \mathrm{E}(X_n))(X_1 - \mathrm{E}(X_1))] & \cdots & \mathrm{E}[(X_n - \mathrm{E}(X_n))(X_n - \mathrm{E}(X_n))] \end{bmatrix}$$

10.3.2.2. *Correlation matrix*

The correlation matrix is a statistical tool from which we can distinguish the strength of the relationship, if it exists, between any two variables involved in the matrix. The correlation between two variables is called the correlation coefficient and ranges from between -1 and 1 CRS (2018).

The correlation matrix R associated with the matrix Σ is obtained by dividing the (i, j) element of Σ with the term $\text{VarX}_i\text{VarX}_j$.

10.3.2.3. *Extracted eigenvalues – eigenvectors*

In order to determine the number of components in PCA, several techniques have been proposed, all of which rely on the evaluation of the eigenvalues and the corresponding eigenvectors of either Σ or R. The most popular ones are:

– ***Kaiser's rule***

This technique keeps the components with eigenvalues greater than one.

– ***Scree plot***

Catell's technique (Cattell 1966), also known as a scree test. According to this technique, a plot with the eigenvalues sorted from maximum to minimum is created. At some point, the curve connecting the values will create an angle, like an elbow, after which the curve will remain almost straight and parallel to the x-axis. The number of components selected is defined by the number of values that appear up to that elbow.

– ***Proportion of variance explained***

It is sometimes thought that a good factor analysis should explain two-thirds of the variance (King and Jackson 1999).

10.3.2.4. *Component matrix*

The components are a set of uncorrelated vectors that have been created by the following methodology:

Let us denote by C_j the jth component, λ_j the corresponding eigenvalue and v_{ij} the corresponding eigenvector, $i = 1, 2, ..., n$, $j = 1, 2, ..., m$ where n is the number of observations and m is the total number of original variables/covariates. Hence, C_j is defined as:

$$C_j = -\sqrt{\lambda_j}\, v_{ij} = -\sqrt{\lambda_j} \begin{bmatrix} v_{1j} \\ v_{2j} \\ \vdots \\ v_{nj} \end{bmatrix},$$

$\forall\, i = 1, ..., n, \;\; j = 1, ..., m,$

Each new variable $Z_j,\ j = 1, ..., m$ is a linear combination of the components matrix and the original dataset matrix. Indeed, let us denote by

X_{ij} : the elements of the original dataset matrix,
C_{jj} : the elements of the component matrix and
Z_{ij} : the elements of the new variable matrix.

Then, the Z_{ij} element of the vector $Z_j = (Z_{1j}, ..., Z_{mj})'$ is:

$$Z_{ij} = \sum_{j=1}^{m} X_{ij} C_{jj} \qquad \forall\, i = 1, ..., n, j = 1, ..., m.$$

10.4. Application

In this section, we will use the methodology discussed in the previous sections for the modeling of PPE.

10.4.1. *The modeling of PPE*

The modeling of PPE, according to the relevant theory (Hickman 1968; Homburg 2000; Diamonds 2001; Holzmann 2009), is based on 20 factors which are most likely related and possibly affect it. For this work, 20 European countries were selected based on the completeness of the available data, mostly derived through Knoema and OECD. The data which is annual, covers the period 2001–2015. Note that at the time of this work, the data for 2016–2018 was not fully available. Based on the available data, three individual datasets were created corresponding to the time-periods 2001–2005, 2006–2010 and 2011–2015, representing, respectively, a crisis irrelevant, a pre-crisis and a post-crisis period. The value of each variable for each five-year time period is taken to be equal to the average of all annual values of the specific variable within the specific time period. The selected countries and explanatory variables, in alphabetical order, are given in Tables 10.1 and 10.2, respectively.

Austria	France	Latvia	Republic of Slovenia
Belgium	Germany	Netherlands	Spain
Czech Republic	Greece	Poland	Sweden
Denmark	Iceland	Portugal	Switzerland
Finland	Italy	Slovak Republic	United Kingdom

Table 10.1. *Selected European countries based on the completeness of the selected variables*

Compensation of employees	Gross domestic product (GDP)	Median age of population	Total household savings
Consumer price index (CPI)	Imports of goods and services	Net number of migration flows	Total household spendings
Current account balance (CAB)	Inflation	Net number of births	Total labor force
Demographic dependency	Investments	Private sector debt	Total savings rate
Exports of goods and services	Long-term interest rates	Short-term interest rates	Unemployment rate

Table 10.2. *Selected covariates*

Some of the variables in Table 10.2 are directly related to PPE, such as inflation, CAB, GDP and unemployment rate, while others are indirectly related. The purpose of this analysis is to identify the variables/factors that affect the PPE.

Beale *et al.*'s technique is the first dimension reduction technique that was used. Table 10.3 presents the results of Beale *et al.*'s (1967) method for each of the three datasets examined.

Based on the above discussion and taking into consideration the significance of *GDP*, we arrived at the conclusion that, irrespective of the time period, only the variables extracted from the original data before *GDP* must be excluded from further analysis, while all other variables should remain and be considered for the next step of the reduction process. As a result, the variables *Exports of goods and services*, *Total household spendings*, *Short-term interest rates*, *Total household savings* and *Total savings rate* have been extracted from all three datasets.

The reduction process continues, with the implementation of the PCA using the following 15 variables (see Table 10.4) for each time period.

The implementation of PCA results in the full 15 principal components for each time period, with the corresponding eigenvalues ranging from almost eight to nearly zero. Based on the overall results and the fact that the loss of important information of the model was not desirable, we conclude that the first seven components should be kept regardless of the eigenvalues because they contain a considerable amount of the total information/variability (approximately 75%). The described variability played a key role in the decision since the intention was to keep that many components so that a considerable proportion of the original variability will be described by the components chosen.

Dataset	2001–2005	2006–2010	2011–2015
1st extraction	Total household savings	Exports of goods and services	Exports of goods and services
2nd extraction	Exports of goods and services	Short-term interest rates	Total household spendings
3rd extraction	Total household spendings	GDP	GDP
4th extraction	Total savings rate	Number of births	Total labor force
5th extraction	GDP	Total household spendings	Short-term interest rates
6th extraction	Imports of goods and services	Total savings rate	Private sector debt
7th extraction	Number of Births	Imports of goods and services	Imports of goods and services
8th extraction	Total labor force	Long-term interest rates	Total savings rate
9th extraction	Short-term interest rates	Private sector debt	Inflation rate
10th extraction	Inflation rate	Median age of population	Net number of migrants
11th extraction	Median age of population	Demographic dependency	Median age of population
12th extraction	Net number of migrants	Inflation rate	Total household savings
13th extraction	Long-term interest rates	Total household savings	Investments
14th extraction	Investments	Net number of migrants	Unemployment rate
15th extraction	Demographic dependency	Compensation of employees	Demographic dependency
16th extraction	Compensation of employees	Investments	Compensation of employees
17th extraction	CPI		

Table 10.3. *Beale et al.*

Compensation of employees	GDP	Long-term interest rates	Private sector debt
CPI	Imports of goods and services	Median age of population	Total labor force
CAB	Inflation	Net number of migration flows	Unemployment rate
Demographic dependency	Investments	Net number of births	

Table 10.4. *Remaining covariates after Beale et al.'s technique*

NOTE.– *In order to determine which variables are significant in each component, the following empirical "rule" was followed. For the top two components, the variables for which the absolute value of the associated coefficient is at least equal to 0.8 are kept as significant. For the remaining five components, only one variable is kept as important, the one that has the highest absolute value among all (due to the low variability explanation by the component). Although there is no specific rule, a value of around 0.8 is considered to be satisfactory in retaining a sufficient amount of information (Jolliffe 2002).*

Table 10.5 presents the most significant variables based on the components (coefficients) as a result of the PCA method, for all three datasets examined.

Dataset	2001–2005	2006–2010	2011–2015
1st Component	GDP (.96)	GDP (.96)	GDP (.98)
	Imports of goods and services (.94)	Imports of goods and services (.93)	Imports of goods and services (.95)
	Inflation (.89)	Inflation (.94)	Inflation (.76)
	Investments (.79)	Investments (.83)	Investments (.81)
			Net number of migration flows (.95)
	Number of births (.90)	Number of births (.90)	Number of births (.90)
	Private sector debt (.93)	Private sector debt (.95)	Private sector debt (.87)
	Total labor force (.91)	Total labor force (.93)	Total labor force (.92)
2nd Component	Median age of population (-.70)	Median age of population (-.71)	Unemployment rate (.81)
	Long-term interest rates (.75)		
3rd Component	CAB (.70)	CAB (.65)	CAB (-.71)
4th Component	Unemployment rate (-.80)	Unemployment rate (-.54)	CAB (-.46)
5th Component	CPI (-.45)	Demographic dependency (.46)	Compensation of employees (-.42)
6th Component	Compensation of employees (-.52)	Compensation of employees (-.60)	Compensation of employees (-.46)
7th Component	Investments (-.33)	Investments (-.40)	Long-term interest rates (.34)

Table 10.5. *Principal component analysis – the seven primary components*

The first component, denoted by Z_1 in all three datasets, holds roughly 50% of the total variation of the dataset, while the second one, denoted by Z_2, holds at most, 15% of it. The rest of the components contain the remaining percentage of variation (see the end of section 3). By construction, the first component is considered to be the most important, in which the analysis is primarily based on. Having said that, we observed in the above analysis that in all three datasets the variables that were significant in every component were almost always the same, with the variable playing the primary role and having the most influence in each of the three sets being GDP.

However, it should be pointed out that there is one important exception. Indeed, in the third time-period, the net number of migration flows has found to be significant in the first component. This variable might have an impact on the modeling process that was possibly not as important in the past as it is in this particular time period. This can be due to two very important events that have begun to emerge in Europe since 2010, the European Migrant Crisis (Lendaro 2016; Garcia-Zamor 2018) and the Spanish, Icelandic, Portuguese and Greek Economic Crisis (Gibson *et al.* 2014).

After dimensionality reduction, the regression analysis (Scheffe 1999; Anderson 2009; Sheather 2009) is implemented, using the seven components of PCA from Table 10.5 as independent variables and the logarithm of PPE as the dependent variable. The intention is to identify the significance of each component (independent covariates) and obtain an "ideal" model for the PPE for descriptive, as well as predictive purposes. The logarithm transformation was decided to be used in order to achieve the linearity between the dependent variable and each independent variable as well as the homoscedasticity of the residuals.

Table 10.6 presents the regression results for all of models examined. Note that the overall model is the average model of all three time-based periods under examination.

	2001–2005		**2006–2010**		**2011–-2015**		**Overall model**	
	Test	**Graph**	**Test**	**Graph**	**Test**	**Graph**	**Test**	**Graph**
R^2	87%		86%		70%		68%	
Adj R^2	82%		78%		53%		57%	
F-statistic	Reject H_0		Reject H_0		Reject H_0		Reject H_0	
T-statistic Significance	All		All		None		None	
Normality Assumption	✓*	✗	✓*	✗	✓**	✗	✓**	✗
Homoscedasticity assumption		✓		✓				✓
Autocorrelation assumption	✓	✓	✓	✓	✓	✓	✓	✓
Independence assumption	✓		✓		✓		✓	
Linearity assumption		✓		✓		✓		✓
Multicollinearity assumption	High		High		High		High	

Table 10.6. *Regression analysis results*

In Table 10.6, the assumptions have been checked by both statistical (**Test**) and graphical (**Graph**) methods. Furthermore, note the following:

– All results were interpreted at $\alpha = 5\%$.

– ✓ means that the null hypothesis failed to be rejected.

– ✗ means that the null hypothesis is rejected.

– The existence of * means that we reject this null hypothesis at $\alpha = 5\%$, but the value is near 0.01, and if we choose $\alpha = 1\%$, then the null hypothesis can fail to be rejected.

– The existence of ** means that we reject this null hypothesis in $\alpha = 5\%$, but in the case of $\alpha = 1\%$, the null hypothesis would fail to be rejected.

Observe that over time, the selected model is becoming less profitable, which means that as time passes by, new factors that have had no big impact in the previous years and are not included in the existing model, nowadays, play an important role in the formation of the PPE. As it was mentioned above, it is recommended that the phenomena of the European Debt Crisis and the European Migrant Crisis must be analyzed and be included as variables in future cases of modeling of PPE.

Although it was attempted to address the problem of multicollinearity through dimension reduction techniques, it seems that it continues to exist in the model. However, it has diminished visibly, even if it is still high, in contrast to the original raw data. The first consequence of its existence in the model process can be seen in the analysis of variance. In the third model, as well as in the overall model, the F-test comes to the conclusion that at least one variable is statistically significant, while at the same time the T-test suggests that none of the variables are, which as mentioned above is a phenomenon that occurs in the presence of multicollinearity.

A second consequence of the existence of multicollinearity is the overfitting of the model. High R-squared values, especially in the 2001–2005 and 2006–2010 datasets, suggest that the results might be misleading due to the multicollinearity.

In order to try to achieve the maximum reduction, or even the elimination of the remaining multicollinearity of the model, a transformation in the original covariates was made. The standardization of the explanatory variables was performed through the next formula:

$$X_{ij} = \frac{X_{ij} - \mu_{X_j}}{\sigma_{X_j}}$$

where μ and σ are the mean and standard deviation of X, respectively.

Using the transformed dataset and repeating the Beale *et al.* (1967) and PCA techniques, we came to the same conclusions as before with relatively minor changes that do not affect the overall interpretation of the covariates.

10.4.2. *Concluding remarks*

Concerning the results of the regression, the implementation appears improved. The assumptions remain the same, while multicollinearity is non-existent in all cases under examination. The F-test and T-test come into full agreement, with the first indicating that at least one factor is statistically significant and the second showing that at least Z_1 or Z_2 are important in all datasets. Also, R-squared now ranges from almost 38% to 57%, which confirms the suggested hypothesis that due to multicollinearity and the complexity of the model, overfitting lurks.

This simple transformation seems to manage the handling of the multicollinearity of factors, although based on theory, PCA was expected to succeed in doing so. Based on literature studies, we can use the correlation matrix instead of covariance matrix, especially when factors have a different unit of measurement, in order to apply the PCA. The new Z variables were correlated, although all the procedures that followed were correct. The conclusion of this work concerned the credibility of PCA, a well-known and established technique that has been used for a long period of time, when it comes to big data. From this study, we can see that when a large amount of information (observations) is used, in this case around 6,000 observations, techniques like PCA and similar ones do not perform well and their reliability is diminishing. This is due to the fact that when those techniques were developed, the accessibility to very large databases had nothing to do with the accessibility that exists nowadays, with the ever-increasing development of technology contributing dramatically to the growth of available data for modeling purposes. This work aimed to reveal the problems, such as multicollinearity, that exist when it comes to Big Data Analysis and how to overcome them simply when well-established techniques fail to perform.

10.5. Acknowledgments

This work was completed as part of the activities of the Laboratory of Statistics and Data Analysis of the University of the Aegean LabSTADA.

10.6. References

Anderson, D.W. (2009). *An Introduction to Multivariate Statistical Analysis*. John Wiley & Sons, Hoboken, New Jersey.

Artemiou, A. and Li, B. (2009). On principal components and regression: A statistical explanation of a natural phenomenon. *Statistica Sinica*, 19, 1557–1565.

Artemiou, A. and Li, B. (2013). Predictive power of principal components for single-index model and sufficient dimension reduction. *Journal of Multivariate Analysis*, 119, 176–184.

Barr, N. and Diamond, P.A. (2006). The economics of pensions. *Oxford Review of Economic Policy*, 22(1), 15–39.

Beale, E.M.L., Kendall, M.G., Mann, D.W. (1967). The discarding of variables in multivariate analysis. *Biometrika*, 54(3/4), 357–366.

Belsley, D.A. (1991). A guide to using the collinearity diagnostics. *Computer Science in Economics and Management*, 4(1), 33–50.

Bonoli, G. (2003). Two worlds of pension reform in Western Europe. *Comparative Politics*, 35(4), 399–416.

Bonoli, G. and Shinkawa, T. (2005). *Ageing and Pension Reform Around the World: Evidence from Eleven Countries*. Edward Elgar, Cheltenham.

Cattell, R. (1966). The Scree test for the number of factors. *Multivariate Behavioral Research*, 1(2), 245–276.

Cheever, E. (2020). Eigenvalues and eigenvectors [Online]. Available at: http://lpsa.swarthmore.edu/MtrxVibe/EigMat/MatrixEigen.html.

Creative Research Systems (2016). Correlation [Online]. Available at: https://www.surveysystem.com/correlation.htm [Accessed 27 March 2018].

Diamonds, P.A. (2001). Towards an optimal social security design. Working Paper No. 4, CeRP, Turin.

Farrell, J. and Shoag, D. (2017). Risky choices: Simulating public pension funding stress with realistic shocks. Working Paper No. 37, Hutchins Center.

Franco, D., Marino, M.-R., Zotteri, S. (2006). Pension expenditure projections, pension liabilities and European Union Fiscal Rules. *SSRN* [Online]. Available at: https://ssrn.com/abstract=2005199.

de la Fuente, A. (2011). A simple model of aggregate pension expenditure. Working Paper No. 553, Barcelona Graduate School of Economics.

Garcia-Zamor, J.C. (2018). The European migrant and refugee crisis. *Ethical Dilemmas of Migration*. PAGG Springer, Wiesbaden.

Geary, R.C. and Leser, C.E.V. (1968). Significance tests in multiple regression. *The American Statistician*, 22(1), 20–21.

Gibson, H.-D., Palivos, T., Tavlas, G.-S. (2014). The crisis in the Euro Area: An analytic overview. *Journal of Macroeconomics*, 39(2), 233–239.

Hair, J.F., Black, W.C., Babin, B.J., Anderson, R.E. (2010). *Advanced Diagnostics for Multiple Regression: A Supplement to Multivariate Data Analysis*. Pearson Prentice Hall Publishing, Upper Saddle River.

Hickman, J.C. (1968). Funding theories for social insurance. *Casualty Actuarial Society*, 55, 303–311.

Holzmann, R. (2009). *Aging Population, Pension Funds, and Financial Markets: Regional Perspectives and Global Challenges for Central, Easter, and Southern Europe*. World Bank, Washington, D.C.

Homburg, S. (2000). Compulsory savings in the welfare state. *Journal of Public Economics*, 77(2), 233–239.

Hotelling, H. (1933). Analysis of a complex of statistical variables into principal components. *Journal of Educational Psychology*, 24, 417–441, 498–520.

Hotelling, H. (1936). Relations between two sets of variates. *Biometrika*, 28(3/4), 321–377.

IBM (N/A). What is big data analytics? [Online]. Available at: https://www.ibm.com/analytics/hadoop/big-data-analytics.

Jolliffe, I.T. (1972). Discarding variables in a principal component analysis I: Artificial data. *Journal of the Royal Statistical Society. Series C (Applied Statistics)*, 21(2), 160–173.

Jolliffe, I.T. (2002). *Principal Components Analysis*, 2nd edition. Springer, New York.

King, J.R. and Jackson, D.A. (1999). Variable selection in large environmental data sets using principal components analysis. *Environmetrics*, 10, 67–77.

Knoema (N/A). Database [Online]. Available at: https://knoema.com.

Laboratory of Statistics and Data Analysis (2019). Laboratory of Statistics and Data Analysis [Online]. Available at: http://actuarweb.aegean.gr/labstada/index.html.

Lachowska, M. and Myck, M. (2018). The effect of public pension wealth on saving and expenditure. *American Economic Journal: Economic Policy*, 10(3), 284–308.

Largey, A. and Spencer, J.E. (1996). F- and T-tests in multiple regression: The possibility of "conflicting" outcomes. *Journal of the Royal Statistical Society. Series D (The Statistician)*, 45(1), 105–109.

Lendaro, A. (2016). A European migrant crisis? Some thoughts on mediterranean borders. *Studies in Ethnicity and Nationalism*, 16(1), 148–157.

Marcinkiewicz, E. and Chybalski, F. (2014). How to measure and compare pension expenditures in cross-country analyses? Some methodological remarks. *International Journal of Business and Management*, 2(4), 43–59.

Marcinkiewicz, E. and Chybalski, F. (2016). A new proposal of pension regimes typology: Empirical analysis of the OECD countries. *ENRSP Conference*, Poland.

McCullagh, P. and Nelder, J.A. (1989). *Generalized Linear Models*. Chapman Hall, London.

Organisation for Economic Co-operation and Development (OECD) (2020). Featured charts [Online]. Available at: https://data.oecd.org.

Organisation for Economic Co-operation and Development (OECD) (2019). Pension expenditures [Online]. Available at: https://data.oecd.org/socialexp/pension-spending.htm.

Pearson, K. (1901). On lines and planes of closest fit to systems of points in space. *The London, Edinburgh, and Dublin Philosophical Magazine and Journal of Science*, 2(11), 559–572.

Pereira, J., Karam, P., Muir, D., Tuladhar, A. (2010). Macroeconomic effects of public pension reforms. Working Paper No. 10/297, International Monetary Fund.

Scheffe, H. (1999). *The Analysis of Variance*. John Wiley & Sons, New Media.

Sheather, S. (2009). *A Modern Approach to Regression with R*. Springer Science & Business Media, Cham.

Smallman, L., Artemiou, A., Morgan, J. (2018). Sparse generalised principal component analysis. *Pattern Recognition*, 83, 443–455.

Wasserman, L. (2004). *All of Statistics: A Course in Statistical Inference*. Springer, New York.

Wikipedia (2020). Multicollinearity [Online]. Available at: https://en.wikipedia.org/wiki/Multicollinearity.

11

Weak Signals in High-Dimensional Poisson Regression Models

We addressed parameter estimation in the context of high-dimensional sparse Poisson regression models, in which the number of predictors exceeds the sample size. Generally, predictor screening via penalized maximum likelihood methods is required to provide the sparsity structure in the parsimonious model, before applying post-screening parameter estimation via maximum likelihood based on such a model. The major problem is that the use of different screening methods produces different sparsity structures, usually of unknown correctness. This may produce either overfitted or underfitted models, making post-screening maximum likelihood estimation based on these models inefficient. We therefore proposed post-screening estimation based on linear shrinkage, pretest and Stein-type shrinkage strategies to address inefficient maximum likelihood estimation based on the parsimonious models from the screening stage of unknown of appropriateness. Through Monte Carlo simulations, with unknown correctness in the predictor screening stage, the proposed estimators were shown to be significantly more efficient than the classical maximum likelihood estimators.

11.1. Introduction

The Poisson regression model is widely applied for predicting count response in fields including medicine, biotechnology and quality control; see Myers *et al.* (2012) or Agresti (2015) for more details. Maximum likelihood (ML) is a well-known statistical method for estimating unknown parameters in the model. Unfortunately, it cannot be directly used in high-dimensional regimes, where the number of predictors/features (p) exceeds the sample size (n).

Chapter written by Orawan REANGSEPHET, Supranee LISAWADI and Syed Ejaz AHMED.

We are primarily interested in analyzing high-dimensional count data ($p > n$), which is attracting growing attention. Though there are many candidate predictors, some may be relatively unimportant in predicting the response and should be eliminated from the model. When analyzing the high-dimensional Poisson regression model, a predictor screening method is required to screen out these unimportant predictors. This provides a parsimonious model with only important predictors used for post-screening parameter inference and predicting the response of interest.

The big problem is that a wide range of different predictor screening methods results in different sparse structures for parsimonious models. Following Ahmed and Yüzbaşı (2016), some methods may result in a lower-dimensional parsimonious model than others. Furthermore, these results generally provide parsimonious models with uncertainty of appropriateness. After the predictor screening stage, we consequently may encounter underfitting when the screening method screens out too many important predictors, or overfitting if too many predictors having very weak or no signals, considered as noise, are kept in the model. These raise questions about the reliability and correctness of post-screening ML estimation, based on the models having different sparsity structures and unknown appropriateness.

In this work, we investigated whether the performance of the different screening methods using penalized ML methods (PMLMs) performed well for varying-coefficient high-dimensional Poisson regression models. We then proposed robust post-screening parameter estimation based on linear shrinkage, preliminary test and Stein-type shrinkage strategies. The goal was to improve classical post-screening ML estimation based on parsimonious models which are seriously affected by underfitting or overfitting. These strategies have been applied in different contexts of parameter estimation by many researchers, for example, Yüzbaşı *et al.* (2017), Li *et al.* (2018), Reangsephet *et al.* (2018) and Karbalaee *et al.* (2019).

The rest of this chapter is organized as follows. The statistical background of the Poisson regression model and its ML estimation are presented in section 11.2. The methodologies for parameter estimation in high-dimensional regimes are proposed in section 11.3. Monte Carlo simulations were carried out to investigate the performance of the proposed methods under varying-coefficient models and the results are presented and discussed in section 11.4. Section 11.5 presents our conclusions.

11.2. Statistical background

Given a sample of size n, let y_i be a count response variable and $\vec{x}_i = (1, x_{i1}, x_{i2}, ..., x_{ip})^{\mathrm{T}} \in \mathbb{R}^{p+1}$ be a vector of p collected predictors for $i = 1, 2, ..., n$, where the notation ^{T} denotes the transpose of the vector. The probability function of the Poisson regression model for the ith observation is given by

$$f(y_i|\vec{x}_i; \vec{\beta}) = \exp(-\mu_i)\mu_i^{y_i}/y_i!, \quad y_i = 0, 1, 2, ..., \qquad [11.1]$$

with $\mu_i = \exp(\vec{x}_i^{\mathrm{T}}\vec{\beta})$, ensuring $\mu_i > 0$, as the mean parameter. Here, $\vec{\beta} = (\beta_0, \beta_1, \beta_2, ..., \beta_p)^{\mathrm{T}} \in \mathbb{R}^{p+1}$ is the vector of unknown regression coefficients and the notation ^{T} indicates the transpose of the vector or matrix.

Under the assumption of independent observations, the log-likelihood function is given by

$$\ell(\vec{\beta}) = \sum_{i=1}^{n} \left(y_i \vec{x}_i^{\mathrm{T}} \vec{\beta} - \exp(\vec{x}_i^{\mathrm{T}} \vec{\beta}) - \ln y_i! \right). \qquad [11.2]$$

The ML estimator of parameter vector $\vec{\beta}$ is obtained as the solution to:

$$\frac{\partial \ell(\vec{\beta})}{\partial \ell \vec{\beta}} = \sum_{i=1}^{n} \left(y_i \vec{x}_i - \exp(\vec{x}_i^{\mathrm{T}} \vec{\beta}) \vec{x}_i \right) = \vec{0}_p, \qquad [11.3]$$

which is not a closed-form solution. The Newton–Raphson iteration method is used to obtain the ML estimates of $\vec{\beta}$. Following Santos and Neves (2008), under certain regularity conditions, the asymptotic properties of the ML estimator of $\vec{\beta}$ are consistency and asymptotic normality with covariance matrix: $1/\left(\sum_{i=1}^{n} \exp(\vec{x}_i^{\mathrm{T}} \vec{\beta}) \vec{x}_i \vec{x}_i^{\mathrm{T}}\right)$.

11.3. Methodologies

11.3.1. *Predictor screening methods*

We start by proposing PMLMs for effectively screening out the unimportant predictors from the initial model, making the model parsimonious. In general, the PMLM imposes a penalty function $P_\lambda(\vec{\beta})$ on the regression coefficients in the ML estimation and is obtained by maximizing the function $M_\lambda(\vec{\beta})$:

$$M_\lambda(\vec{\beta}) = \sum_{i=1}^{n} \left(y_i \vec{x}_i^{\mathrm{T}} \vec{\beta} - \exp(\vec{x}_i^{\mathrm{T}} \vec{\beta}) - \ln y_i! \right) - \lambda P_\lambda(\vec{\beta}), \quad \lambda > 0. \qquad [11.4]$$

Here, λ is called the sparse controller, which controls the strength of the penalty function. Its value can be obtained using k-fold cross-validation.

In this work, we introduce two commonly used PMLMs: least absolute shrinkage and selection operator (LASSO) and adaptive LASSO. The LASSO was proposed by Park and Hastie (2007) and its penalty function is defined as

$$P_\lambda(\vec{\beta}) = \sum_{j=1}^{p} |\beta_j|. \qquad [11.5]$$

Following Zou (2006), the penalty function of adaptive LASSO is given by

$$P_\lambda(\vec{\beta}) = \sum_{j=1}^{p} |\beta_j|\tau_j. \qquad [11.6]$$

Here, for $z > 0$ and $j = 1, 2, ..., p$, $\tau_j = |\widetilde{\beta}_j|^{-z}$ is the adaptive weight, used for penalizing coefficients with different weights, and $\widetilde{\beta}_j$ is the initial estimate of β_j, which is obtained through the ridge PMLM with penalty function $P_\lambda(\vec{\beta}) = \sum_{j=1}^{p} \beta_j^2$; see Wahid *et al.* (2007).

Adaptive LASSO generally selects fewer predictors as important than LASSO (Ahmed and Yüzbaşı 2016). Suppose that LASSO selects $p_1 + p_2$ predictors, whereas only p_1 predictors are selected by adaptive LASSO, such that $p_1 + p_2 < p$. After screening the predictors and then reordering the coefficients, there are two competing parsimonious Poisson regression models:

1. **LASSO-based:** $f(y_i|\vec{x}_i; \vec{\beta}) = \dfrac{\exp(-\mu_i)\mu_i^{y_i}}{y_i!}$; $\mu_i = \exp(\vec{x}_{i1}^{\mathrm{T}}\vec{\beta}_1 + \vec{x}_{i2}^{\mathrm{T}}\vec{\beta}_2)$. [11.7]

2. **Adaptive LASSO-based:** $f(y_i|\vec{x}_i; \vec{\beta}) = \dfrac{\exp(-\mu_i)\mu_i^{y_i}}{y_i!}$; $\mu_i = \exp(\vec{x}_{i1}^{\mathrm{T}}\vec{\beta}_1)$. [11.8]

Here, $\vec{\beta}_1 = (\beta_0, \beta_1, \beta_2, ..., \beta_{p_1})^{\mathrm{T}} \in \mathbb{R}^{p_1+1}$ is a vector of unknown coefficients related to p_1 predictors $\vec{x}_{i1} = (1, x_{i1}, x_{i2}, ..., x_{ip_1})^{\mathrm{T}} \in \mathbb{R}^{p_1+1}$ selected by both methods, and $\vec{\beta}_2 = (\beta_{p_1+1}, \beta_{p_1+2}, ..., \beta_{p_1+p_2})^{\mathrm{T}} \in \mathbb{R}^{p_2}$ is a vector of unknown coefficients related to p_2 predictors $\vec{x}_{i2} = (x_{i(p_1+1)}, x_{i(p_1+2)}, ..., x_{i(p_1+p_2)})^{\mathrm{T}} \in \mathbb{R}^{p_2}$ selected only by LASSO.

It should be noted that either LASSO or adaptive LASSO methods may work well in screening important predictors, but not in all situations. Following Ahmed and Yüzbaşı (2016), it is possible that LASSO may cause a parsimonious model to overfit, as too many confounding predictors having slightly or no signals are retained, so that $\vec{\beta}_2 = (0, 0, ..., 0)^{\mathrm{T}} = \vec{0}_{p_2}$. Conversely, underfitting may occur if adaptive LASSO eliminates too many important predictors having strong or moderate signals, indicating that information $\vec{\beta}_2 = \vec{0}_{p_2}$ is false. This will degrade the performance of post-screening ML estimation based on the resulting models. We overcome this problem by proposing robust post-screening estimation for estimating the parameter vector $\vec{\beta}_{\text{post}} = (\vec{\beta}_1^{\mathrm{T}}, \vec{\beta}_2^{\mathrm{T}})^{\mathrm{T}}$. This is presented in the next section.

11.3.2. *Post-screening parameter estimation methods*

First, we have the two alternative ML estimators of parameter vector $\vec{\beta}_{\text{post}} = (\vec{\beta}_1^{\mathrm{T}}, \vec{\beta}_2^{\mathrm{T}})^{\mathrm{T}}$ from the results of the predictor screening stage. One is that the ML estimator based on model [11.7] is called overfitted (OF), denoted as $\hat{\vec{\beta}}^{\text{OF}}_{\text{post}} = (\beta^{\text{OF}}_1, \beta^{\text{OF}}_2, ..., \beta^{\text{OF}}_{p_1+p_2})^{\text{T}}$. The other is that the ML estimator based on model [11.8], in which $\vec{\beta}_2$ is restricted to $\vec{0}_{p_2}$, is called underfitted (UF), denoted as $\hat{\vec{\beta}}^{\text{UF}}_{\text{post}} = (\beta^{\text{UF}}_1, \beta^{\text{UF}}_2, ..., \beta^{\text{UF}}_{p_1}, \vec{0}^{\text{T}}_{p_2})^{\text{T}}$.

We now propose the novel post-screening estimators of $\vec{\beta}_{\text{post}}$, designed to handle the uncertainty of predictor screening results. The idea is to optimally combine $\hat{\vec{\beta}}^{\text{OF}}_{\text{post}}$ and $\hat{\vec{\beta}}^{\text{UF}}_{\text{post}}$. We first propose the linear shrinkage (LS) estimator given by

$$\hat{\vec{\beta}}^{\text{LS}}_{\text{post}} = \hat{\vec{\beta}}^{\text{OF}}_{\text{post}} - c(\hat{\vec{\beta}}^{\text{OF}}_{\text{post}} - \hat{\vec{\beta}}^{\text{UF}}_{\text{post}}), \quad 0 \leq c \leq 1. \qquad [11.9]$$

This estimator effectively shrinks $\hat{\vec{\beta}}^{\text{OF}}_{\text{post}}$ towards $\hat{\vec{\beta}}^{\text{UF}}_{\text{post}}$ through the confidence level in the information $\vec{\beta}_2 = \vec{0}_{p_2}$ (c) to control the uncertainty in the predictor screening stage. The value of c can be set by the judgment of experimenter or by minimizing the mean square error of the estimator; see Ahmed (2014) or Reangsephet *et al.* (2018) for some insights. However, the value of c is conservatively set as 0.50.

Next, the pretest (PT) estimator suitably decides to select $\hat{\vec{\beta}}^{\text{OF}}_{\text{post}}$ or $\hat{\vec{\beta}}^{\text{UF}}_{\text{post}}$ as the reasonable estimator using the PT strategy of Bancroft (1944). This strategy tests whether the information $\vec{\beta}_2 = \vec{0}_{p_2}$ is correct or not. Hence, the PT estimator can be written as

$$\hat{\vec{\beta}}^{\text{PT}}_{\text{post}} = I(T_n > \chi^2_{p_2,\alpha})\hat{\vec{\beta}}^{\text{OF}}_{\text{post}} + I(T_n \leq \chi^2_{p_2,\alpha})\hat{\vec{\beta}}^{\text{UF}}_{\text{post}}. \qquad [11.10]$$

Here, $I(.)$ is an indicator function, T_n is the likelihood ratio test statistic given as Eq. [11.11], widely used for testing $H_0 : \vec{\beta}_2 = \vec{0}_{p_2}$ against $H_1 : \vec{\beta}_2 \neq \vec{0}_{p_2}$, and $\chi^2_{p_2,\alpha}$ is the critical value of the χ^2 distribution with p_2 d.f. at a significance level of α

$$T_n = 2\left(\ln L(\hat{\vec{\beta}}^{\text{OF}}_{\text{post}}; y_1, y_2, ..., y_n) - \ln L(\hat{\vec{\beta}}^{\text{UF}}_{\text{post}}; y_1, y_2, ..., y_n)\right). \qquad [11.11]$$

Given H_0 to be true, T_n converges in distribution to the χ^2 distribution with p_2 d.f. as $n \to \infty$. The result of $\hat{\vec{\beta}}^{\text{PT}}_{\text{post}}$ is $\hat{\vec{\beta}}^{\text{OF}}_{\text{post}}$ if H_0 is rejected and $\hat{\vec{\beta}}^{\text{UF}}_{\text{post}}$ otherwise. We present the shrinkage pretest (SP) estimator to avoid the use of two extreme estimators ($\hat{\vec{\beta}}^{\text{OF}}_{\text{post}}$ or $\hat{\vec{\beta}}^{\text{UF}}_{\text{post}}$). The SP estimator is defined as

$$\hat{\vec{\beta}}^{\text{SP}}_{\text{post}} = I(T_n > \chi^2_{p_2,\alpha})\hat{\vec{\beta}}^{\text{OF}}_{\text{post}} + I(T_n \leq \chi^2_{p_2,\alpha})\hat{\vec{\beta}}^{\text{LS}}_{\text{post}}. \qquad [11.12]$$

Discussions of the superiority of $\overset{\frown\text{SP}}{\vec{\beta}}_{\text{post}}$ over $\overset{\frown\text{PT}}{\vec{\beta}}_{\text{post}}$ can be found in Reangsephet *et al.* (2018).

Lastly, we propose the post-screening minimax estimator using the Stein-type shrinkage strategy of Stein (1956) and James and Stein (1992). This is called the shrinkage (S) estimator and is given as

$$\overset{\frown\text{S}}{\vec{\beta}}_{\text{post}} = \overset{\frown\text{UF}}{\vec{\beta}}_{\text{post}} + \left(1 - \frac{p_2 - 2}{T_n}\right)(\overset{\frown\text{OF}}{\vec{\beta}}_{\text{post}} - \overset{\frown\text{UF}}{\vec{\beta}}_{\text{post}}), \quad p_2 \geq 3. \qquad [11.13]$$

The shortcoming of $\overset{\frown\text{S}}{\vec{\beta}}_{\text{post}}$ is that overshrinking, which leads to misinterpretation of coefficients in the resulting model, occurs when $p_2 - 2 > T_n$. To tackle this, we introduce the positive-part shrinkage (PS) estimator, which is the robust version of $\overset{\frown\text{S}}{\vec{\beta}}_{\text{post}}$:

$$\overset{\frown\text{PS}}{\vec{\beta}}_{\text{post}} = \overset{\frown\text{UF}}{\vec{\beta}}_{\text{post}} + \left(1 - \frac{p_2 - 2}{T_n}\right)^{+}(\overset{\frown\text{OF}}{\vec{\beta}}_{\text{post}} - \overset{\frown\text{UF}}{\vec{\beta}}_{\text{post}}), \quad p_2 \geq 3. \qquad [11.14]$$

Here, $\left(1 - \frac{p_2-2}{T_n}\right)^{+} = \max\left\{0, 1 - \frac{p_2-2}{T_n}\right\}$.

The utility of the last two estimators is that their performance does not depend either on c or α, unlike LS, PT and SP estimators. We next numerically compared the performance of the proposed methods via a Monte Carlo simulation under different design models.

11.4. Numerical studies

11.4.1. *Simulation settings and performance criteria*

We are now in the era of high-dimensional data ($p > n$). Considering the varying-coefficient Poisson regression model with sample size n and initial predictors p, the p predictors generally can be characterized as p_s strong, p_w weak–moderate and p_n no signals, such that $p_s + p_w + p_n = p$, following Ahmed and Ahmed and Yüzbaşı (2016) and Li *et al.* (2018). Without loss of generality, we omit the intercept term ($\beta_0 = 0$) and then generate the Poisson response as follows:

$$y_i = \exp(\vec{x}_i^{\text{T}}\vec{\beta}).$$

Here, the predictor $\vec{x}_i = (x_{i1}, x_{i2}, ..., x_{ip})^{\text{T}} \in \mathbb{R}^p$ was, respectively, constructed from the standardized p-dimensional multivariate normal distribution, for $i = 1, 2, ..., n$.

We assumed the sparsity on the coefficient vector $\vec{\beta} = (\beta_1, \beta_2, ..., \beta_p)^{\mathrm{T}}$ to be known and rearranged the coefficients of $\vec{\beta}$ following the strength of signals. Hence, the true coefficient vector was set as $\vec{\beta} = (\vec{\beta}_s^{\mathrm{T}}, \vec{\beta}_w^{\mathrm{T}}, \vec{\beta}_n^{\mathrm{T}})^{\mathrm{T}}$, where $\vec{\beta}_s = (3, 3, ..., 3)^{\mathrm{T}} \in \mathbb{R}^{p_s}$, $\vec{\beta}_w = (w, w, ..., w)^{\mathrm{T}} \in \mathbb{R}^{p_w}$ and $\vec{\beta}_n = (0, 0, ..., 0)^{\mathrm{T}} \in \mathbb{R}^{p_n}$ were, respectively, the coefficient vector of predictors having strong, weak–moderate and no signals. All non-zero coefficients were randomly assigned to have either positive or negative signs.

The values of w were set as 0.01, 0.05, 0.50, 1.00 and 1.50 to study whether the performance of methods was affected by changing the very weak signal to moderate. In this study, we considered two cases of (n, p_s, p_w, p_n): case 1: $(n, p_s, p_w, p_n) = (100, 5, 45, 150)$ and case 2: $(n, p_s, p_w, p_n) = (150, 5, 75, 300)$, satisfying the usual assumptions $p_s \leq p_w < n$ and $p_n > n$. The values of sparse controller λ of PMLMs were computed by 10-fold cross-validation. The values of parameters c and α were conservatively set as $c = 0.50$ and $\alpha = 0.05$. Each configuration was run over 1,000 times to get a stable result, implemented in the R program (R Core Team (2018)).

The effectiveness of LASSO and adaptive LASSO for screening the predictors in the initial stage was investigated using the selection percentage criterion. For comparison of post-screening parameter estimation performance, the common criterion was the simulated relative efficiency (SRE):

$$\mathrm{SRE}(\widehat{\vec{\beta}}_{\text{post}}^{\text{OF}} : \widehat{\vec{\beta}}_{\text{post}}^{*}) = \mathrm{MSE}(\widehat{\vec{\beta}}_{\text{post}}^{\text{OF}})/\mathrm{MSE}(\widehat{\vec{\beta}}_{\text{post}}^{*}), \qquad [11.15]$$

where

$$\mathrm{MSE}(\widehat{\vec{\beta}}_{\text{post}}^{*}) = \sum_{j=1}^{1,000} (\widehat{\vec{\beta}}_{\text{post}}^{*(j)} - \vec{\beta}_{\text{post}})^{\mathrm{T}}(\widehat{\vec{\beta}}_{\text{post}}^{*(j)} - \vec{\beta}_{\text{post}})/1,000. \qquad [11.16]$$

Here, MSE $(\widehat{\vec{\beta}}_{\text{post}}^{*})$ is the mean square estimation error of $\widehat{\vec{\beta}}_{\text{post}}^{*}$, which is one of the proposed post-screening estimators, and the notation $^{(j)}$ indicates the jth simulation run. An SRE $(\widehat{\vec{\beta}}_{\text{post}}^{\text{OF}} : \widehat{\vec{\beta}}_{\text{post}}^{*})$ greater than one indicates the superiority of $\widehat{\vec{\beta}}_{\text{post}}^{*}$ over $\widehat{\vec{\beta}}_{\text{post}}^{\text{OF}}$, used as a benchmark.

11.4.2. *Results*

In the predictor screening stage, the LASSO-based and adaptive LASSO-based selection percentage of predictors for each signal level are reported in Tables 11.1 and 11.2, and those of each predictor are graphically shown in Figures 11.1–11.10.

Method	w	Selection percentage of predictor (%)		
		Strong	Weak to moderate	No
LASSO	0.01	100.000	7.898	5.189
	0.05	99.940	13.891	11.281
	0.50	86.280	26.242	16.581
	1.00	61.200	34.593	21.227
	1.50	50.860	40.513	29.464
Adaptive LASSO	0.01	99.960	2.536	1.777
	0.05	96.160	7.360	5.924
	0.50	71.640	16.571	8.769
	1.00	51.960	23.993	13.224
	1.50	39.480	27.398	19.539

Table 11.1. *LASSO-based and adaptive LASSO-based selection percentage of predictors with strong, weak–moderate and no signals for case 1*

Method	w	Selection percentage of predictor (%)		
		Strong	Weak to moderate	No
LASSO	0.01	100.000	14.212	10.277
	0.05	99.880	21.384	17.881
	0.50	91.980	30.657	21.984
	1.00	65.220	40.012	26.277
	1.50	60.100	45.327	33.397
Adaptive LASSO	0.01	99.780	3.537	2.544
	0.05	96.520	11.352	8.727
	0.50	83.540	16.847	11.500
	1.00	56.280	25.847	15.050
	1.50	48.220	31.928	21.147

Table 11.2. *LASSO-based and adaptive LASSO-based selection percentage of predictors with strong, weak–moderate and no signals for case 2*

Overall, adaptive LASSO screened out more predictors than LASSO. Its screening results introduced a lower-dimensional parsimonious model. As w increased, the selection percentage of predictors with strong signals decreased for both methods, in contrast to that of predictors with other signals. LASSO showed higher performance than adaptive LASSO in selecting predictors with strong signals. Unfortunately, it also kept more predictors with no signals.

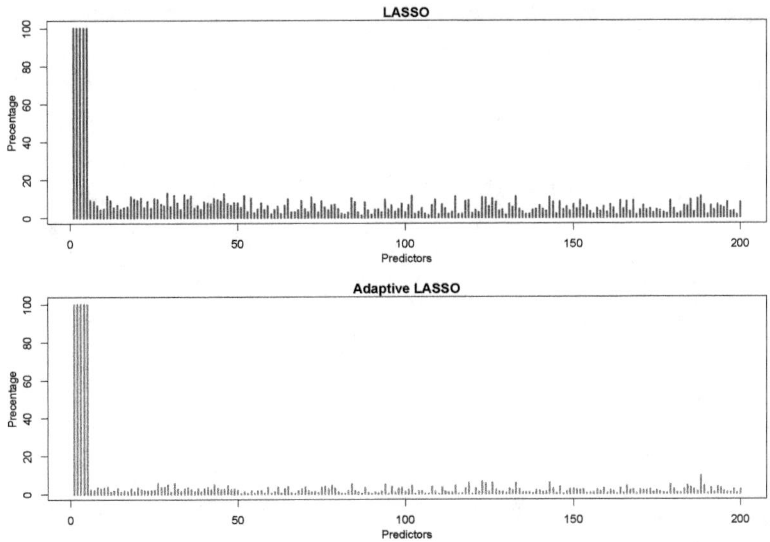

Figure 11.1. *Comparison of the percentage of each predictor selected by LASSO and adaptive LASSO when* $w = 0.01$ *for case 1. For a color version of this figure, see www.iste.co.uk/dimotikalis/analysis1.zip*

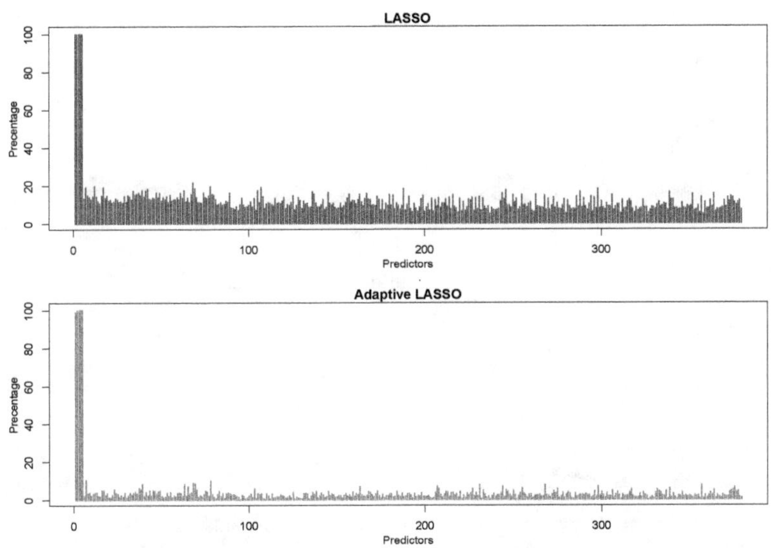

Figure 11.2. *Comparison of the percentage of each predictor selected by LASSO and adaptive LASSO when* $w = 0.01$ *for case 2. For a color version of this figure, see www.iste.co.uk/dimotikalis/analysis1.zip*

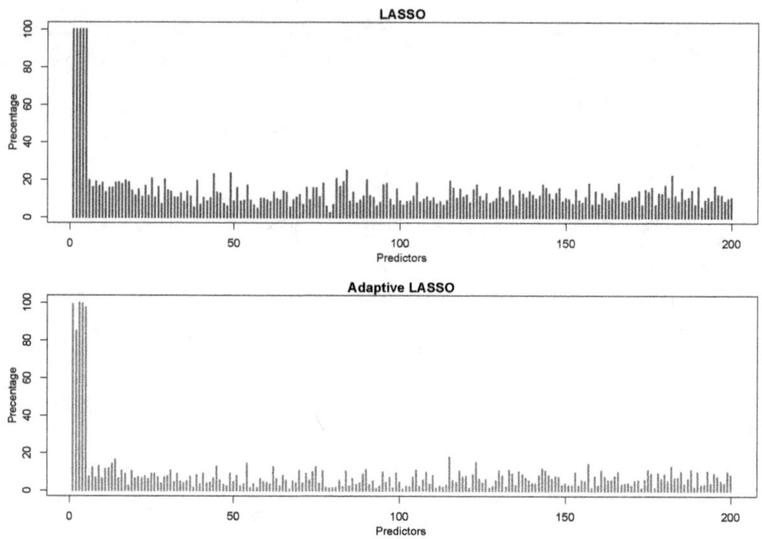

Figure 11.3. *Comparison of the percentage of each predictor selected by LASSO and adaptive LASSO when $w = 0.05$ for case 1. For a color version of this figure, see www.iste.co.uk/dimotikalis/analysis1.zip*

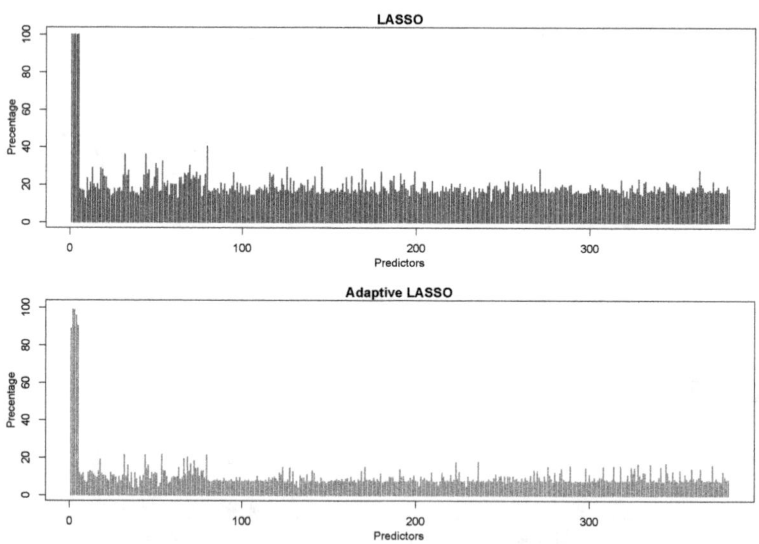

Figure 11.4. *Comparison of the percentage of each predictor selected by LASSO and adaptive LASSO when $w = 0.05$ for case 2. For a color version of this figure, see www.iste.co.uk/dimotikalis/analysis1.zip*

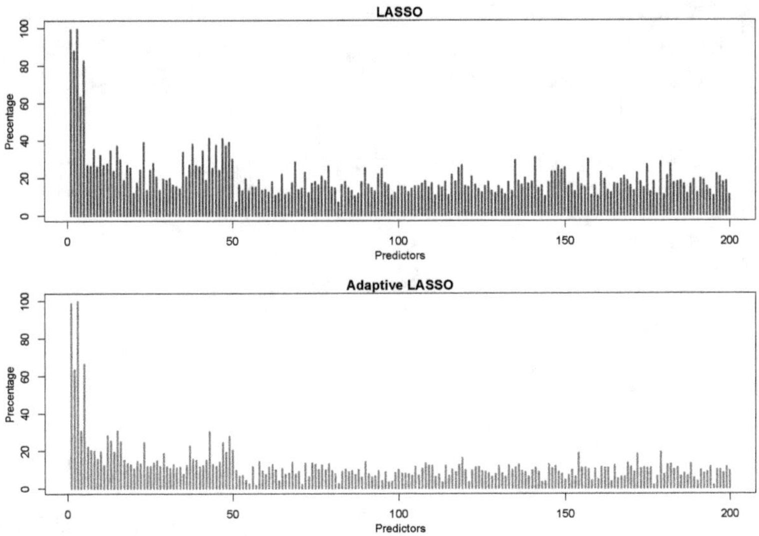

Figure 11.5. *Comparison of the percentage of each predictor selected by LASSO and adaptive LASSO when $w = 0.50$ for case 1. For a color version of this figure, see www.iste.co.uk/dimotikalis/analysis1.zip*

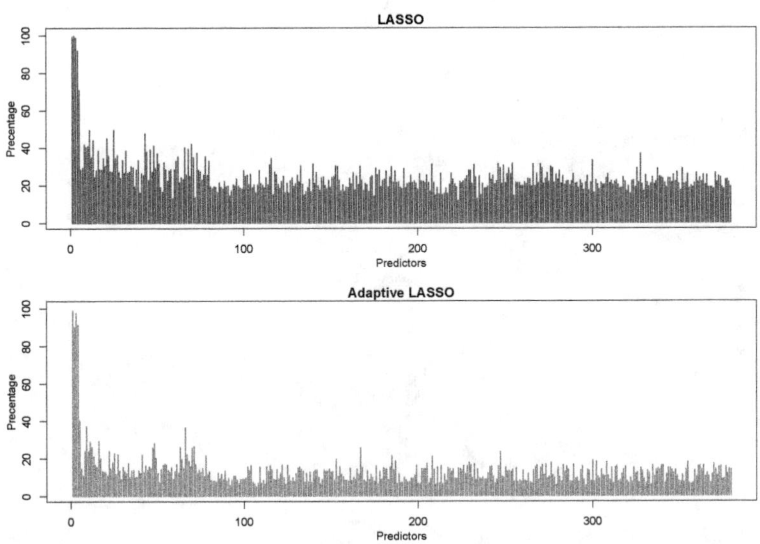

Figure 11.6. *Comparison of the percentage of each predictor selected by LASSO and adaptive LASSO when $w = 0.50$ for case 2. For a color version of this figure, see www.iste.co.uk/dimotikalis/analysis1.zip*

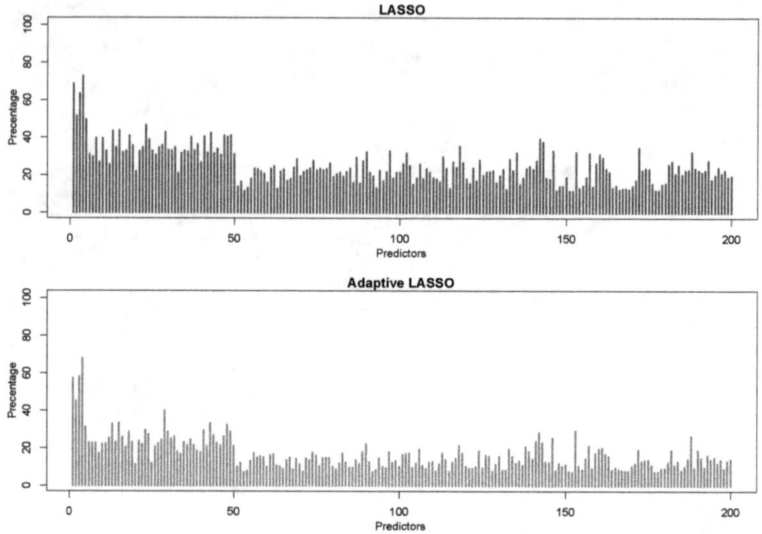

Figure 11.7. *Comparison of the percentage of each predictor selected by LASSO and adaptive LASSO when $w = 1.00$ for case 1. For a color version of this figure, see www.iste.co.uk/dimotikalis/analysis1.zip*

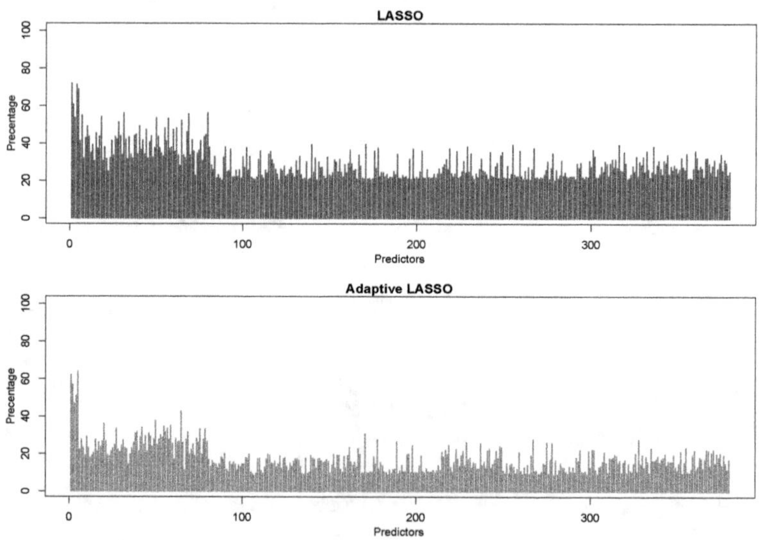

Figure 11.8. *Comparison of the percentage of each predictor selected by LASSO and adaptive LASSO when $w = 1.00$ for case 2. For a color version of this figure, see www.iste.co.uk/dimotikalis/analysis1.zip*

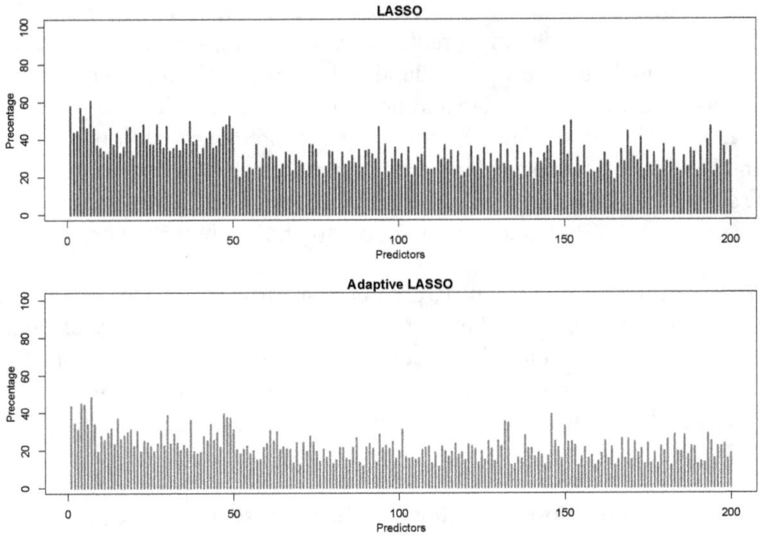

Figure 11.9. *Comparison of the percentage of each predictor selected by LASSO and adaptive LASSO when* $w = 1.50$ *for case 1. For a color version of this figure, see www.iste.co.uk/dimotikalis/analysis1.zip*

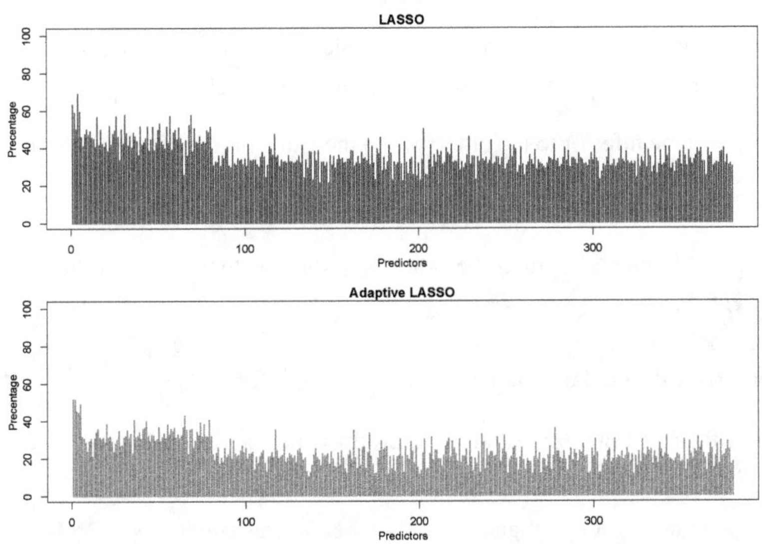

Figure 11.10. *Comparison of the percentage of each predictor selected by LASSO and adaptive LASSO when* $w = 1.50$ *for case 2. For a color version of this figure, see www.iste.co.uk/dimotikalis/analysis1.zip*

When considering the magnitude of a weak–moderate signal w, the performance of the two methods in selecting predictors with as important strong signals was comparable for small w. However, adaptive LASSO performed well in screening out predictors with weak–moderate and no signals, as their selection percentage was very small. When w became large, indicating that predictors with this signals were significant, LASSO outperformed in selecting predictors with both strong and weak–moderate signals, even though it still screened out fewer predictors with no signals. In contrast, adaptive LASSO tended to screen out too many important predictors.

It is noteworthy that, for small w, predictors with weak–moderate signals are of little importance for predicting the response, and should be entirely excluded from the resulting parsimonious model. In contrast, they become significant and should be included in the model when w becomes large. For this reason, LASSO resulted in an overfitted model for small w, whereas adaptive LASSO-based screening resulted in underfitted model when w was large.

Next, we applied the post-screening estimators presented in the previous section, which are a combination of the ML estimators based on LASSO-based and adaptive LASSO-based parsimonious models for estimating $\vec{\beta}_{\text{post}}$. The SRE results of the estimators are reported in Tables 11.3 and 11.4. The key findings were as follows:

– All estimators achieved highest performance when $w = 0.01$, and their performance deteriorated as w increased;

– When w was small, $\hat{\vec{\beta}}^{\text{UF}}_{\text{post}}$ was the most efficient, indicating that adaptive LASSO produced a more appropriate parsimonious model, in which $\vec{\beta}_2 = \vec{0}_{p_2}$ was true;

– $\hat{\vec{\beta}}^{\text{UF}}_{\text{post}}$ became inferior to the others when the value of w increased. This confirmed that the LASSO-based model was more appropriate as very weak signals became moderate, whereas adaptive LASSO caused underfitting in which $\vec{\beta}_2 = \vec{0}_{p_2}$ was false;

– $\hat{\vec{\beta}}^{\text{LS}}_{\text{post}}$ had higher SRE than the others for some values of w, as did $\hat{\vec{\beta}}^{\text{PT}}_{\text{post}}$, $\hat{\vec{\beta}}^{\text{SP}}_{\text{post}}$, $\hat{\vec{\beta}}^{\text{S}}_{\text{post}}$ and $\hat{\vec{\beta}}^{\text{PS}}_{\text{post}}$;

– For moderate and large values of w, $\hat{\vec{\beta}}^{\text{LS}}_{\text{post}}$ outperformed $\hat{\vec{\beta}}^{\text{UF}}_{\text{post}}$;

– The SREs of both $\hat{\vec{\beta}}^{\text{PT}}_{\text{post}}$ and $\hat{\vec{\beta}}^{\text{SP}}_{\text{post}}$ dropped below one as w increased, before increasing again to one;

– $\hat{\vec{\beta}}^{\text{SP}}_{\text{post}}$ was less efficient than $\hat{\vec{\beta}}^{\text{PT}}_{\text{post}}$ only when w was small;

– The SRE of $\hat{\vec{\beta}}^{\text{PS}}_{\text{post}}$ was higher than or equal to that of $\hat{\vec{\beta}}^{\text{S}}_{\text{post}}$;

– $\hat{\vec{\beta}}^{\text{PS}}_{\text{post}}$ and $\hat{\vec{\beta}}^{\text{S}}_{\text{post}}$ consistently outperformed $\hat{\vec{\beta}}^{\text{OF}}_{\text{post}}$, with their SREs greater than one.

w	Estimator					
	UF	LS	PT	SP	SE	PSE
0.01	3.113	1.864	2.354	1.733	2.014	2.092
0.05	2.337	1.772	2.168	1.626	2.001	2.058
0.50	1.259	1.399	1.029	1.055	1.238	1.239
1.00	0.851	0.906	0.964	0.996	1.033	1.034
1.50	0.622	0.881	1.000	1.000	1.010	1.012

Table 11.3. *Simulated relative efficiency results of $\hat{\vec{\beta}}^{UF}_{post}$, $\hat{\vec{\beta}}^{LS}_{post}$, $\hat{\vec{\beta}}^{PT}_{post}$, $\hat{\vec{\beta}}^{SP}_{post}$, $\hat{\vec{\beta}}^{S}_{post}$ and $\hat{\vec{\beta}}^{PS}_{post}$ with respect to $\hat{\vec{\beta}}^{OF}_{post}$ for case 1*

w	Estimator					
	UF	LS	PT	SP	SE	PSE
0.01	3.976	2.097	2.529	2.083	2.379	2.381
0.05	2.839	2.080	2.425	1.652	2.152	2.252
0.50	1.391	1.303	1.164	1.177	1.289	1.304
1.00	0.885	1.001	0.989	0.999	1.041	1.042
1.50	0.733	0.982	1.000	1.000	1.024	1.024

Table 11.4. *Simulated relative efficiency results of $\hat{\vec{\beta}}^{UF}_{post}$, $\hat{\vec{\beta}}^{LS}_{post}$, $\hat{\vec{\beta}}^{PT}_{post}$, $\hat{\vec{\beta}}^{SP}_{post}$, $\hat{\vec{\beta}}^{S}_{post}$ and $\hat{\vec{\beta}}^{PS}_{post}$ with respect to $\hat{\vec{\beta}}^{OF}_{post}$ for case 2*

This results confirmed that the proposed post-screening estimators were robust even when the parsimonious models were inappropriate, making two initial competitors, $\hat{\vec{\beta}}^{OF}_{post}$ and $\hat{\vec{\beta}}^{UF}_{post}$, inefficient. Our SRE results were strongly consistent with those of previous studies including (Ahmed 2014; Ahmed and Yüzbaşı 2017; Reangsephet *et al.* 2018).

11.5. Conclusion

In this work, we studied parameter estimation for sparse Poisson regression models in high-dimensional regimes. We used LASSO and adaptive LASSO as the predictor screening methodologies and proposed the improved estimation based on linear shrinkage, PT, and Stein-type shrinkage strategies for post-screening parameter inference. Via Monte Carlo simulations, the predictor screening performance of LASSO and adaptive LASSO was compared, measured from the selection percentage, and the performance of the proposed post-screening estimators was evaluated, measured from the SRE.

The results showed that the sets of important predictors obtained from LASSO and adaptive LASSO screening were different, introducing different parsimonious models induced by different sparsity structures. Neither LASSO nor adaptive LASSO provided appropriate parsimony in all situations. However, adaptive LASSO was shown to screen a smaller number of significant predictors. LASSO introduced overfitting when there were many predictors with very weak signals, which should be considered as noise when predicting the response. Adaptive LASSO produced underfitting when very weak signals became moderate, eliminating these significant signals.

After the predictor screening stage, inappropriate screening consequently resulted in the ML estimators based on the screening results becoming unreliable and ineffective. Regardless of whether predictor screening results were trustworthy or not, the use of estimators based on a Stein-type shrinkage strategy is suggested. However, when the condition of Stein-type shrinkage-based estimators ($p_2 \geq 3$) is not satisfied, we recommend using PT-based estimators, given their superior performance.

11.6. Acknowledgments

The research of Professor S. Ejaz Ahmed was partially supported by the Natural Sciences and Engineering Research Council of Canada and Bualuang ASEAN Chair Professorship, Thammasat University. Orawan Reangsephet and Supranee Lisawadi gratefully acknowledge the financial support provided by the Faculty of Science and Technology, Thammasat University.

11.7. References

Agresti, A. (2015). *Foundations of Linear and Generalized Linear Models*. John Wiley and Sons, New Jersey.

Ahmed, S.E. (2014). *Penalty, Shrinkage and Pretest Strategies: Variable Selection and Estimation*. Springer, New York.

Ahmed, S.E. and Yüzbaşı, B. (2016). Big data analytics: Integrating penalty strategies. *International Journal of Management Science and Engineering Management*, 11(2), 105–115.

Ahmed, S.E. and Yüzbaşı, B. (2017). High dimensional data analysis: Integrating submodels. *Big and Complex Data Analysis*, 285–304.

Bancroft, T.A. (1944). On biases in estimation due to the use of preliminary tests of significance. *The Annals of Mathematical Statistics*, 15(2), 190–204.

James, W. and Stein, C. (1992). Estimation with quadratic loss. *Breakthroughs in Statistics*. Springer, New York.

Karbalaee, M.H., Tabatabaey, S.M.M., Arashi, M. (2019). On the preliminary test generalized Liu estimator with series of stochastic restrictions. *Journal of The Iranian Statistical Society*, 18(1), 113–131.

Li, Y., Hong, H.G., Ahmed, S.E., Li, Y. (2018). Weak signals in high-dimensional regression: Detection, estimation and prediction. *Applied Stochastic Models in Business and Industry*, 35(2), 283–298.

Myers, R.H., Montgomery, D.C., Vining, G.G., Robinson, T.J. (2012). *Generalized Linear Models: With Applications in Engineering and the Sciences*. John Wiley and Sons, Hoboken, New Jersey.

Park, M.Y. and Hastie, T. (2007). L1 regularization path algorithm for generalized linear models. *Journal of the Royal Statistical Society: Series B (Statistical Methodology)*, 69(4), 659–677.

R Core Team (2018). *R: A Language and Environment for Statistical Computing*. R Foundation for Statistical Computing, Vienna, Austria.

Reangsephet, O., Lisawadi, S., Ahmed, S.E. (2018). Improving estimation of regression parameters in negative binomial regression model. *International Conference on Management Science and Engineering Management*, 265–275.

Santos, J.A. and Neves, M.M. (2008). A local maximum likelihood estimator for Poisson regression. *Metrika*, 68(3), 257–270.

Stein, C. (1956). The admissibility of Hotelling's -test. *The Annals of Mathematical Statistics*, 27(3), 616–623.

Wahid, A., Khan, D.M., Hussain, I. (2017). Robust adaptive lasso method for parameter's estimation and variable selection in high-dimensional sparse models. *PloS One*, 12(8), e0183518.

Yüzbaşı, B., Ahmed, S.E., Aydın, D. (2017). Ridge-type pretest and shrinkage estimations in partially linear models. *Statistical Papers*, 1–30.

Zou, H. (2006). The adaptive lasso and its oracle properties. *Journal of the American Statistical Association*, 101(476), 1418–1429.

12

Groundwater Level Forecasting for Water Resource Management

The actual increase of groundwater demand, which, in the Mediterranean regions, depends on several natural (e.g. climate change) and human factors (e.g. agriculture, domestic and industrial use), is the cause of depauperating in both the quantity and quality of the groundwater resource. Analysis of the groundwater level time series data and predicting future trends could be an alternative way, with respect to local numerical groundwater models, to manage water use in large areas, promoting sustainable development and identifying causes of water level decline. In order to forecast groundwater trends and show the usefulness of such methodology for decision-makers, we have studied a small area located on the Tuscan coast (Italy). Results of the integrated time series analysis not only give information about future hydrologic trends, but can also be useful in understanding possible climate change and related effects in hydrologic system.

12.1. Introduction

Groundwater resource management is one of the most important pillars for civil, agricultural and industrial development. In these contexts, groundwater overexploitation is one of the main societal challenges, in particular for coastal areas where overexploitation may induce fast impoverishment and also affect groundwater quality due to seawater intrusion (Alfarrah and Walraevens 2018). Moreover, overexploitation not only impacts groundwater quality and quantity, but also energy consumption, and consequently CO_2 emissions (Pfeiffer and Lin 2012; Vetter *et al.* 2017), since deeper water levels require higher energy inputs to be abstracted. This could eventually create a vicious cycle, since overexploitation is opposed by the

Chapter written by Andrea ZIRULIA, Alessio BARBAGLI and Enrico GUASTALDI.

natural groundwater recharge processes, which are heavy related to drought events and climate change scenarios (Giorgi and Lionello 2008; Gemitzi and Lakshmi 2018). Time series analysis of groundwater levels can be a good starting point to study the overexploitation phenomenon and to detect tendency, behavior and causes of the water level drawdown (Gong *et al.* 2018). In this work, we analyze water table level time series related to four boreholes (Steccaia, Bibbona, Castagneto Carducci, San Vincenzo) located on the Etruscan Coast (from the Cecina River to San Vincenzo; Tuscany, Italy), in order to forecast future trends using ARIMA and SARIMA models (Moraes *et al.* 2018). In particular, the specific objectives are:

– time series decomposition to characterize trends, seasonality, errors and verify the autocorrelation between values;

– ARIMA and SARIMA model computation, in which the most appropriate model is selected on the base of the Corrected Akaike quality index Information Criterion (Hurvich and Tsai 1989) (AICc), and the root mean squared error (RMSE) is used to evaluate the model accuracy;

– model implementation and short-term forecasts (2019–2021) with equal confidence levels between 80% and 95%.

12.2. Materials and methods

12.2.1. *Study area*

The coastal Aquifer of the Etruscan Coast (between the Cecina River and San Vincenzo 12.1) is formed by a sequence of permeable gravel and sand layers separated by low-permeability silty-clayey deposits (Cerrina *et al.* 2010). The aquifer bottom is represented by clayey deposits in the northern and central areas and by low-permeability Ligurian units in the southern area. From a morphological point of view, the study area can be divided into two zones: a zone with a low-to-medium steep layout and an altitude between a few tens of meters and 500–550 meters a.m.s.l; a broad zone, mostly flat, with a constant slope towards the sea. The steeper zone characterizes the eastern part of the Etruscan Coast where the hills crown a vast flat area. Furthermore, several water course incisions characterize the hilly zone mainly flowing from East to West.

12.2.2. *Forecast method*

Hydroclimatic time series analysis can be well represented thanks to the use of ARIMA models, which explain auto-regression and moving average processes useful for modeling stationary series (Garima and Bhawna 2017; Chen *et al.* 2018). The whole procedure is an iterative and analytical process used to identify, estimate and verify a forecast ARIMA model. In addition, the seasonal ARIMA model (SARIMA) was studied to define seasonal components in addictions (P, D, Q), in order to represent seasonal and periodic components for water level fluctuation behaviors.

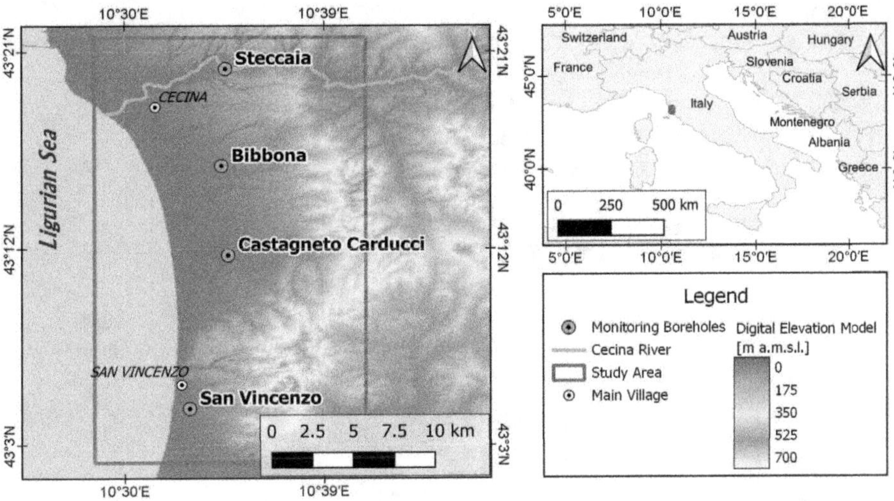

Figure 12.1. *Study area and observed borehole location. For a color version of this figure, see www.iste.co.uk/dimotikalis/analysis1.zip*

An outlier analysis was carried out by the Box–Cox transformation in order to identify their values and estimate their replacements. The stationary of the series was verified instead by using the Dickey–Fuller test, while a seasonal time series decomposition was done using Loess (STL) procedures (Cleveland *et al.* 1990) for additive time series through weighted polynomial regression. Finally, autocorrelation (ACF) and partial autocorrelation (PACF) analysis were useful in evaluating the ARIMA model values (Khorasani *et al.* 2016).

For each time series, the best model was identified by comparing the AICc with different parameterization degrees found in ACF and PACF analysis (Hurvich and Tsai 1989).

Since residual analysis provides important information on groundwater level variability not associated with cyclical and seasonal systematic phenomena (i.e. residuals should be random), the residuals were studied using the Ljung–Box test. Finally, prediction forecast intervals at 80% and 95% were set and the best model (between ARIMA and SARIMA) was selected according to the lowest RMSE. The RMSE value is calculated from a training set of measures based on the difference between f(x) and fitted-(f(x)).

12.3. Results

Monthly water level data (Table 12.1) was downloaded from the Tuscan Environmental Agency's database SIR and analyzed through R software (Hyndman *et al.* 2019).

Borehole	Start	End
Steccaia	March 2005	March 2019
Bibbona	December 2010	March 2019
C. Carducci	December 2010	March 2019
San Vincenzo	November 2010	March 2019

Table 12.1. *Time series length*

First, outlier analysis and recomputation were carried out (Table 12.2) according to the Box–Cox method.

Borehole	Outliers
Steccaia	Yes
Bibbona	No
C. Carducci	No
San Vincenzo	No

Table 12.2. *Box–Cox results to find outliers in time series*

The STL decomposition analysis revealed a general decrease in monthly time series. The sites of Bibbona, Castagneto Carducci and San Vincenzo data show clear negative trends, while the Steccaia borehole appears to be relating to no-trend behavior, in which groundwater level seasonal fluctuations seem to be constant in time (Figures 12.2–12.5).

The Dickey–Fuller check results are shown in Table 12.3 and non-stationary time series were transformed to stationary with a differentiation process.

Borehole	Time series	p-value
Steccaia	Stationary	0.01
Bibbona	Non-stationary	0.63
C. Carducci	Non-stationary	0.07
San Vincenzo	Non-stationary	0.19

Table 12.3. *Dickey–Fuller test results*

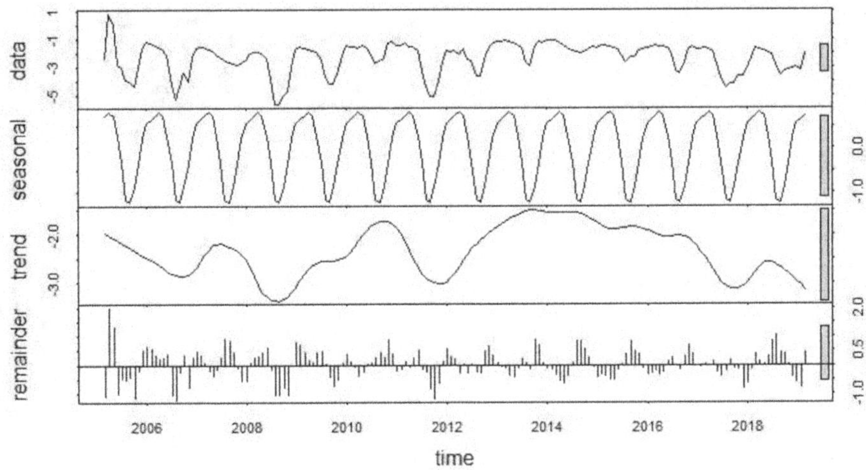

Figure 12.2. *Steccaia borehole STL decomposition*

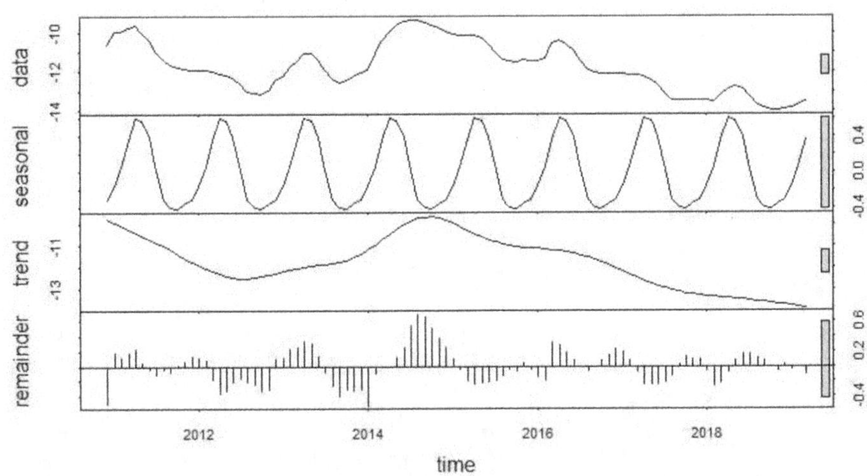

Figure 12.3. *Bibbona borehole STL decomposition*

An automatic model with the "auto.arima" function (Hyndman *et al.* 2019) was computed in order to obtain the best SARIMA model with a minimum AICc and MLE index. In Table 12.5, these model values are represented for all of the time series.

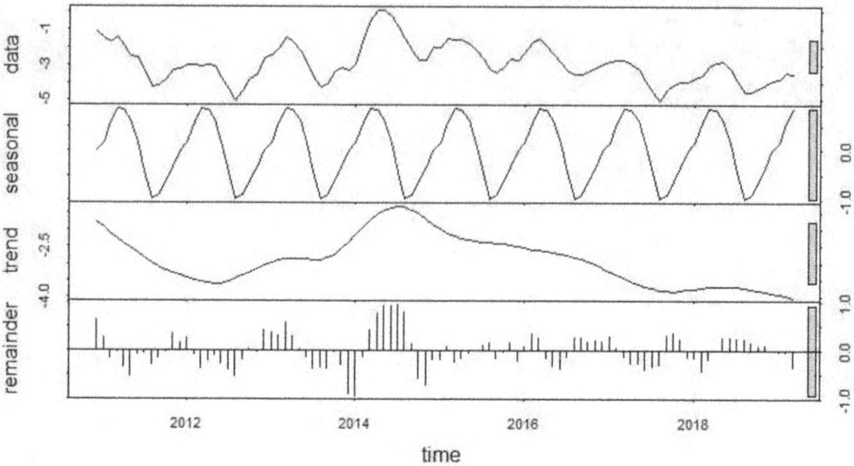

Figure 12.4. *Castagneto Carducci borehole STL decomposition*

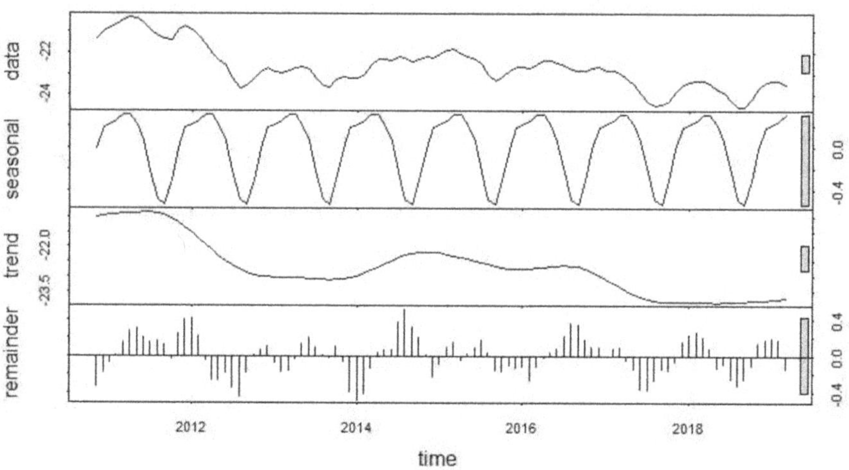

Figure 12.5. *San Vincenzo borehole STL decomposition*

The ARIMA model was, instead, set manually with the ACF and PACF analysis (Figure 12.6) according to lower AICc values. The residual analysis of fitted values was performed to model the evaluation, and the statistical results ($\alpha = 0.05$) are shown in Table 12.4. Finally, the ACF residual distribution with residual histogram confirms normal assumption for ARIMA models (Figure 12.7).

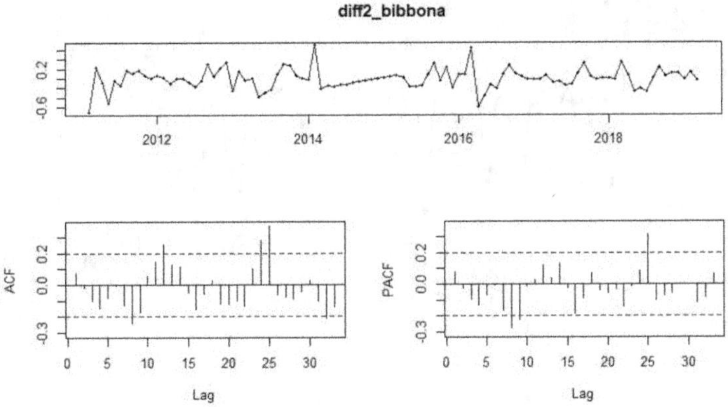

Figure 12.6. *ACF and PACF analysis to ARIMA parameter evaluation in Bibbona borehole*

Borehole	ARIMA model	p-value
Steccaia	5,0,5	0.11
Bibbona	5,2,3	0.10
C. Carducci	3,2,5	0.45
San Vincenzo	2,1,4	0.35

Table 12.4. *ARIMA model and residual normality analysis*

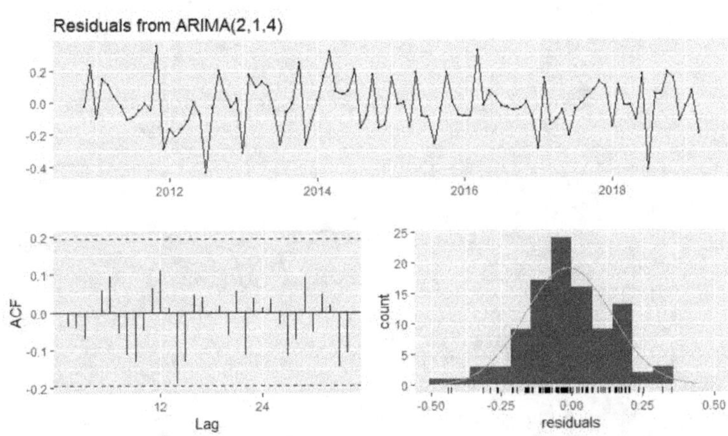

Figure 12.7. *San Vincenzo borehole residual analysis. For a color version of this figure, see www.iste.co.uk/dimotikalis/analysis1.zip*

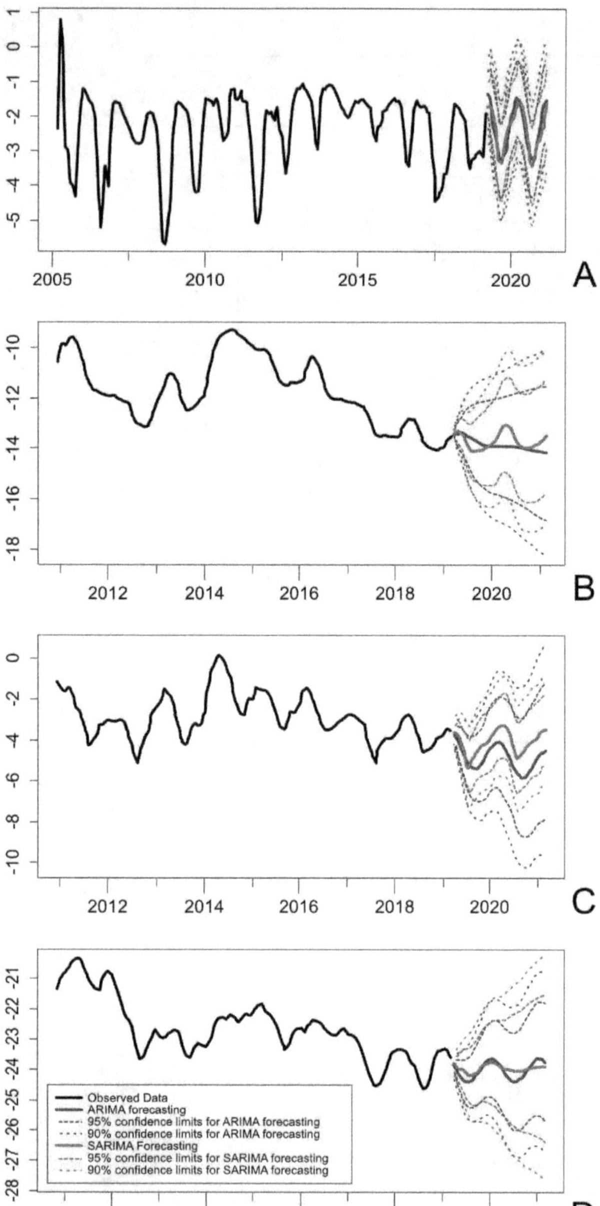

Figure 12.8. *ARIMA (blue) and SARIMA (red) groundwater level (m) forecast in Steccaia (A), Bibbona (B), Castagneto Carducci (C) and San Vincenzo (D) boreholes. For a color version of this figure, see www.iste.co.uk/dimotikalis/analysis1.zip*

Borehole	SARIMA model
Steccaia	(2,0,0)(2,1,2)12
Bibbona	(2,0,1)(1,1,0)12
Castagneto Carducci	(2,0,0)(1,1,0)12
San Vincenzo	(2,1,0)(1,0,0)12

Table 12.5. *Best SARIMA model, for each time series, obtained with the "auto.arima" function*

Borehole	SARIMA model	ARIMA
Steccaia	0.5000	0.4796
Bibbona	0.1769	0.1856
Castagneto Carducci	0.3189	0.3132
San Vincenzo	0.1593	0.1520

Table 12.6. *RMSE summary of the ARIMA and SARIMA models*

Borehole	Month	Forecast value (m)	Real value (m)	Delta value (cm)
Steccaia	April	-1.34	-1.89	0.55
	May	-1.34	-1.89	0.55
	June	-1.34	-1.89	0.55
Bibbona	April	-13.36	-13.63	0.27
	May	-13.32	-13.73	0.41
	June	-13.46	-13.91	0.45
C. Carducci	April	-3.73	-3.60	0.13
	May	-4.04	-3.73	0.31
	June	-4.46	-4.10	0.36
S. Vincenzo	April	-23.84	-23.74	0.10
	May	-24.10	-24.03	0.08
	June	-24.28	-24.60	0.32

Table 12.7. *Comparison between monthly forecast values and real monthly values for April, May and June 2019*

The modeled groundwater levels shown in Figure 12.8 were used to forecast the following time period (April 2019 to March 2021). The model accuracy based on RMSE values is reported in Table 12.6. Table 12.7 shows monthly average forecast values compared to the observed values from the monitoring boreholes with relative differences expressed in centimeters (April, May and June 2019). In particular, we focused the comparison on these forecasted months because, at the time, only these months were available to verify real values with forecasted values. These values show a deviation between 0.08 and 0.55 centimeters for the four boreholes examined, and a general groundwater level decrement, also according to seasonality. Only the Steccaia time series shows stable levels for the observed and forecasted periods. However, the differences between Steccaia and the other boreholes could be explained by the proximity of the Cecina River (i.e. a more stable source of groundwater recharge) to Steccaia.

12.4. Conclusion

Finally, a time series analysis for groundwater water table level conditions, performed for the large Tuscan coastal zone, is presented as a comparative study of ARIMA and SARIMA models to evaluate water table level fluctuations. The fluctuations are caused by changes in precipitation between seasons and years. So, seasonal models are more appropriate to explain hydrological phenomena.

The designated models SARIMA and ARIMA were verified in terms of best information criterion (AICc) and model accuracy (RMSE). In our case, by comparing four auto-setting SARIMA models and their relative four ARIMA manual-setting models, we achieved similar forecast errors with a very small difference between RMSE values. In particular, three ARIMA models are slightly better compared to their respective SARIMA models. This fact could suggest that a manual setting, with a more accurate parameter model evaluation, leads to a better result. Consequently, by applying a manual setting to the SARIMA parameters, this small difference can be improved, giving a better representation of seasonality phenomena that an auto-setting cannot catch.

The time series decomposition based on monthly level data reveals negative trends (possible overexploitation) at Bibbona, Castagneto Carducci and San Vincenzo. This behavior could generally continue in the coming years (as we can see from forecast scenarios), while groundwater level fluctuations at Steccaia could remain constant over the next few years.

Time series analysis and forecasts applied to groundwater resources could not only provide a useful indication for regulatory agencies and water utilities to detect overexploitation, develop specific local policies and use more efficiently available funds, but also to develop and complete further studies on the possible effect that climate change has on natural groundwater recharge phenomena.

12.5. References

Alfarrah, N. and Walraevens, K. (2018). Groundwater overexploitation and seawater intrusion in coastal areas of arid and semi-arid regions. *Water*, 10, 143 [Online]. Available at: https://doi.org/10.3390/10020143.

Cerrina Feroni, A., Da Prato, S., Doveri, M., Ellero, A., Lelli, M., Marini, L., Masetti, G., Nisi, B., Raco, B. (2010). Corpi idrici sotterranei della Val di Cecina. *Memorie descrittive della Carta Geologica d'Italia*, ISPRA, Rome.

Chen P., Niu, A., Liu, D., Jiang, W., Ma, B. (2018). Time series forecasting of temperatures using SARIMA: An example from Nanjing. *IOP Conference Series: Materials Science and Engineering*, 394(5).

Cleveland R.B., Cleveland, W.S., McRae, J.E., Terpenning, I. (1990). STL: A seasonal-trend decomposition procedure based on loess. *Journal of Official Statistics*, 6, 3–73.

Garima J. and Bhawna, M. (2017). A study of time series models ARIMA and ETS [Online]. Available at: https://ssrn.com/abstract=2898968 or http://dx.doi.org/10.2139/ssrn.2898968.

Gemitzi, A. and Lakshmi, V. (2018). Evaluating renewable groundwater stress with GRACE data in Greece. *Groundwater*, 56(3), 501–514 [Online]. Available at: https://doi.org/10.1111/gwat.12591.

Giorgi, F. and Lionello, P. (2008). Climate change projections for the Mediterranean region. *Global and Planetary Change*, 63(2), 90–104 [Online]. Available at: https://doi.org/10.1016/j.gloplacha.2007.09.005.

Gong, Y., Wang, Z., Xu, G., Zhang, Z. (2018). A comparative study of groundwater level forecasting using data-driven models based on ensemble empirical mode decomposition. *Water*, 10, 730 [Online]. Available at: https://doi.org/10.3390/w10060730.

Hurvich, C.M. and Tsai, C.L. (1989). Regression and time series model selection in small samples. *Biometrika*, 76(2), 297–307.

Hyndman R., Athanasopoulos, G., Bergmeir, C., Caceres, G., Chhay, L., O'Hara-Wild, M., Petropoulos, F., Razbash, S., Wang, E., Yasmeen, F. (2019). Forecast: Forecasting functions for time series and linear models [Online]. Available at: http://pkg.robjhyndman.com/forecast.

Khorasani, M., Ehteshami, M., Ghadimi, H., Salari, M. (2016). Simulation and analysis of temporal changes of groundwater depth using time series modeling. *Modeling Earth Systems and Environment*, 2, 90 [Online]. Available at: https://doi.org/10.1007/s40808-016-0164-0.

Moraes Takafuji, E.H., da Rocha, M.M., Manzione, R.L. (2018). Groundwater level prediction/forecasting and assessment of uncertainty using SGS and ARIMA models: A case study in the Bauru Aquifer system (Brazil). *Natural Resources Research* [Online]. Available at: https://doi.org/10.1007/s11053-018-9403-6.

Pfeiffer, L. and Lin, C.Y.C. (2012). Groundwater pumping and spatial externalities in agriculture. *Journal of Environmental Economics and Management*, 64(1), 16–30 [Online]. Available at: https://doi.org/10.1016/j.jeem.2012.03.003.

SIR (N/A) Settore Idrologico e Geologico Regionale [Online]. Available at: http://www.sir.toscana.it.

Vetter, S.H., Sapkota, T.B., Hillier, J., Stirling, C.M., Macdiarmid, J.I., Aleksandrowicz, L., Green, R., Joy, E.J.M., Dangour, A.D., Smith, P. (2017). Greenhouse gas emissions from agricultural food production to supply Indian diets: Implications for climate change mitigation. *Agriculture, Ecosystems and Environment*, 237, 234–241 [Online]. Available at: https://doi.org/10.1016/j.agee.2016.12.024.

13

Phase I Non-parametric Control Charts for Individual Observations: A Selective Review and Some Results

Parametric control charts are based on assumptions of a specific form for the underlying process distribution. One challenge, commonly encountered in non-manufacturing processes, is that the underlying process distribution of the quality characteristic(s) significantly deviates from normality and is usually not known. Therefore, statistical properties of the most commonly used parametric control charts are highly affected and their performance generally deteriorates. There are many applications in non-manufacturing operations, where there is insufficient information to justify such assumptions and a normal transformation of the observations is feasible; however, this is at the expense of the interpretability of the analysis. To this end, non-parametric control charts with a minimal set of distributional assumption requirements are in high demand, and such distribution-free charts have been proposed in recent years. However, most of the existing non-parametric control charts are designed for Phase II monitoring. Moreover, little has been done in developing non-parametric Phase I control charts, especially for individual observations that are prevalent in non-manufacturing applications. In this contribution, we bring a selective review forward to 2020, discuss the main ideas that shaped the field of univariate Phase I non-parametric process monitoring (focusing on individual observations) and relate them to the recently developed control charts in existing literature. The charts were found to have stable in-control properties, be robust against a broad class of out-of-control scenarios and process distributions, and be surprisingly efficient in comparison with their parametric counterparts.

Chapter written by Christina PARPOULA.

13.1. Introduction

Statistical process control (SPC) includes a wide range of statistical procedures and problem-solving techniques with powerful applications in diverse scientific areas. Control charts form the core of SPC for monitoring the characteristics of a process over time (Montgomery 2013) and are perhaps the most widely used techniques in practice. The SPC chart applications are divided into two phases. In Phase I (the initial phase), a retrospective analysis is performed on historical data for characterizing the in-control (IC) state. More specifically, in Phase I, the control charts are used to access the stability of the process, determine the process distribution and estimate the population parameters that are required for setting up the Phase II control charts. Phase I analysis constitutes the most important stage of SPC since the success of ongoing prospective process monitoring in Phase II critically depends on the success of Phase I stage. The importance of an effective Phase I analysis using appropriate methods and metrics is highlighted in several reviews of Phase I control charting such as those, among others, of Chakraborti *et al.* (2009) and Jones-Farmer *et al.* (2014).

13.1.1. *Background*

For many years, SPC chart applications have typically been found in industrial manufacturing operations. However, in the last few decades, control chart applications have begun to popularize them in non-manufacturing industries, with applications in areas such as healthcare monitoring, applied behavior analysis, environment monitoring, banking and insurance, etc. (Parpoula and Karagrigoriou 2020).

As these non-manufacturing applications continue to spread, new research issues and challenges inevitably arise, which require developing more effective control charts, especially designed for different contexts and analysis purposes. One such challenge commonly encountered in non-manufacturing processes is that the underlying process distribution of the quality characteristic(s) being studied significantly deviates from normality and is usually unknown. Thus, statistical properties of the most commonly used parametric charts (based on an assumption of a particular underlying process distribution, such as the normal one) are highly affected and their performance generally deteriorates (their false alarm rates are inflated, often significantly, when the underlying assumptions are not met). Besides, in the early Phase I stage, it is often not possible to know much about the underlying distribution; hence, a specific distributional assumption (such as normality) cannot be reasonably justified (a discussion about this issue can be found, among others, in Woodall (2000) and Coelho *et al.* (2015)). It is therefore important to develop control charting techniques that are either more robust or do not require the assumption of normality of the quality characteristic(s) being monitored (Ning *et al.* 2015).

Given these concerns, non-parametric control charts can provide a useful and robust alternative to the practitioner and appear to be ideal candidates for expanding

Phase I applications into broader scientific areas. Several authors, for example, Woodall and Montgomery (1999), Woodall (2000) and Ning *et al.* (2015) have recognized the need to develop non-parametric control charts. One of the major advantages of the non-parametric control charts is that we can control the false alarm probability (FAP) without assuming the form and the shape of the underlying process distribution. This appealing property makes the non-parametric control charts particularly suitable for Phase I applications, since if the IC properties of a chart are not stable and robust, its further application in Phase II becomes somewhat meaningless as pointed out by Coelho *et al.* (2015).

For a comprehensive review of the development of Phase I and Phase II non-parametric control charts for univariate and multivariate process monitoring up until 2020, the interested reader may refer to Chakraborti and Graham (2019). In this contribution, our goal is not to present the vast fast-growing non-parametric control chart literature in an exhaustive fashion, but rather to present the main ideas that have shaped the field of univariate Phase I non-parametric process monitoring (focusing on individual observations) and to relate them to recently developed control charts existing in the literature.

13.1.2. *Univariate non-parametric process monitoring*

A lot of work has been done in univariate non-parametric SPC over the last few years, and numerous control charts have appeared in statistical and quality engineering literature. Most existing univariate non-parametric control charts use either charting statistics that are based on ranking/ordering information of the observations across different time points, or some even more recent work dealt with this problem from a change-point model (CPM) perspective, developing univariate non-parametric control charts using rank or likelihood ratio-based statistics. For a review of these approaches, the interested reader may refer, among others, to Ning *et al.* (2015) and Qiu (2019) and references therein.

It is worth noting that the vast majority of the existing univariate non-parametric control charts are designed for prospective Phase II process monitoring, assuming the availability of an IC reference sample from a Phase I analysis. As such, they are not suitable for retrospective analysis in Phase I control. Despite the widely recognized importance of Phase I control in SPC, univariate non-parametric Phase I control charts have not received enough attention; hence, further work in this field would be welcome, as also highlighted by Chakraborti and Graham (2019).

One such charting mechanism is the non-parametric Phase I Shewhart-type chart for subgroup location, called the mean-rank chart, proposed by Jones-Farmer *et al.* (2009). This chart is based on the well-known Kruskal–Wallis test, and the authors proposed ranking each observation in each subgroup with respect to the entire sample and calculating the standardized average of the ranks in each subgroup test. Graham *et al.* (2010) proposed a non-parametric Phase I median chart. This chart is based on the subgroup precedence counts from the pooled median. A head-to-head performance comparison of the mean-rank and median charts was performed by Coelho *et al.* (2015). While these charts are useful, since both of them are distribution-free, and are effective at detecting shifts in the location of the process and establishing control in Phase I, one potential limitation is that they are not directly usable with individual data, and there are constraints on the subgroup size and the number of subgroups, as pointed out by Chakraborti and Graham (2019).

Along these lines, Hawkins and Deng (2010) and Ross and Adams (2012) (while focusing on Phase II monitoring) briefly discussed univariate non-parametric Phase I control charts based on a CPM for individual observations. Hawkins and Deng (2010) used a two-sample Mann–Whitney (MW) test to detect a possible change in the location parameter, whereas Ross and Adams (2012) suggested two control charts based on the comparison between two empirical cumulative distribution functions by either the Cramer–von Mises (CvM) or Kolmogorov–Smirnov (KS)-type test statistic. Moreover, Ross (2015) provided an R package which contains several implementations of the CPM framework in both batch (Phase I) and sequential (Phase II) settings, where the sequences may contain either a single or multiple change-points.

Capizzi and Masarotto (2013) introduced a new distribution-free strategy for detecting shifts in process location and/or scale in Phase I, based on a recursive segmentation and permutation (RS_P) approach, which leads to an effective procedure both in terms of maintaining an FAP and out-of-control (OC) performance, according to the derived simulation results. Ning *et al.* (2015) proposed a non-parametric Phase I control chart for monitoring the process mean with individual observations. The proposed chart is based on a CPM formulation and on the empirical likelihood ratio (ELR) test, and was found to compare favorably with other recently developed charts for individual observations in terms of signal probability. Parpoula and Karagrigoriou (2020) constructed adjusted generalized likelihood ratio test-based control charts for detecting a step shift either in the mean and/or variance of a sequence of individual observations. Note that the resulting control charts, with permutation-based control limits, were distribution-free. Recently, Parpoula (2020), following only the “level” part of the RS_P approach described by Capizzi and Masarotto (2013), presented an effective Phase I distribution-free procedure which is directly implementable with individual data. Specifically, the test statistics and the possible change-points are computed based on an algorithmic procedure, which is recursive in the sense that each change-point is conditionally selected on the results of the previous stages and is based on a segmentation approach in the sense that the algorithm transforms individual time series data into multiple segments between change-points. Thus, it reflects the individual time series behavior in each generated segment.

Conclusively, little has been done to develop univariate Phase I non-parametric control charts, and there is a need for more work in this field. Specifically, more research needs to be done regarding Phase I analysis for small subgroup sizes and even for individual observations because, although traditional SPC applications of control charts involve sub-grouped data, recent advances have led to more and more instances where individual measurements are collected over time (Coelho *et al.* 2015). Later in this chapter, an extensive simulation study is conducted comparing existing Phase I parametric and non-parametric control charts – directly usable with individual data – in order to demonstrate their properties, advantages and disadvantages as well as their efficacy under different distributions and shift patterns.

The rest of the chapter is organized as follows. In section 13.2, we briefly formulate the problem from a CPM perspective on which all control charts examined in this study are based. Section 13.3 is devoted to comparing the performance of competing Phase I parametric and non-parametric control charts for individual observations. Finally, section 13.4 gives a number of concluding remarks and possible future research along the same lines.

13.2. Problem formulation

As mentioned earlier, there are many practical manufacturing as well as non-manufacturing applications in which only individual observations can be collected for SPC implementation (Ning *et al.* 2015). Let X represent the quality characteristic to be monitored, whose distribution, denoted by F, is unknown. Let $E(X) = \mu$ and Var$(X) = \sigma^2$ denote the mean and variance of X, respectively. When the process is IC, we assume that $E(X) = \mu_0$ and Var$(X) = \sigma_0^2$, where both μ_0 and σ_0^2 are unknown. Here, our focus is on monitoring the process mean μ in a Phase I application; we therefore further assume that the process variance remains at σ_0^2 during the course of the Phase I application (Ning *et al.* 2015).

The *a posteriori* (or offline) problem of change-point detection arises when the series of observations is complete at the time to process it. We only consider in this work the *a posteriori* problem; hence, our goal consists of recovering the configuration of change-points using the whole observed series. Let x_1, x_2, ..., x_m be m independent random observations from F, collected for the purpose of a Phase I application. That is, each x_i, $i = 1, 2, \ldots, m$, represents the only individual observation sampled from F at the ith sampling period. Under this framework, the conventional Shewhart chart (Shewhart 1939), i.e. the X-chart, plots each x_i against predetermined control limits, and the chart signals when any of the x_i's plots are outside of the control limits (Ning *et al.* 2015).

We then formulate the problem from a CPM perspective on which all Phase I parametric and non-parametric control charts examined in this study are based. When the process is IC (stable), we assume the individual observations (x_i, $i = 1, 2, \dots, m$) to be independent and drawn from an unknown but common cumulative distribution function (c.d.f.), $F_0(x)$, whereas when the process is OC (unstable), the observations can be thought to be drawn by the following general non-parametric multiple CPM:

$$x_i \sim \begin{cases} F_0(x) & \text{if } 0 < i \le \tau_1, \\ F_1(x) & \text{if } \tau_1 < i \le \tau_2, \\ \vdots & \vdots \\ F_k(x) & \text{if } \tau_k < i \le m, \end{cases} \qquad [13.1]$$

where $0 < \tau_1 < \tau_2 < \dots < \tau_k < m$ represent k change-points and $F_r(\cdot)$, $r = 0, \dots, k$, are unknown c.d.f. for x which, at one or several times, may shift in position. The shift times are also assumed to be unknown.

Note here that model [13.1] includes a wide variety of OC situations, i.e. processes which are not stable everywhere. It can describe processes subject to *step shifts* described, for example, by

$$x_i \sim \begin{cases} F_0(x) & \text{if } 0 < i \le \tau_1, \\ F_1(x) & \text{if } \tau_1 < i \le \tau_2, \\ F_2(x) & \text{if } \tau_2 < i \le m, \end{cases} \qquad [13.2]$$

for some unknown τ_1, τ_2, $F_0(x)$, $F_1(x)$, $F_2(x)$; *transient shifts* described, for example, by

$$x_i \sim \begin{cases} F_0(x) & \text{if } 0 < i \le \tau_1, \\ F_1(x) & \text{if } \tau_1 < i \le \tau_2, \\ F_0(x) & \text{if } \tau_2 < i \le m, \end{cases} \qquad [13.3]$$

for some unknown τ_1, τ_2, $F_0(x)$, $F_1(x)$; and even *isolated shifts* described, for example, by

$$x_i \sim \begin{cases} F_0(x) & \text{if } i \neq \tau, \\ F_1(x) & \text{if } i = \tau. \end{cases} \qquad [13.4]$$

for some unknown τ, $F_0(x)$, $F_1(x)$.

Since we are interested in performing Phase I analysis, we test/verify the hypotheses H_0, where the process was IC $\forall\ i, i = 1, \dots, m$ ($k = 0$) versus H_1, $\exists\ i, i = 1, \dots, m$, where process was OC ($k > 0$). This hypothesis testing system requires the specification of a nominal FAP which is the probability of at least one false alarm out of m Phase I observations, where a false alarm is defined as the event that a single charting statistic plot on or outside the control limit(s) when the process is IC. Therefore, choosing an acceptable FAP value, say α, we test the stability over time of the level parameter. Here, the typical difference between parametric and non-parametric tests takes place. A parametric test can guarantee a prescribed FAP only if $F_0(\cdot)$ belongs to a particular family of probability distribution (such as normal), whereas a non-parametric procedure enables practitioners control the FAP without any knowledge of the specific distribution from which individual observations are drawn.

13.3. A comparative study

In this section, an extensive simulation study is conducted comparing existing Phase I parametric and non-parametric control charts for individual observations under the CPM framework. All of the considered charts try to detect level changes and are able to detect both increases and decreases in the parameter monitored. We compare both the IC and OC performances of the Phase I control charts. The IC performance indicates how robust the chart is with respect to the nominal FAP value. A chart with attained FAP values very close to the nominal value is favored. On the other hand, the OC comparison involves comparing the probabilities of alarm (at least one signal) under some OC condition when the charts have the same (or roughly the same) nominal FAP value. The chart with the highest probability of at least one signal under the OC condition is favored.

13.3.1. *The existing methodologies*

We investigate here the performance of existing Phase I parametric and non-parametric counterparts for detecting a single or multiple mean shifts in a sequence of individual observations. In the context of a CPM, we consider:

1) the conventional Shewhart (1939) (X-chart): The sample standard deviation is computed as $\frac{\overline{MR}}{d_2}$, where $\overline{MR}$ is the average of moving ranges of length 2, and $d_2 = 1.128$ is the adjusting constant for estimating the population standard deviation using the average of sample ranges of size 2. Note that the observations are standardized before being plotted on the X-chart;

2) *adjusted generalized likelihood ratio* (AdjGLR): The control statistic is based on the generalized likelihood ratio (GLR) test, as in Sullivan and Woodall (1996), computed under a Gaussian assumption. However, the control limits are here computed by permutation so that the desired FAP is guaranteed. The resulting AdjGLR test statistic-based chart (with permutation-based control limits) is distribution-free and is used to detect location shifts in a stream with a (possibly unknown) non-Gaussian distribution;

3) *rank-based adjusted generalized likelihood ratio* (RAdjGLR): Identical to AdjGLR, except that a preliminary rank transformation of the original data is now used (to improve the performance in the case of nonnormal data). The resulting RAdjGLR

test statistic-based chart (with permutation-based control limits) is distribution-free and is also used to detect location shifts in a stream with a (possibly unknown) non-Gaussian distribution. The interested reader may refer to Parpoula and Karagrigoriou (2020) for more detail regarding the AdjGLR and RAdjGLR test statistic-based charts;

4) *recursive segmentation and permutation* (RS_P): The "level" part of the RS_P procedure, as in Capizzi and Masarotto (2013), directly implementable with individual data as in Parpoula (2020), used to detect location shifts in a stream with a (possibly unknown) non-Gaussian distribution;

5) *Mann–Whitney* (MW): MW test statistic-based chart, as in Hawkins and Deng (2010), used to detect location shifts in a stream with a (possibly unknown) non-Gaussian distribution;

6) *Cramer–von Mises* (CvM): CvM test statistic-based chart, as in Ross and Adams (2012), used to detect arbitrary changes in a stream with a (possibly unknown) non-Gaussian distribution;

7) *Kolmogorov–Smirnov* (KS): KS test statistic-based chart, as in Ross and Adams (2012), used to detect arbitrary changes in a stream with a (possibly unknown) non-Gaussian distribution;

8) *empirical likelihood ratio* (ELR): ELR test statistic-based chart, as in Ning *et al.* (2015), used to detect location shifts in a stream with a (possibly unknown) non-Gaussian distribution.

13.3.2. *Simulation settings*

We follow similar simulation settings as in Ning *et al.* (2015). Performance is evaluated using the probability of giving an alarm that has been estimated by simulations using Monte Carlo replications. We used 300,000 Monte Carlo replications for all considered cases. As for the ELR test statistic-based chart, 10,000 replications were used, as in Ning *et al.* (2015), to sustain reasonable simulation execution times. We only consider one change-point location in the simulation study for the sake of simplicity. Under the assumptions of normally distributed data and known IC parameters, the Shewhart control chart based on 3-sigma limits yields an FAP_0 of 0.0027. However, we set here the IC signal probability (nominal FAP_0) to be 0.005 to be more lenient with the demands placed on the control chart. We consider two sample sizes $m = 50$ and $m = 100$. Similar conclusions have been obtained using other values of m but we omit them to conserve space. We considered the following IC distribution functions:

– *normal*, standard normal distribution;

– *exponential*, negative exponential distribution of mean equal to 1;

– *Student*, Student's t distribution with 3 degrees of freedom (t_3).

The exponential and t_3 distributions are chosen to represent the cases of continuous skewed and longer-tailed distributions. The exponential is a positive asymmetric distribution with support $[0, \infty]$, whereas Student is a symmetric distribution with heavy tails.

It is important to note here that the control limits used in the X-chart are specifically chosen so that the signal probability is approximately 0.005 when the IC process follows a standard normal distribution. However, when the IC process follows a nonnormal distribution, the actual IC signal probability is likely to be different if the same control limits are used. The X-chart is a parametric method assuming a normal IC process; however, in theory, it can be designed to achieve a prescribed FAP under any known IC distribution. Nevertheless, due to the fact that it is difficult to have precise information on the underlying distribution and distributional assumptions cannot be verified before checking the stability of the process, as explained by Woodall (2000) and highlighted by Capizzi and Masarotto (2013), we consider here only control limits based on the normal IC distribution.

Similar to the existing literature in which Phase I control charts are evaluated and compared, we consider the typical OC scenario of *a step change in the process mean.* In practical applications, we do not know in advance when the transition between an IC and OC state occurs. Thus, the positions of the change times were assumed to be random, and were chosen according to the following three patterns:

Pattern I: The OC period is longer than the IC period (40 OC observations and 10 IC observations for $m = 50$, 80 OC observations and 20 IC observations for $m = 100$);

Pattern II: The IC period is as long as the OC period (25 OC observations and 25 IC observations for $m = 50$, 50 OC observations and 50 IC observations for $m = 100$);

Pattern III: The IC period is longer than the OC period (10 OC observations and 40 IC observations for $m = 50$, 20 OC observations and 80 IC observations for $m = 100$).

Thus, for $m = 50$, the change-point locations considered are i= 10, 25 and 40. As for $m = 100$, the change-point locations are set at i= 20, 50 and 80.

For each of these Patterns (I–III), we simulate and compare the OC signal probabilities of the competing charts for a range of process mean shifts, as characterized by $\mu_1 = \mu_0 + \delta\sigma_0$, where μ_0 and σ_0 are the IC mean and standard deviation, respectively, of a given distribution. The OC shift configurations considered here are of size $\delta = 0.25, 0.50, 1.00, 1.50, 2.00, 3.00$ (in units of standard deviations).

REMARK 13.1.– *The presence of outlying observations in the Phase I sample is not examined in this study. Detecting isolated shifts in a non-parametric framework is possible only when the data is gathered in the form of subgroups. In a sequence of individual observations (as in our case), an isolated shift due to an isolated outlier cannot be detected without additional information about the shape of the distribution (Capizzi and Masarotto 2013). Moreover, existing non-parametric Phase I charts have been proven ineffective in detecting the presence of outlying observations in the Phase I sample (Ning et al. 2015).*

13.3.3. *Simulation-study results*

For a given sample size, we proceed below with the simulation study, in order to evaluate the performance of the existing Phase I parametric and non-parametric control charts for individual observations, in terms of their IC and OC performances.

Normal Distribution								
m	X-chart	AdjGLR	RAdjGLR	RS_P	MW	CvM	KS	ELR
50	0.00502	0.00634	0.00616	0.00500	0.00496	0.00480	0.00480	0.00454
100	0.00504	0.00667	0.00649	0.00532	0.00499	0.00490	0.00510	0.00489
Exponential Distribution								
m	X-chart	AdjGLR	RAdjGLR	RS_P	MW	CvM	KS	ELR
50	0.42520	0.00585	0.00630	0.00573	0.00510	0.00498	0.00498	0.00892
100	0.65570	0.00882	0.00625	0.00572	0.00480	0.00493	0.00493	0.00833
Student Distribution								
m	X-chart	AdjGLR	RAdjGLR	RS_P	MW	CvM	KS	ELR
50	0.39300	0.00555	0.00642	0.00518	0.00504	0.00491	0.00493	0.00346
100	0.63920	0.01006	0.00655	0.00514	0.00501	0.00503	0.00511	0.00387

Table 13.1. *Attained FAP values for different IC distributions (desired* $FAP_0 = 0.005$*)*

13.3.3.1. *IC performance*

Table 13.1 shows that:

(i) The X-chart attains the specified FAP values only in the case of normally distributed observations (i.e. under its design conditions). Thus, the lack of IC robustness of the Phase I X-chart should be a serious concern for its implementation in practice. In general, the X-chart should be used only when the shape of the distribution is known.

(ii) The AdjGLR and RAdjGLR test statistic-based charts can be used without prior information on the IC distribution; however, their attained FAP values are generally larger than the nominal FAP and may not be preferred.

(iii) The RS_P and MW test statistic-based charts were found to perform well and to approximately guarantee the prescribed FAP in all considered cases.

The simulated IC signal probabilities for the CvM and KS test statistic-based charts are very close to but slightly more conservative than 0.005. Statistically speaking, a conservative method is one that provides less chance of rejecting the null hypothesis in comparison to some other method or some pre-defined standard.

(iv) The simulated IC signal probabilities for the ELR test statistic-based chart are close to but generally more conservative than 0.005 (except for the exponential distribution where the IC signal probabilities are larger than the nominal level at 0.005).

13.3.3.2. *OC performance*

The simulated OC signal probabilities when the process mean incurs a step change are summarized in Figures 13.1–13.3 under *normal* distribution; Figures 13.4–13.6 under *exponential* distribution; Figures 13.7–13.9 under *Student* t_3 distribution, for Patterns I–III, respectively. Note that Figures 13.1–13.9 present the OC signal probabilities for a given sample size ($m = 50$ and $m = 100$), as a function of δ, which is a measure of the shift size.

Figures 13.1–13.9 show that:

(i) The "best" chart is different under different distributions and positions of change times (Patterns I–III).

(ii) For any given control chart, the signal probability is generally higher if the sample size is larger (except for the X-chart).

(iii) For a step change in process mean the signal probability is generally higher if the change-point occurs in the middle of the Phase I sample (Pattern II) than early (Pattern I) or later (Pattern III) in the sample.

(iv) The conventional Phase I X-chart is not as effective in detecting OC processes not only when the process distribution is non-normal but also when the process follows a normal distribution.

(v) Of the existing non-parametric Phase I control charts, the AdjGLR test statistic-based chart seems to be the best overall performing chart under normal distribution, whereas the RAdjGLR, RS_P, MW, CvM, KS and ELR test statistic-based charts satisfactorily perform under a non-normal distribution.

(vi) Of the existing non-parametric Phase I control charts, the RS_P, MW and ELR test statistic-based charts seem to be the best overall performing charts in almost all of the OC scenarios considered regardless of the process distribution.

(vii) In practical Phase I applications, the exact distribution of the process is usually not readily available. The results summarized in Figures 13.1–13.9 indicate that RAdjGLR, RS_P, MW, CvM, KS and ELR test statistic-based charts are effective non-parametric Phase I control charts, since they provide reasonably good detecting power overall, under various OC scenarios and process distributions.

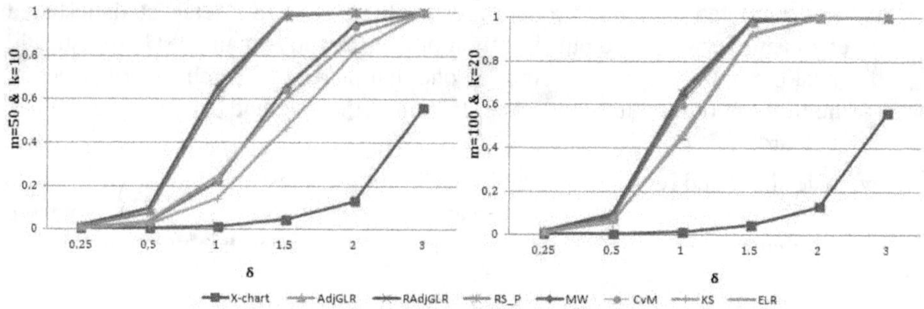

Figure 13.1. *The OC signal probabilities for IC Normal distribution (**Pattern I**). For a color version of this figure, see www.iste.co.uk/dimotikalis/analysis1.zip*

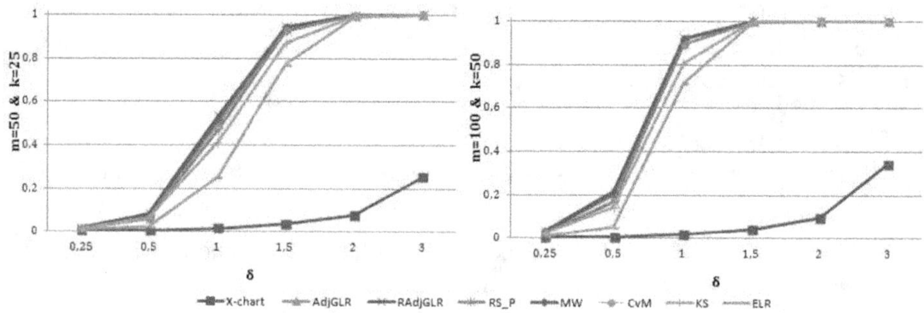

Figure 13.2. *The OC signal probabilities for IC Normal distribution (**Pattern II**). For a color version of this figure, see www.iste.co.uk/dimotikalis/analysis1.zip*

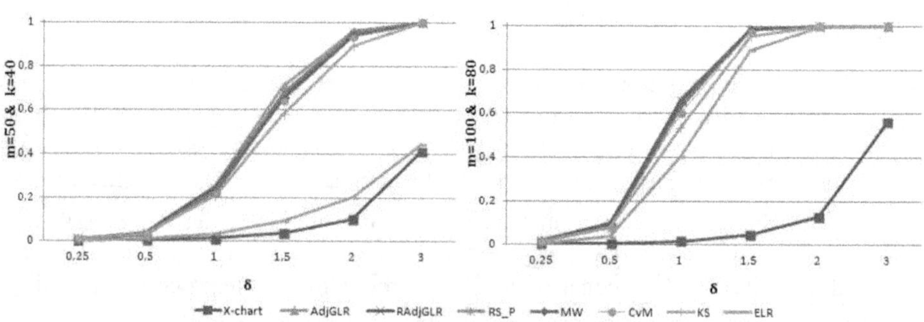

Figure 13.3. *The OC signal probabilities for IC Normal distribution (**Pattern III**). For a color version of this figure, see www.iste.co.uk/dimotikalis/analysis1.zip*

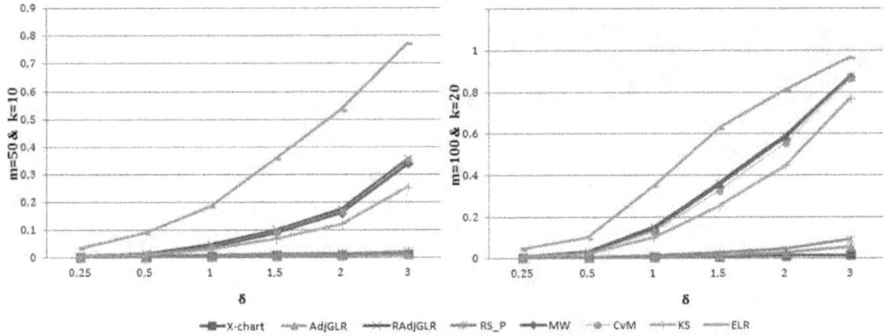

Figure 13.4. *The OC signal probabilities for IC Exponential distribution (**Pattern I**). For a color version of this figure, see www.iste.co.uk/dimotikalis/analysis1.zip*

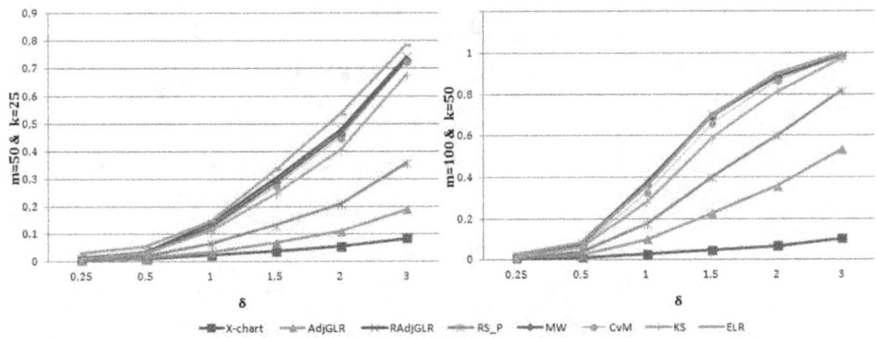

Figure 13.5. *The OC signal probabilities for IC Exponential distribution (**Pattern II**). For a color version of this figure, see www.iste.co.uk/dimotikalis/analysis1.zip*

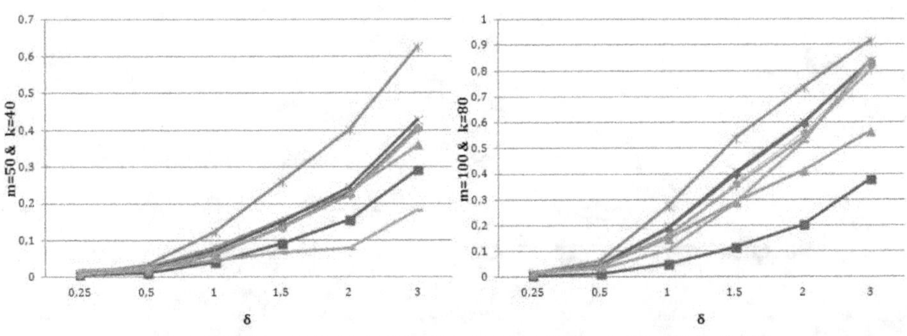

Figure 13.6. *The OC signal probabilities for IC Exponential distribution (**Pattern III**). For a color version of this figure, see www.iste.co.uk/dimotikalis/analysis1.zip*

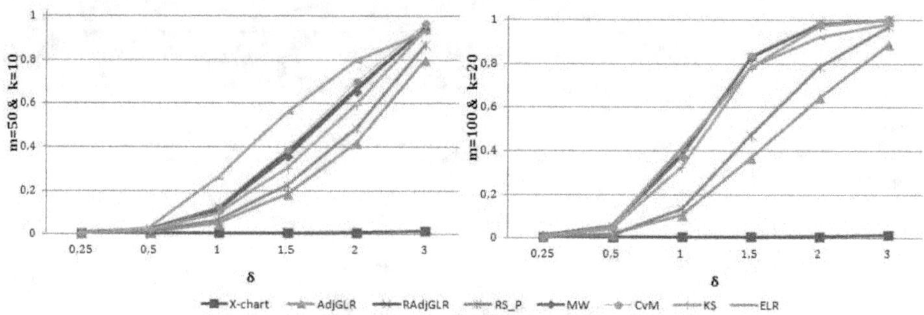

Figure 13.7. *The OC signal probabilities for IC Student distribution (**Pattern I**). For a color version of this figure, see www.iste.co.uk/dimotikalis/analysis1.zip*

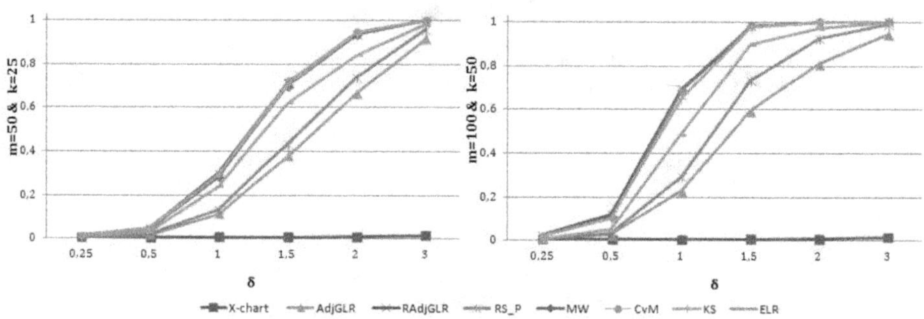

Figure 13.8. *The OC signal probabilities for IC Student distribution (**Pattern II**). For a color version of this figure, see www.iste.co.uk/dimotikalis/analysis1.zip*

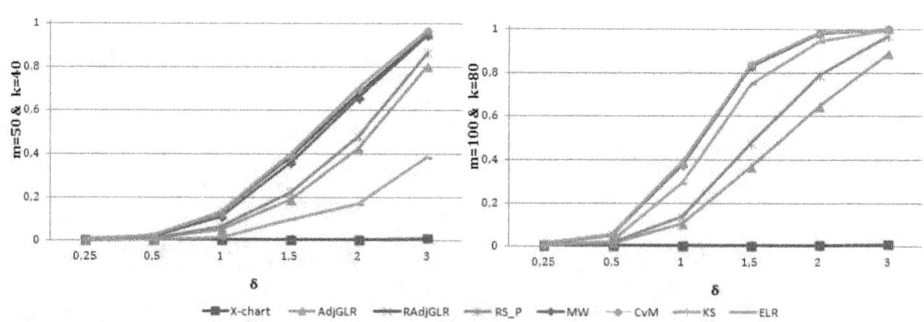

Figure 13.9. *The OC signal probabilities for IC Student distribution (**Pattern III**). For a color version of this figure, see www.iste.co.uk/dimotikalis/analysis1.zip*

13.4. Concluding remarks

In this chapter, we have examined the performance of existing Phase I parametric and non-parametric control charting techniques (based on a CPM formulation) for monitoring the process mean with individual observations. The IC and OC performances of these control schemes were thoroughly investigated using Monte Carlo replications, considering a broad class of different distributions and shift patterns. The AdjGLR and RAdjGLR test statistic-based charts can be used without prior information on the IC distribution; however, their attained FAP values were generally larger than the nominal FAP and may not be preferred. The simulated IC signal probabilities for the CvM, KS and ELR test statistic-based charts were close to but generally more conservative than the nominal FAP. The RS_P and MW test statistic-based charts were found to perform well and to guarantee approximately the prescribed FAP in all considered cases. When the process is normal, it is "surprising" that the X-chart is even less effective as its non-parametric counterparts. The OC signal probabilities of the X-chart under normal processes (as shown in Figures 13.1–13.3) should give practitioners some caution as to whether the X-chart is an effective Phase I control chart even under normal processes, when the process mean and standard deviation have to be estimated using Phase I data. As mentioned earlier, in practice, the process distribution may not be easy to determine, especially if a large Phase I sample is not available. If the prevalent concern in Phase I is that the process mean may incur a sustained shift, the derived results indicate that RS_P, MW and ELR test statistic-based charts are effective non-parametric Phase I control charts, with a satisfactory and robust performance in a variety of OC scenarios and process distributions. It is worth noting that the CPM approach is not limited to test the process mean. It would be worthwhile to study how the same CPM approach can be used to develop such non-parametric Phase I control charts with individual observations for monitoring process variability, a topic that has also received very little attention in relevant literature.

13.5. References

Capizzi, G. and Masarotto, G. (2013). Phase I distribution-free analysis of univariate data. *J. Qual. Technol.*, 45, 273–284.

Chakraborti, S. and Graham, M.A. (2019). Nonparametric (distribution-free) control charts: An updated overview and some results. *Qual. Eng.*, 31, 523–544.

Chakraborti, S., Human, S.W., Graham, M.A. (2009). Phase I statistical process control charts: An overview and some results. *Qual. Eng.*, 21, 52–62.

Coelho, M.L.I., Chakraborti, S., Graham, M.A. (2015). A comparison of Phase I control charts. *S. Afr. J. Ind. Eng.*, 26, 178–190.

Graham, M.A., Human, S.W., Chakraborti, S. (2010). A phase I nonparametric Shewhart-type control chart based on the median. *J. Appl. Stat.*, 37, 1795–1813.

Hawkins, D.M. and Deng, Q. (2010). A nonparametric change-point control chart. *J. Qual. Technol.*, 42, 165–173.

Jones-Farmer, L.A., Jordan, V., Champ, C.W. (2009). Distribution-free Phase I control charts for subgroup location. *J. Qual. Technol.*, 41, 304–316.

Jones-Farmer, L.A., Woodall, W.H., Steiner, S.H., Champ, C.W. (2014). An overview of Phase I analysis for process improvement and monitoring. *J. Qual. Technol.*, 46, 265–280.

Montgomery, D.C. (2013). *Introduction to Statistical Quality Control*, 7th edition. John Wiley and Sons, Hoboken, New Jersey.

Ning, W., Yeh, A.B., Wu, X., Wang, B. (2015). A nonparametric Phase I control chart for individual observations based on empirical likelihood ratio. *Qual. Reliab. Eng. Int.*, 31, 37–55.

Parpoula, C. (2020). A distribution-free control charting technique based on change-point analysis for outbreak detection (submitted).

Parpoula, C. and Karagrigoriou, A. (2020). On change-point analysis-based distribution-free control charts with Phase I applications. In *Distribution-free Methods for Statistical Process Monitoring and Control*, Koutras M., Triantafyllou I. (eds). Springer International Publishing, Cham.

Qiu, P. (2019). Some recent studies in statistical process control. In *Statistical Quality Technologies*, Lio, Y., Ng, H., Tsai, T.R., Chen, D.G. (eds). ICSA Book Series in Statistics, Springer, Cham.

Ross, G.J. (2015). Parametric and nonparametric sequential change detection in R: The cpm package. *J. Stat. Softw.*, 66, 1–20.

Ross, G.J. and Adams, N.M. (2012). Two nonparametric control charts for detecting arbitrary distribution changes. *J. Qual. Technol.*, 44, 102–116.

Shewhart, W.A. (1939). *Statistical Method from the Viewpoint of Quality Control*. Dover Publications, New York.

Sullivan, J.H. and Woodall, W.H. (1996). A control chart for preliminary analysis of individual observations. *J. Qual. Technol.*, 28, 265–278.

Woodall, W.H. (2000). Controversies and contradictions in statistical process control. *J. Qual. Technol.*, 32, 341–350.

Woodall, W.H. and Montgomery, D.C. (1999). Research issues and ideas in statistical process control. *J. Qual. Technol.*, 31, 376–386.

14

On Divergence and Dissimilarity Measures for Multiple Time Series

Divergence and dissimilarity measures play an important role in mathematical statistics and statistical data analysis. This work is devoted to a review on the most popular divergence and dissimilarity measures, as well as on some new advanced measures associated with multiple time series data.

14.1. Introduction

An issue of fundamental importance in Statistics is the investigation of Information Measures, which constitutes a broad class of measures that include, among others, the divergence and dissimilarity measures. These measures are used to measure the extent of information contained in the data and/or the divergence or dissimilarity between two populations, functions or data sets. Traditionally, the measures of information are classified into four main categories, namely divergence-type, entropy-type, Fisher-type and Bayesian-type (see Vonta and Karagrigoriou 2011).

Measures of divergence between two probability distributions have a more than 100 year long history initiated by Pearson, Mahalanobis, Lévy and Kolmogorov. Among the most popular measures of divergence are the Kullback–Leibler measure of divergence (Kullback and Leibler 1951) and the Csiszar's φ-divergence family of measures (Csiszár 1963; Ali and Silvey 1966). Cressie and Read (1984) attempted to provide a unified analysis by introducing the so-called power divergence family of statistics that involved an index and is used in goodness-of-fit tests, primarily for multinomial distributions. The Cressie and Read family includes a number of well-known measures like Pearson's χ^2 measure and the classical loglikelihood ratio

Chapter written by Konstantinos MAKRIS, Alex KARAGRIGORIOU and Ilia VONTA.

statistic. In recent years, the BHHJ divergence measure was introduced by Basu *et al.* (1998) as a robust estimating procedure. In 2010, the generalized BHHJ family of measures was proposed by Mattheou *et al.* (2009) for hypothesis testing purposes. The BHHJ family, like the Cressie and Read family, relies on an index, which controls the trade-off between robustness and efficiency.

It should be pointed out that measures of divergence play a significant role in statistical inference and have several applications. Measures of dissimilarity are used for estimating purposes (Toma 2008, 2009), for the construction of test statistics for goodness-of-fit tests (e.g. Huber-Carol *et al.* 2002; Zhang 2002; Meselidis and Karagrigoriou 2020) or for the construction of model selection criteria. It should be noted, for instance, that the classical Kullback–Leibler measure has been used for the development of various criteria like the famous Akaike Information Criterion (e.g. Cavanaugh 2004; Shang 2008). Applications of divergence and dissimilarity measures can also be found in biosciences like biomedicine and biostatistics. Indeed, the existence of censoring schemes in survival modeling makes the modeling a very challenging problem. For related references, see Gail and Ware (1979) who studied grouped censored survival, and Akritas (1988) or Chen *et al.* (2004) who proposed Pearson-type goodness-of-fit tests.

In the same setting, we frequently encounter the concept of similarity between time series. A similarity or dissimilarity measure between two (or more) time series is used to investigate the common behavior of the series involved. Among the techniques that appear frequently in practice are classical mathematical measures or metrics (like the Euclidean distance), special data transformations or even algorithmic procedures.

In this chapter, we focus on divergence and dissimilarity measures, explore their capabilities and discuss the so-called MKN dimensionless coefficients which are focusing on the K largest ($M = 1$) or lowest ($M = 0$) ordered values of $N(N \geq 2)$ time series. Sections 14.2 and 14.3 are devoted to standard mathematical and divergence measures. Section 14.4 deals with a class of advanced measures, the so-called MKN dissimilarity measures, and provides some illustrative examples.

14.2. Classical measures

As indicated earlier, the evaluation of similarity or closeness of two or more frequently, complex sets of data, requires a proper measure of similarity. There is often subjectivity in choosing a distance measure. In general, the distance d between two random vectors $X = (x_1, ..., x_p)$ and $Y = (y_1, ..., y_p)$ satisfies the classical properties of:

- **Positivity**, i.e. $d(X, Y) \geq 0 \; with \; d(X, Y) = 0 \; if \; and \; only \; if \; X = Y$
- **Symmetry**, i.e. $d(X, Y) = d(Y, X)$
- **Triangle inequality**, i.e. $d(X, Y) = d(Y, V) + d(X, V)$ for any vector V.

Some of the most popular classical measures that are used to quantify the distance between two datasets are presented below. Observe that not all measures necessarily satisfy all of properties mentioned above. In fact, it should be emphasized that for statistical purposes, only the first of the above three properties is mandatory as long as the distance between two objects or functions is non-negative.

— **Euclidean distance** is defined by

$$d(X,Y) = |X - Y| = \sqrt{\sum_{i=1}^{p} (x_i - y_i)^2} = \sqrt{(X - Y)'(X - Y)}.$$

The Euclidean distance highly depends on the measurement units, and by changing the scale, we obtain a different measure of distance. Also, large absolute values and outliers have a much greater impact and often determine the magnitude of the distance. Finally, note that the distance ignores statistical properties and characteristics such as the variability involved in each variable. A way to avoid these deficiencies is a proper standardization of each variable involved in the analysis.

— **City-block distance** is defined by

$$d_C(X,Y) = \sum_{i=1}^{p} |x_i - y_i|$$

and is used in the presence of outliers since the absolute value downweights the effect of extreme observations, as compared to the square used in the Euclidean distance.

— **Minkowski distance**, given by

$$d_M(X,Y) = \left(\sum_{i=1}^{p} |x_i - y_i|^v \right)^{1/v}, \ v \geq 1$$

is a generalization of the Euclidean distance which reduces the effect of outlying observations.

— **Chebyshev distance**

$$d_T(X,Y) = max|x_i - y_i|\ , i = 1, ..., p$$

greatly depends on differences in the measurement scale of variables.

— **Czekanowski coefficient** is defined by

$$d(X,Y) = 1 - \frac{\sum_{i=1}^{p} |x_i - y_i|}{\sum_{i=1}^{p} (x_i + y_i)}$$

— **Mahalanobis distance** is given by

$$D_{Mh}(X,Y) = (X - Y)' \Sigma^{-1} (X - Y)$$

where Σ^{-1} is the inverse of the variance–covariance matrix Σ. The Mahalanobis distance is constructed based on statistical characteristics and in particular, by taking into consideration the variances and covariances through the Σ matrix. It is often used to remove the multicollinearity among variables.

14.3. Divergence measures

Measures of information are powerful statistical tools with diverse applicability. In this section, we will focus on a specific type of information measure, known as the measure of discrepancy (distance or divergence) between the distributions of two variables X and Y with pdf's f and g. Furthermore, note that such measures are used to evaluate the discrepancy between (i) the distribution of X as deduced from an available set of data and (ii) the distribution of a hypothesized distribution that is believed to be the generating mechanism that produced the set of data at hand. Such general measures are applicable in Reliability Theory for new as well used items, which are appropriate for measuring the discrepancy between the tail-part of the involved distributions. In such cases, we could provide ways to quantify the divergence between the residual lives as well as the past lifetimes which are frequently encountered in Reliability Theory and are associated with the tail heaviness of the distributions.

For historical reasons, we first present Shannon's entropy (Shannon 1948) given by

$$I^S(X) \equiv I^S(f) = -\int f \ln f d\mu = E_f[-\ln f],$$

where X is a random variable with density function $f(x)$ and μ is a probability measure on $\mathbb{R}$. Shannon's entropy was introduced and used during World War II in Communication Engineering. Shannon derived the discrete version of $I^S(f)$, where f is a probability mass function and named it *entropy* because of its similarity with thermodynamics entropy. The discrete version is defined by analogy. For a finite number of points, Shannon's entropy measures the expected information of a signal transferred without noise from a source X, with density $f(x)$, and is related to the Kullback–Leibler divergence (Kullback and Leibler 1951) through the following expression:

$$I^S(f) = I^S(h) - I_X^{KL}(f, h)$$

where h is the density of the uniform distribution, and the Kullback–Leibler divergence between two densities $f(x)$ and $g(x)$ is given by

$$I_X^{KL}(f, g) = \int f \ln(f/g) d\mu = E_f[\ln(f/g)]. \quad [14.1]$$

Many generalizations of Shannon's entropy were hereupon introduced. Rényi's (1961) entropy, as extended by Liese and Vajda (1987), is given by

$$I^{R_{lv},a}(X) \equiv I^{R_{lv},a}(f) = \frac{1}{a\,(a-1)} \ln E_f\left[f\right]^{a-1}, \; a \neq 0, 1.$$

For more details about entropy measures, the reader is referred to Mathai and Rathie (1975), Nadarajah and Zografos (2003) and Zografos and Nadarajah (2005).

A measure of divergence is used as a way to evaluate the distance (divergence) between any two functions f and g associated with the variables X and Y. Among the most popular measures of divergence are the Kullback–Leibler measure of divergence given in [14.1] and Csiszar's φ-divergence family of measures (Csiszar 1963; Ali and Silvey 1966) given by

$$I_{f,g}^{\varphi} = \int_0^{\infty} g(x)\varphi\left(\frac{f(x)}{g(x)}\right) dx. \qquad [14.2]$$

The class of Csiszar's measures includes a number of widely used measures that can be recovered with appropriate choices of the function φ. When the function φ is defined by

$$\varphi(u) = u \log u \;\text{ or }\; \varphi(u) = u \log u + 1 - u, \qquad [14.3]$$

the above measure reduces to the Kullback–Leibler measure given in [14.1]. If

$$\varphi(u) = \frac{1}{2}(1-u)^2, \qquad [14.4]$$

Csiszar's measure yields Pearson's chi-square divergence. If

$$\varphi(u) \doteq \varphi_1(u) = \left(u^{a+1} - u - a(u-1)\right)/(a(a+1)) \qquad [14.5]$$

we obtain the Cressie and Read power divergence (Cressie and Read 1984), $a \neq 0, -1$.

Another function that is usually considered is

$$\varphi(u) \doteq \varphi_2(u) = 1 - (1 + \frac{1}{a})u + \frac{u^{1+a}}{a}, \quad a \neq 0. \qquad [14.6]$$

This function is associated with the BHHJ power divergence (Basu *et al.* 1998), while it is a special case of the BHHJ family of divergence measures proposed by Mattheou *et al.* (2009)

$$I_X^{BHHJ} = \int g^{1+\alpha}(x)\,\varphi\left(\frac{f(x)}{g(x)}\right) dx, \quad \alpha \geq 0. \qquad [14.7]$$

Appropriately chosen functions $\varphi(\cdot)$ give rise to the special measures mentioned above, while for $\alpha = 0$, the BHHJ family reduces to Csiszar's family. For more details on divergence measures, see Cavanaugh (2004), Pardo (2006), Toma (2009) and Toma and Broniatowski (2011). For robust inference based on divergence measures specifically, see Basu *et al.* (2011) and a recent paper by Patra *et al.* (2012) on the power divergence and the density power divergence families.

14.4. Dissimilarity measures for ordered data

Let us consider two time series of equal length $\{X_i^1\}_{i=1}^n$ and $\{X_i^2\}_{i=1}^n$ and assume that $X_{(i)}^j$ represents the ith ordered observation of the jth dataset with $j = 1, 2$. Note that the discussion will be based on two datasets but the generalization to N with $N > 2$ is straightforward. Finally, K represents the number of consecutive extreme values (either highest or lowest) used for the analysis. For instance, the $K = 4$ highest values of a series $\{X_i\}_{i=1}^n$ are $X_{(n)}, X_{(n-1)}, X_{(n-2)}, X_{(n-3)}$.

The purpose of dissimilarity measures to be discussed in this section is to examine whether the K highest values of the datasets appear on the same time points or not. If they do, the dissimilarity will be equal to 0, indicating that the datasets are identical. The bigger the difference, the bigger the dissimilarity. Recall that in what follows, we focus on two time series. For the extension to N series, one should replace the index KM by KNM, where $M = 0$ or 1 depending on whether the K minimum or the K maximum values are examined. Naturally, if $K = n$, then the entire dataset is used and the index M becomes redundant.

14.4.1. *Standard dissimilarity measures*

The first measure we consider is the $K - similarity$ measure between X_1 and X_2 denoted by V[KM] and defined by

$$V_{X_1,X_2}[KM] = \frac{1}{K}\sum_{i=1}^{K}\left(X_{(i)}^1 - X_{(i)}^2\right)^2. \qquad [14.8]$$

Observe that instead of comparing the corresponding observed values, we compare the corresponding ordered values. If the measure is equal to zero, this means that the two series have the exact same values but may possibly occurr on different time points, since the i_{th} larger values of X_1 and X_2 may coincide without necessarily being recorded at the exact same (corresponding) time point. If we are instead interested in examining whether the shapes of the series are similar or not, with not necessarily the same magnitude, we could focus on the time points of occurrence of the ordered observations. As a typical example, consider a seasonal flu that could occur at the exact same period in two consecutive years but with different magnitudes. Thus, if we wish to explore the repetition in time of such an annual time series, we may consider the time points instead of the actual values. Let T_i^j represent the time point of the ith highest (lowest) observation of the jth dataset with $j = 1, 2$. Hence, T_i^j is the time point of the $X_{(i)}^j$ ordered observation. Then, the $K - time - similarity$ measure is defined by

$$V_{1,2}[KM] = \frac{1}{K} \sum_{i=1}^{K} \left(T_i^1 - T_i^2\right)^2. \qquad [14.9]$$

Observe the importance of the above proposal, which indicates that two series are assumed to be (fully) similar if the corresponding i^{th} highest values of the two series occurred at exactly the same time point, irrespectively of the magnitude of the associated observed values of the two series. The measure could be evaluated for any value of K, $K = 1, 2, \ldots, n$.

The following properties are straightforward:

- $V_{j,j}[KM] = \frac{1}{K} \sum_{i=1}^{K} \left(T_i^j - T_i^j\right)^2 \equiv V_{X_j,X_j}[KM] = 0,\ j = 1, 2.$
- $V_{1,2}[KM] = V_{2,1}[KM]$
- $V_{X_1,X_2}[KM] = V_{X_2,X_1}[KM]$
- $V_{i,j}[KNM] = V_{j,i}[KNM],\ i, j = 1, 2, \ldots, N$
- $V_{X_i,X_j}[KNM],\ i, j = 1, 2, \ldots, N.$

In cases such as the seasonal flu or the behavior of financial markets, we may investigate whether the similarity occurs shifted by a certain time period between the series. Indeed, in such cases, the occurrence of the same pattern may occur with a time delay in one market, compared with the other, or from one season to the next. In a case like this, the proper measure could be defined by

$$V_{1,2}^{m}[KM] = \frac{1}{K} \sum_{i=1}^{K} \left(T_i^1 - T_{i+m}^2\right)^2 \qquad [14.10]$$

where m could be any integer (even a negative one). Obviously for $m = 0$, we return back to the original definition.

It is also possible to combine the time points together with the actual observations (equations [14.9] and [14.10]) in order to make it possible to distinguish between series that are similar for just one of the two measures:

$$V[KM] = V_{X_1,X_2}[KM] + V_{1,2}[KM]. \qquad [14.11]$$

Finally, one may place weights w_1 and w_2 on the two components of the above measure. Then, the measure becomes

$$V^{w}[KM] = w_1 V_{X_1,X_2}[KM] + w_2 V_{1,2}[KM] \qquad [14.12]$$

where $w_1, w_2 \geq 0$ with $w_1 + w_2 = 1$. As expected, the original individual measures $V_{X_1,X_2}[KM]$ and $V_{1,2}[KM]$ are obtained for $w_1 = 1, w_2 = 0$ and $w_1 = 0, w_2 = 1$, respectively.

Two related measures which do not compare the times of the K ordered values one by one but instead compare either the summation or the product of the times of the K ordered values provided below. More specifically, the $K - Stime - variation$ is defined by

$$V_{1,2}^S[KM] = \frac{\sum_{i=1}^K T_i^1}{\sum_{i=1}^K T_i^2} \qquad [14.13]$$

while the $K - \prod time - variation$ is given by

$$V_{1,2}^{\prod}[KM] = \frac{\prod_{i=1}^K T_i^1}{\prod_{i=1}^K T_i^2}. \qquad [14.14]$$

REMARK 14.1.– *Observe that the above two special measures may produce misleading results because it is possible that the K extreme times may be the same without being in one-to-one correspondence. Indeed, assume that the five highest observed daily values of two series of length* $n = 40$ *occur at the same time points, say* $T_{36} - T_{40}$, *but not all with the same order. For instance,* $T_{36}^1 = T_{40}^2$, *say equal to 30 (30th day) and* $T_{40}^1 = T_{36}^2$, *say equal to 20 (20th day) with the other three being identical. Hence, naturally* $V_{1,2}^S[KM] = V_{1,2}^{\prod}[KM] = 1$. *On the other hand,* $V_{1,2}[KM] = 200$.

REMARK 14.2.– *Consider another example where all but one of the five highest observations occur at exactly the same time points, with just one occurring on extremely different time points, say, for instance that,* $T_{36}^1 = 5$ *and* $T_{36}^2 = 40$. *In such a case, the Euclidean type measure* $V_{1,2}[KM]$ *will be quite large* $(35^2 = 1225)$, *while the summation type measure will be equal to* $(40 + A)/(5 + A) < 8$ *irrespectively of the value of A. Similarly for the product measure.*

As a consequence of the above Remarks, we conclude that the previously defined measures may be unstable depending on the special features of the K extreme values of the series involved. In the next subsection, we will present an alternative, stable dimensionless measure which will not experience the same disadvantages as the above measures.

14.4.2. *Advanced dissimilarity measures*

In this section, we present the dimensionless KM measure which attempts to overcome the defects associated with standard measures presented in the previous subsection. The basic idea is the use of a measure of central tendency among either the K observed ordered values or the corresponding time points where these values occur. These measures were introduced in Makris (2017) for experimental data for two wind turbines, and later by the same author (Makris 2018) for incidence rate data of influenza like illness (ILI).

Let $X^j_{(i)}$, $i = 1, 2, \dots, n$ be the ordered observations of the j series, $j = 1, 2$. Let T^j_i also be the time point associated with the ith ordered observation of the jth series.

DEFINITION 14.1.– *The* $K-dimensionless-coefficient$ *for a series* X *is denoted by* $D_X[KM]$ *and is defined by*

$$D_X[KM] = \begin{cases} \frac{Average\{|X_{(i)}|, i=n-K+1,\dots,n\}}{Average\{|X_{(i)}|, i=1,\dots,n\}}, & if M = 0 \\ \\ \frac{Average\{|X_{(i)}|, i=1,\dots,K\}}{Average\{|X_{(i)}|, i=1,\dots,n\}}, & if M = 1. \end{cases} \qquad [14.15]$$

It should be noted that the absolute value is used to handle cases with negative values. Also observe that for $K = n$ the entire series is used and the coefficient is equal to 1.

REMARK 14.3.– *The series* X *will be used as a reference series and the* $K - dimensionless - coefficient$ *for any other series, say* $D_Y[KM]$ *of the series* Y*, will be calculated on the exact same positions of the* K *extreme observations of the reference series* X.

It is worth mentioning that the similarity of two series based on the above $K - dimensionless - coefficient$ does not imply that the K extreme ordered values of the reference series coincide with the corresponding K extreme values of the second series. Instead, the similarity indicates that the *average* of the K ordered values of the two series coincides. It should also be pointed out that extreme values within the K extreme ordered observations will considerably affect the average used in the coefficient. The investigator may consider using the median, if such a case is observed or expected to occur.

The fraction

$$D_{X,Y}[KM] = \frac{D_X[KM]}{D_Y[KM]} \qquad [14.16]$$

could be viewed as the $K - dimensionless - measure$ between the series X and Y. In such a case, a convenient $K - dimensionless - measure$ matrix could be constructed for visual inspection. The matrix will be extremely useful, especially if several series are involved. The matrix for three series is presented as in Table 14.1.

X	Y	Z	Reference
1	$D_Y[KM]/D_X[KM]$	$D_Z[KM]/D_X[KM]$	X
$D_X[KM]/D_Y[KM]$	1	$D_Z[KM]/D_Y[KM]$	Y
$D_X[KM]/D_Z[KM]$	$D_Y[KM]/D_Z[KM]$	1	Z

Table 14.1. *K-dimensionless-measure matrix for three series X, Y and Z*

Observe that Table 14.1 provides the measure given in [14.17] with each of the series as a reference series.

Some basic properties of the proposed coefficient are given below:

– $D_X[nM] = 1$ for both $M = 0$ and $M = 1$

– $D_Y[KM] \leq D_X[KM]$ for any series Y with X being the reference series.

– $D_Y[KM] = D_X[KM]$ for a specific K means that the series are similar in the sense of Definition 1

– $D_X[K_1M] \leq D_X[K_2M]$ for $K_1 > K_2$

– $\frac{D_X[KM]}{D_X[KM]} = 1$

– $D_{X,Y}[KM] = \frac{D_X[KM]}{D_Y[KM]} \leq 1.$

Finally, as in the previous section, we may focus on the time points associated with the K extreme ordered observations. Then, the $K - time - dimensionless - coefficient$ of the reference series X will be given by

$$D_1[KM] = \begin{cases} \frac{Average\{|T_i|, i=n-K+1,\ldots,n\}}{Average\{|T_i|, i=1,\ldots,n\}}, & ifM = 0 \\ \\ \frac{Average\{|T_i|, i=1,\ldots,K\}}{Average\{|T_i|, i=1,\ldots,n\}}, & ifM = 1. \end{cases} \qquad [14.17]$$

Note that in the above case, the denominator for the reference series coincides with the denominator of any other series to be compared with X. Hence, the $K - time - dimensionless - measure$ between X and Y can be simply defined by

$$D_{1,2}[KM] = \begin{cases} \frac{Average\{|T_i^1|, i=n-K+1,\ldots,n\}}{Average\{|T_i^2|, i=n-K+1,\ldots,n\}}, & ifM = 0 \\ \\ \frac{Average\{|T_i^1|, i=1,\ldots,K\}}{Average\{|T_i^2|, i=1,\ldots,K\}}, & ifM = 1. \end{cases} \qquad [14.18]$$

We close this section with a matrix of the $K - dimensionless - measure$ between three series, based on weekly data of influence like illness (ili) cases for the three periods 2011–2012, 2012–2013 and 2013–2014 for Greece. Note that the data only refers to the weekly cases for the 40th week of each year up to the 8th week of the following year, which is the typical influenz period. The results were obtained from the National Public Health Organization (NPHO) of Greece.

Observe the weak similarity between series Y and Z, when Z is the reference series. Also observe a strong similarity between X and Z, when both Z and X are the reference series. The latter observation indicates that the average of the K higher values of X is very similar to the average of the K higher values of Z.

X	Y	Z	Reference
1	0.8563	0.9630	X
0.8858	1	0.8758	Y
0.9645	0.7107	1	Z

Table 14.2. *K-dimensionless-measure matrix for three ili periods for Greece*

14.5. Conclusion

This work is devoted to dissimilarity and divergence measures, with emphasis on comparing time series of equal length. A number of measures have been discussed which appear to be attractive, and could be useful in a number of applications in various fields such as biosciences, reliability theory and in general, in phenomena occuring periodically. The research is ongoing, and further results will be presented in the near future.

14.6. References

Akritas, M.G. (1988). Pearson-type goodness-of-fit tests: The univariate case. *J. Amer. Statist. Assoc.*, 83, 222–230.

Ali, S.M. and Silvey, S.D. (1963). A general class of coefficients of divergence of one distribution from another. *J. Roy. Statist. Soc. B*, 28, 131–142.

Basu, A., Harris, I.R., Hjort, N.L., Jones, M.C. (1998). Robust and efficient estimation by minimising a density power divergence. *Biometrika*, 85, 549–559.

Basu, A., Shioya, H., Park, C. (2011). *Statistical Inference: The Minimum Distance Approach.* Chapman & Hall/CRC, Boca Raton.

Cavanaugh, J.E. (2004). Criteria for linear model selection based on Kullback's symmetric divergence. *Aust. N. Z. J. Stat.*, 46, 257–274.

Chen H.S., Lai, K., Ying, Z. (2004). Goodness of fit tests and minimum power divergence estimators for survival data. *Statistica Sinica*, 14, 231–248.

Cressie, N. and Read, T.R.C. (1984). Multinomial goodness-of-fit tests. *J. Roy. Stat. Soc. B*, 5, 440–454.

Csiszar, I. (1963). Eine Informationstheoretische Ungleichung und ihre Anwendung auf den Bewis der Ergodizitat on Markhoffschen Ketten. *Publication of the Mathematical Institute of the Hungarian Academy of Sciences*, 8, 84–108.

Gail, M.H. and Ware, J.H. (1979). Comparing observed life table data with a known survival curve in the presence of random censorship. *Biometrics*, 35, 385–391.

Huber-Carol, C., Balakrishnan, N., Nikulin, M.S., Mesbah, M. (2002). *Goodness-of-fit Tests and Model Validity*. Birkhäuser, Boston.

Kullback, S. and Leibler, R. (1951). On information and sufficiency. *Ann. Math. Stat.*, 22, 79–86.

Liese, F. and Vajda, I. (1987). *Convex Statistical Distances*. Teubner, Leipzig.

Makris, K. (2017). Statistical analysis of random waves in SPAR-type and TLP-type wind turbines (in Greek). MSc. Thesis, National Technical University of Athens, Greece.

Makris, K. (2018). Statistical analysis of epidemiological time series data (in Greek). MSc. Thesis, National Technical University of Athens, Greece.

Mathai, A. and Rathie, P.N. (1975). *Basic Concepts in Information Theory*. John Wiley and Sons, New York.

Mattheou, K., Lee, S., Karagrigoriou, A. (2009). A model selection criterion based on the BHHJ measure of divergence. *J. Statist. Plann. Infer.*, 139, 128–135.

Meselidis, C. and Karagrigoriou, A. (2020). Statistical inference for multinomial populations based on a double index family of test statistics. *J. Statist. Comput. and Simul.*, 90(10), 1773–1792.

Nadarajah, S. and Zografos, K. (2003). Formulas for Renyi information and related measures for univariate distributions. *Infor. Sci.*, 155, 118–119.

Pardo, L. (2006). *Statistical Inference Based on Divergence Measures*. Chapman and Hall/CRC, Boca Raton.

Patra, S., Maji, A., Basu, A., Pardo, L. (2013). The power divergence and the density power divergence families: The mathematical connection. *Sankhya B*, 75, 16–28.

Renyi, A. (1961). On measures of entropy and information. *Proc. of the Fourth Berkeley Symposium on Mathematical Statistics and Probability*, 1, 547–561.

Shang, J. (2008). Selection criteria based on Monte Carlo simulation and cross validation in mixed models. *Far East J. Theor. Statist.*, 25, 51–72.

Shannon, C.E. (1948). A mathematical theory of communication. *Bell Sys. Tech. J.*, 27, 379–423.

Toma, A. (2008). Minimum Hellinger distance estimators for multivariate distributions from the Johnson system, *J. Statist. Plan. Infer.*, 138, 803–816.

Toma, A. (2009). Optimal robust M-estimators using divergences. *Stat. Prob. Lett.*, 79, 1–5.

Toma, A. and Broniatowski, M. (2011). Dual divergence estimators and tests: Robustness results. *J. Multivariate Anal.*, 102(1), 20–36.

Vonta, F. and Karagrigoriou, A. (2011). Information measures in biostatistics and reliability. In *Mathematical and Statistical Models and Methods in Reliability*, Rykov, V.V., Balakrishnan, N., Nikulin (eds). Birkhauser, Boston.

Zhang, J. (2002). Powerful goddness-of-fit tests based on likelihood ratio. *J. R. Stat. Soc. Ser. B*, 64(2), 281–294.

Zografos, K. and Nadarajah, S. (2005). Survival exponential entropies. *IEEE Trans. Inform. Theory*, 51, 1239–1246.

List of Authors

Benard ABOLA
Division of Applied Mathematics
Mälardalen University
Västerås
Sweden

Syed Ejaz AHMED
Department of Mathematics and Statistics
Brock University
Ontario
Canada

Collins ANGUZU
Department of Mathematics
Makerere University
Kampala
Uganda

Alessio BARBAGLI
Department of Physics and Earth Science
University of Ferrara
Italy

Pitos Seleka BIGANDA
Department of Mathematics
University of Dar-es-Salaam
Tanzania

Jose BLANCHET
Department of Management Science and Engineering
Stanford University
USA

Dominique DESBOIS
UMR 0210 Économie Publique, INRAE-AgroParisTech
Université Paris-Saclay
France

Yannis DIMOTIKALIS
Department of Management Science and Technology
Hellenic Mediterranean University
Heraklion
Crete
Greece

Christopher ENGSTRÖM
Division of Applied Mathematics
Mälardalen University
Västerås
Sweden

Gwenael GATTO
Department of Mathematics and Statistics
Hunter College
New York
USA

Franciszek GRABSKI
Department of Mathematics and Physics
Polish Naval Academy
Gdynia
Poland

Enrico GUASTALDI
Geoexplorer Impresa Sociale S.r.l.
San Giovanni Valdarno
Italy

Olympia HADJILIADIS
Department of Mathematics and Statistics
Hunter College
New York
USA

Fei HE
Industrial Engineering and Operations Research
Columbia University
New York
USA

Zhangyi HU
Department of Earth and Environmental Science
University of Michigan
Ann Arbor
USA

Godwin KAKUBA
Department of Mathematics
Makerere University
Kampala
Uganda

Yang KANG
Department of Statistics
Columbia University
New York
USA

Alex KARAGRIGORIOU
Department of Statistics and Actuarial-Financial Mathematics
University of the Aegean
Samos
Greece

Supranee LISAWADI
Department of Mathematics and Statistics
Thammasat University
Pathum Thani
Thailand

Konstantinos MAKRIS
Department of Mathematics
School of Mathematical and Physical Sciences
National Technical University of Athens
Greece

John Magero MANGO
Department of Mathematics
Makerere University
Kampala
Uganda

Kimon NTOTSIS
Department of Statistics and
Actuarial-Financial Mathematics
University of the Aegean
Samos
Greece

Christina PARPOULA
Department of Psychology
Panteion University of Social and
Political Sciences
Athens
Greece

Orawan REANGSEPHET
Department of Mathematics and
Statistics
Thammasat University
Pathum Thani
Thailand

Dmitrii SILVESTROV
Department of Mathematics
Stockholm University
Sweden

Sergei SILVESTROV
Division of Applied Mathematics
Mälardalen University
Västerås
Sweden

Christos H. SKIADAS
ManLab
Department of Production
Engineering and Management
Technical University of Crete
Chania
Greece

Anna L. SMITH
Department of Statistics
University of Kentucky
Lexington
USA

Ilia VONTA
Department of Mathematics
School of Mathematical and
Physical Sciences
National Technical University
of Athens
Greece

Jing WU
Department of Statistics
Columbia University
New York
USA

Fan ZHANG
Department of Management Science
and Engineering
Stanford University
USA

Tian ZHENG
Department of Statistics
Columbia University
New York
USA

Andrea ZIRULIA
CGT Center for GeoTechnologies
University of Siena
San Giovanni Valdarno
Italy

Index

A, C

agricultural production cost, 149, 150, 171
asymptotic expansion, 24, 25, 29–32, 35–38, 46, 48, 60
availability, 175, 176, 178, 179, 183–185
centrality, 3, 91–109
climate change, 221, 222, 230
control charts, 233–237, 239, 241–243, 247
correction, 115
coupling, 39–43, 45
CUSUM, 114, 117–121, 123, 125

D, E

damping component, 23, 25–27, 29, 39–43, 45, 49, 50
detection, 113–115, 117, 120, 124–126
dissimilarity measure, 249, 250, 254, 256
distribution-free, 233, 236, 239, 240
distributionally robust optimization, xii, 75
divergence measure, 250, 252–254, 259
ergodic theorem, 26, 39, 42, 43, 45
error, 113, 114, 116
estimate intervals, 171
event times, 129–131, 133–135, 138, 140

G, H, I

graph, 4–6, 8–10, 12–15, 17, 20, 21, 57, 58, 61, 72, 159, 166
groundwater, 221–224, 228–230
hidden Markov model, 114, 120
high-dimensional sparse Poisson regression, 205, 206
individual observations, 233, 235–239, 242, 247
information network, 23–26, 32, 33, 37, 45, 50

K, L

K-dimensionless
- coefficient, 257, 258
- measure, 257–259

likelihood ratio, 235, 236, 239, 240
linear
- programing, 175, 179, 181, 185
- shrinkage, 205, 206, 209, 217

M

Markov chain, 23–30, 32, 36, 39–43, 45, 49, 50, 57–66, 69–72, 163, 176, 179, 182
maximization, 175, 176
metric learning, 77–79, 81–83, 85–87
model checking, 133, 134, 138
Monte Carlo simulations, 205, 206, 210, 217
multicollinearity, 187–191, 198–200
multiple time series, 249

N, O

networks (*see* information network), 3, 4, 57–59, 61–67, 69, 71, 72, 91–93, 95, 97, 100, 104, 107, 108
non-parametric, 233–239, 241–243, 247
optimal transport, 75, 76, 79–81, 89
ordered data, 254

P

PageRank, 3–5, 10–18, 20, 21, 24, 25, 57–66, 69–72, 92, 93, 98, 99, 104, 106, 108
penalized maximum likelihood, 205
perturbation, 57–60, 62, 63, 66, 72
  regular, 28, 32, 36, 39
  singular, 25, 26, 28, 39, 41, 43, 45, 49
Phase I analysis, 234, 235, 237, 238
pig, 149, 150, 154, 162, 165–171
point processes, 130–134
pretest, 205, 209
principal component analysis, 149, 154, 155, 161, 171, 187, 192, 197
public pension expenditures, 187, 188

Q, R

quantile regression, 149, 152, 153, 166, 171
random walk, 3, 21, 91–93, 99, 100, 104, 109
rate of convergence, 24–26, 29, 34, 39–41, 43, 49, 58

S

semi-Markov decision process, 176
signal probability, 236, 240, 241, 243
social network, 129, 130, 133
stationary distribution, 24–30, 40–43
Stein-type shrinkage, 205, 206, 210, 217, 218
symbolic data analysis, 149, 150, 171

T, W

time series, 115, 221–225, 229, 230
triangular array mode, 24, 26, 39, 42, 43, 45
water level, 221, 222

Other titles from

iSTE

in

Innovation, Entrepreneurship and Management

2021

BOBILLIER CHAUMON Marc-Eric
Digital Transformations in the Challenge of Activity and Work: Understanding and Supporting Technological Changes
(Technological Changes and Human Resources Set – Volume 3)

2020

ACH Yves-Alain, RMADI-SAÏD Sandra
Financial Information and Brand Value: Reflections, Challenges and Limitations

ANDREOSSO-O'CALLAGHAN Bernadette, DZEVER Sam, JAUSSAUD Jacques, TAYLOR Robert
Sustainable Development and Energy Transition in Europe and Asia
(Innovation and Technology Set – Volume 9)

BEN SLIMANE Sonia, M'HENNI Hatem
Entrepreneurship and Development: Realities and Future Prospects
(Smart Innovation Set – Volume 30)

CHOUTEAU Marianne, FOREST Joëlle, NGUYEN Céline
Innovation for Society: The P.S.I. Approach
(Smart Innovation Set – Volume 28)

CORON Clotilde
Quantifying Human Resources: Uses and Analysis
(Technological Changes and Human Resources Set – Volume 2)

CORON Clotilde, GILBERT Patrick
Technological Change
(Technological Changes and Human Resources Set – Volume 1)

CERDIN Jean-Luc, PERETTI Jean-Marie
The Success of Apprenticeships: Views of Stakeholders on Training and Learning
(Human Resources Management Set – Volume 3)

DELCHET-COCHET Karen
Circular Economy: From Waste Reduction to Value Creation
(Economic Growth Set – Volume 2)

DIDAY Edwin, GUAN Rong, SAPORTA Gilbert, WANG Huiwen
Advances in Data Science
(Big Data, Artificial Intelligence and Data Analysis Set – Volume 4)

DOS SANTOS PAULINO Victor
Innovation Trends in the Space Industry
(Smart Innovation Set – Volume 25)

GASMI Nacer
Corporate Innovation Strategies: Corporate Social Responsibility and Shared Value Creation
(Smart Innovation Set – Volume 33)

GOGLIN Christian
Emotions and Values in Equity Crowdfunding Investment Choices 1: Transdisciplinary Theoretical Approach

GUILHON Bernard
Venture Capital and the Financing of Innovation
(Innovation Between Risk and Reward Set – Volume 6)

LATOUCHE Pascal
Open Innovation: Human Set-up
(Innovation and Technology Set – Volume 10)

LIMA Marcos
Entrepreneurship and Innovation Education: Frameworks and Tools
(Smart Innovation Set – Volume 32)

MACHADO Carolina, DAVIM J. Paulo
Sustainable Management for Managers and Engineers

MAKRIDES Andreas, KARAGRIGORIOU Alex, SKIADAS Christos H.
Data Analysis and Applications 3: Computational, Classification, Financial, Statistical and Stochastic Methods
(Big Data, Artificial Intelligence and Data Analysis Set – Volume 5)
Data Analysis and Applications 4: Financial Data Analysis and Methods
(Big Data, Artificial Intelligence and Data Analysis Set – Volume 6)

MASSOTTE Pierre, CORSI Patrick
Complex Decision-Making in Economy and Finance

MEUNIER François-Xavier
Dual Innovation Systems: Concepts, Tools and Methods
(Smart Innovation Set – Volume 31)

MICHAUD Thomas
Science Fiction and Innovation Design (Innovation in Engineering and Technology Set – Volume 6)

MONINO Jean-Louis
Data Control: Major Challenge for the Digital Society
(Smart Innovation Set – Volume 29)

MORLAT Clément
Sustainable Productive System: Eco-development versus Sustainable Development
(Smart Innovation Set – Volume 26)

SAULAIS Pierre, ERMINE Jean-Louis
Knowledge Management in Innovative Companies 2: Understanding and Deploying a KM Plan within a Learning Organization
(Smart Innovation Set – Volume 27)

2019

AMENDOLA Mario, GAFFARD Jean-Luc
Disorder and Public Concern Around Globalization

BARBAROUX Pierre
Disruptive Technology and Defence Innovation Ecosystems
(Innovation in Engineering and Technology Set – Volume 5)

DOU Henri, JUILLET Alain, CLERC Philippe
Strategic Intelligence for the Future 1: A New Strategic and Operational Approach
Strategic Intelligence for the Future 2: A New Information Function Approach

FRIKHA Azza
Measurement in Marketing: Operationalization of Latent Constructs

FRIMOUSSE Soufyane
Innovation and Agility in the Digital Age
(Human Resources Management Set – Volume 2)

GAY Claudine, SZOSTAK Bérangère L.
Innovation and Creativity in SMEs: Challenges, Evolutions and Prospects
(Smart Innovation Set – Volume 21)

GORIA Stéphane, HUMBERT Pierre, ROUSSEL Benoît
Information, Knowledge and Agile Creativity
(Smart Innovation Set – Volume 22)

HELLER David
Investment Decision-making Using Optional Models
(Economic Growth Set – Volume 2)

HELLER David, DE CHADIRAC Sylvain, HALAOUI Lana, JOUVET Camille
The Emergence of Start-ups
(Economic Growth Set – Volume 1)

HÉRAUD Jean-Alain, KERR Fiona, BURGER-HELMCHEN Thierry
Creative Management of Complex Systems
(Smart Innovation Set – Volume 19)

LATOUCHE Pascal
Open Innovation: Corporate Incubator
(Innovation and Technology Set – Volume 7)

LEHMANN Paul-Jacques
The Future of the Euro Currency

LEIGNEL Jean-Louis, MÉNAGER Emmanuel, YABLONSKY Serge
Sustainable Enterprise Performance: A Comprehensive Evaluation Method

LIÈVRE Pascal, AUBRY Monique, GAREL Gilles
Management of Extreme Situations: From Polar Expeditions to Exploration-Oriented Organizations

MILLOT Michel
Embarrassment of Product Choices 2: Towards a Society of Well-being

N'GOALA Gilles, PEZ-PÉRARD Virginie, PRIM-ALLAZ Isabelle
Augmented Customer Strategy: CRM in the Digital Age

NIKOLOVA Blagovesta
The RRI Challenge: Responsibilization in a State of Tension with Market Regulation
(Innovation and Responsibility Set – Volume 3)

PELLEGRIN-BOUCHER Estelle, ROY Pierre
Innovation in the Cultural and Creative Industries
(Innovation and Technology Set – Volume 8)

PRIOLON Joël
Financial Markets for Commodities

QUINIOU Matthieu
Blockchain: The Advent of Disintermediation

RAVIX Joël-Thomas, DESCHAMPS Marc
Innovation and Industrial Policies
(Innovation between Risk and Reward Set – Volume 5)

ROGER Alain, VINOT Didier
Skills Management: New Applications, New Questions
(Human Resources Management Set – Volume 1)

SAULAIS Pierre, ERMINE Jean-Louis
Knowledge Management in Innovative Companies 1: Understanding and Deploying a KM Plan within a Learning Organization
(Smart Innovation Set – Volume 23)

SERVAJEAN-HILST Romaric
Co-innovation Dynamics: The Management of Client-Supplier Interactions for Open Innovation
(Smart Innovation Set – Volume 20)

SKIADAS Christos H., BOZEMAN James R.
Data Analysis and Applications 1: Clustering and Regression, Modeling-estimating, Forecasting and Data Mining
(Big Data, Artificial Intelligence and Data Analysis Set – Volume 2)
Data Analysis and Applications 2: Utilization of Results in Europe and Other Topics
(Big Data, Artificial Intelligence and Data Analysis Set – Volume 3)

UZUNIDIS Dimitri
Systemic Innovation: Entrepreneurial Strategies and Market Dynamics

VIGEZZI Michel
World Industrialization: Shared Inventions, Competitive Innovations and Social Dynamics
(Smart Innovation Set – Volume 24)

2018

BURKHARDT Kirsten
Private Equity Firms: Their Role in the Formation of Strategic Alliances

CALLENS Stéphane
Creative Globalization
(Smart Innovation Set – Volume 16)

CASADELLA Vanessa
Innovation Systems in Emerging Economies: MINT – Mexico, Indonesia, Nigeria, Turkey
(Smart Innovation Set – Volume 18)

CHOUTEAU Marianne, FOREST Joëlle, NGUYEN Céline
Science, Technology and Innovation Culture
(Innovation in Engineering and Technology Set – Volume 3)

CORLOSQUET-HABART Marine, JANSSEN Jacques
Big Data for Insurance Companies
(Big Data, Artificial Intelligence and Data Analysis Set – Volume 1)

CROS Françoise
Innovation and Society
(Smart Innovation Set – Volume 15)

DEBREF Romain
Environmental Innovation and Ecodesign: Certainties and Controversies
(Smart Innovation Set – Volume 17)

DOMINGUEZ Noémie
SME Internationalization Strategies: Innovation to Conquer New Markets

ERMINE Jean-Louis
Knowledge Management: The Creative Loop
(Innovation and Technology Set – Volume 5)

GILBERT Patrick, BOBADILLA Natalia, GASTALDI Lise,
LE BOULAIRE Martine, LELEBINA Olga
Innovation, Research and Development Management

IBRAHIMI Mohammed
Mergers & Acquisitions: Theory, Strategy, Finance

LEMAÎTRE Denis
Training Engineers for Innovation

LÉVY Aldo, BEN BOUHENI Faten, AMMI Chantal
Financial Management: USGAAP and IFRS Standards
(Innovation and Technology Set – Volume 6)

MILLOT Michel
Embarrassment of Product Choices 1: How to Consume Differently

PANSERA Mario, OWEN Richard
Innovation and Development: The Politics at the Bottom of the Pyramid
(Innovation and Responsibility Set – Volume 2)

RICHEZ Yves
Corporate Talent Detection and Development

SACHETTI Philippe, ZUPPINGER Thibaud
New Technologies and Branding
(Innovation and Technology Set – Volume 4)

SAMIER Henri
Intuition, Creativity, Innovation

TEMPLE Ludovic, COMPAORÉ SAWADOGO Eveline M.F.W.
Innovation Processes in Agro-Ecological Transitions in Developing Countries
(Innovation in Engineering and Technology Set – Volume 2)

UZUNIDIS Dimitri
Collective Innovation Processes: Principles and Practices
(Innovation in Engineering and Technology Set – Volume 4)

VAN HOOREBEKE Delphine
The Management of Living Beings or Emo-management

2017

AÏT-EL-HADJ Smaïl
The Ongoing Technological System
(Smart Innovation Set – Volume 11)

BAUDRY Marc, DUMONT Béatrice
Patents: Prompting or Restricting Innovation?
(Smart Innovation Set – Volume 12)

BÉRARD Céline, TEYSSIER Christine
Risk Management: Lever for SME Development and Stakeholder Value Creation

CHALENÇON Ludivine
Location Strategies and Value Creation of International Mergers and Acquisitions

CHAUVEL Danièle, BORZILLO Stefano
The Innovative Company: An Ill-defined Object
(Innovation between Risk and Reward Set – Volume 1)

CORSI Patrick
Going Past Limits To Growth

D'ANDRIA Aude, GABARRET Inés
Building 21st Century Entrepreneurship
(Innovation and Technology Set – Volume 2)

DAIDJ Nabyla
Cooperation, Coopetition and Innovation
(Innovation and Technology Set – Volume 3)

FERNEZ-WALCH Sandrine
The Multiple Facets of Innovation Project Management
(Innovation between Risk and Reward Set – Volume 4)

FOREST Joëlle
Creative Rationality and Innovation
(Smart Innovation Set – Volume 14)

GUILHON Bernard
Innovation and Production Ecosystems
(Innovation between Risk and Reward Set – Volume 2)

HAMMOUDI Abdelhakim, DAIDJ Nabyla
Game Theory Approach to Managerial Strategies and Value Creation
(Diverse and Global Perspectives on Value Creation Set – Volume 3)

LALLEMENT Rémi
Intellectual Property and Innovation Protection: New Practices and New Policy Issues
(Innovation between Risk and Reward Set – Volume 3)

LAPERCHE Blandine
Enterprise Knowledge Capital
(Smart Innovation Set – Volume 13)

LEBERT Didier, EL YOUNSI Hafida
International Specialization Dynamics
(Smart Innovation Set – Volume 9)

MAESSCHALCK Marc
Reflexive Governance for Research and Innovative Knowledge
(Responsible Research and Innovation Set – Volume 6)

MASSOTTE Pierre
Ethics in Social Networking and Business 1: Theory, Practice and Current Recommendations
Ethics in Social Networking and Business 2: The Future and Changing Paradigms

MASSOTTE Pierre, CORSI Patrick
Smart Decisions in Complex Systems

MEDINA Mercedes, HERRERO Mónica, URGELLÉS Alicia
Current and Emerging Issues in the Audiovisual Industry
(Diverse and Global Perspectives on Value Creation Set – Volume 1)

MICHAUD Thomas
Innovation, Between Science and Science Fiction
(Smart Innovation Set – Volume 10)

PELLÉ Sophie
Business, Innovation and Responsibility
(Responsible Research and Innovation Set – Volume 7)

SAVIGNAC Emmanuelle
The Gamification of Work: The Use of Games in the Workplace

SUGAHARA Satoshi, DAIDJ Nabyla, USHIO Sumitaka
Value Creation in Management Accounting and Strategic Management: An Integrated Approach
(Diverse and Global Perspectives on Value Creation Set –Volume 2)

UZUNIDIS Dimitri, SAULAIS Pierre
Innovation Engines: Entrepreneurs and Enterprises in a Turbulent World
(Innovation in Engineering and Technology Set – Volume 1)

2016

BARBAROUX Pierre, ATTOUR Amel, SCHENK Eric
Knowledge Management and Innovation
(Smart Innovation Set – Volume 6)

BEN BOUHENI Faten, AMMI Chantal, LEVY Aldo
Banking Governance, Performance And Risk-Taking: Conventional Banks Vs Islamic Banks

BOUTILLIER Sophie, CARRÉ Denis, LEVRATTO Nadine
Entrepreneurial Ecosystems (Smart Innovation Set – Volume 2)

BOUTILLIER Sophie, UZUNIDIS Dimitri
The Entrepreneur (Smart Innovation Set – Volume 8)

BOUVARD Patricia, SUZANNE Hervé
Collective Intelligence Development in Business

GALLAUD Delphine, LAPERCHE Blandine
Circular Economy, Industrial Ecology and Short Supply Chains
(Smart Innovation Set – Volume 4)

GUERRIER Claudine
Security and Privacy in the Digital Era
(Innovation and Technology Set – Volume 1)

MEGHOUAR Hicham
Corporate Takeover Targets

MONINO Jean-Louis, SEDKAOUI Soraya
Big Data, Open Data and Data Development
(Smart Innovation Set – Volume 3)

MOREL Laure, LE ROUX Serge
Fab Labs: Innovative User
(Smart Innovation Set – Volume 5)

PICARD Fabienne, TANGUY Corinne
Innovations and Techno-ecological Transition
(Smart Innovation Set – Volume 7)

2015

CASADELLA Vanessa, LIU Zeting, DIMITRI Uzunidis
Innovation Capabilities and Economic Development in Open Economies
(Smart Innovation Set – Volume 1)

CORSI Patrick, MORIN Dominique
Sequencing Apple's DNA

CORSI Patrick, NEAU Erwan
Innovation Capability Maturity Model

FAIVRE-TAVIGNOT Bénédicte
Social Business and Base of the Pyramid

GODÉ Cécile
Team Coordination in Extreme Environments

MAILLARD Pierre
Competitive Quality and Innovation

MASSOTTE Pierre, CORSI Patrick
Operationalizing Sustainability

MASSOTTE Pierre, CORSI Patrick
Sustainability Calling

2014

DUBÉ Jean, LEGROS Diègo
Spatial Econometrics Using Microdata

LESCA Humbert, LESCA Nicolas
Strategic Decisions and Weak Signals

2013

HABART-CORLOSQUET Marine, JANSSEN Jacques, MANCA Raimondo
VaR Methodology for Non-Gaussian Finance

2012

DAL PONT Jean-Pierre
Process Engineering and Industrial Management

MAILLARD Pierre
Competitive Quality Strategies

POMEROL Jean-Charles
Decision-Making and Action

SZYLAR Christian
UCITS Handbook

2011

LESCA Nicolas
Environmental Scanning and Sustainable Development

LESCA Nicolas, LESCA Humbert
Weak Signals for Strategic Intelligence: Anticipation Tool for Managers

MERCIER-LAURENT Eunika
Innovation Ecosystems

2010

SZYLAR Christian
Risk Management under UCITS III/IV

2009

COHEN Corine
Business Intelligence

ZANINETTI Jean-Marc
Sustainable Development in the USA

2008

CORSI Patrick, DULIEU Mike
The Marketing of Technology Intensive Products and Services

DZEVER Sam, JAUSSAUD Jacques, ANDREOSSO Bernadette
Evolving Corporate Structures and Cultures in Asia: Impact of Globalization

2007

AMMI Chantal
Global Consumer Behavior

2006

BOUGHZALA Imed, ERMINE Jean-Louis
Trends in Enterprise Knowledge Management

CORSI Patrick *et al.*
Innovation Engineering: the Power of Intangible Networks

www.ingramcontent.com/pod-product-compliance
Lightning Source LLC
LaVergne TN
LVHW011749200926
848484LV00003B/660